The IMA Volumes in Mathematics and its Applications

Volume 32

Series Editors
Avner Friedman   Willard Miller, Jr.

Institute for Mathematics and
its Applications
IMA

The **Institute for Mathematics and its Applications** was established by
a grant from the National Science Foundation to the University of Minnesota in
1982. The IMA seeks to encourage the development and study of fresh mathematical concepts and questions of concern to the other sciences by bringing together
mathematicians and scientists from diverse fields in an atmosphere that will stim-
ulate discussion and collaboration.

The IMA Volumes are intended to involve the broader scientific community in
this process.

Avner Friedman, Director
Willard Miller, Jr., Associate Director

* * * * * * * * * *

IMA PROGRAMS

1982-1983	**Statistical and Continuum Approaches to Phase Transition**
1983-1984	**Mathematical Models for the Economics of Decentralized Resource Allocation**
1984-1985	**Continuum Physics and Partial Differential Equations**
1985-1986	**Stochastic Differential Equations and Their Applications**
1986-1987	**Scientific Computation**
1987-1988	**Applied Combinatorics**
1988-1989	**Nonlinear Waves**
1989-1990	**Dynamical Systems and Their Applications**
1990-1991	**Phase Transitions and Free Boundaries**

* * * * * * * * * *

SPRINGER LECTURE NOTES FROM THE IMA:

The Mathematics and Physics of Disordered Media
Editors: Barry Hughes and Barry Ninham
(Lecture Notes in Math., Volume 1035, 1983)

Orienting Polymers
Editor: J.L. Ericksen
(Lecture Notes in Math., Volume 1063, 1984)

New Perspectives in Thermodynamics
Editor: James Serrin
(Springer-Verlag, 1986)

Models of Economic Dynamics
Editor: Hugo Sonnenschein
(Lecture Notes in Econ., Volume 264, 1986)

Richard E. Blahut   Willard Miller, Jr.
Calvin H. Wilcox

Radar and Sonar

Part I

With 15 Figures

Springer-Verlag
New York Berlin Heidelberg London
Paris Tokyo Hong Kong Barcelona

Richard E. Blahut
MS 0600
IBM
Owego, NY 13827
USA

Calvin H. Wilcox
Department of Mathematics
University of Utah
Salt Lake City, UT 84112
USA

Willard Miller, Jr.
Institute for Mathematics and its
  Applications
University of Minnesota
Minneapolis, MN 55455
USA

Series Editors

Avner Friedman
Willard Miller, Jr.
Institute for Mathematics and its
  Applications
University of Minnesota
Minneapolis, MN 55455
USA

Mathematics Subject Classifications: 78A45, 22E70, 43A80

Library of Congress Cataloging-in-Publication Data
Blahut, Richard E.
    Radar and sonar / Richard E. Blahut, Willard Miller, Jr., Calvin
  H. Wilcox.
        p.   cm. — (IMA volumes in mathematics and its applications;
  v. 32)
      Includes bibliographical references.
      ISBN 0-387-97516-0. — ISBN 3-540-97516-0 (EUR)
      1. Radar — Mathematical models.   2. Sonar — Mathematical models.
  I. Miller, Willard.   II. Wilcox, Calvin H. (Calvin Hayden).
  III. Title.   IV. Series.
  TK6580.B5   1991
  621.3848 — dc20   90-27282

Printed on acid-free paper.

Camera-ready copy prepared by the IMA.
Printed and bound by Edwards Brothers, Inc., Ann Arbor, Michigan.
Printed in the United States of America.

9 8 7 6 5 4 3 2 1

ISBN 0-387-97516-0 Springer-Verlag New York Berlin Heidelberg
ISBN 3-540-97516-0 Springer-Verlag Berlin Heidelberg New York

The IMA Volumes
in Mathematics and its Applications

Current Volumes:

CONTENTS

INTRODUCTION

This volume contains the lecture notes from the three sets of tutorial lectures which were given during the first week of the IMA summer program RADAR AND SONAR, June 18 - June 29, 1990. (The second week was devoted to research problems and the proceedings of that part of the program will appear in a second IMA volume.) The first week was run as a summer school with an audience consisting mainly of mathematicians and engineers. The tutorial topics were on mathematics (Topics in Harmonic Analysis with Applications to Radar and Sonar, by Willard Miller, Jr.), on the physical aspects of scattering (Sonar and Radar Echo Structure, by Calvin H. Wilcox), and on the engineering modelling and processing of the phenomena under consideration (Theory of Remote Surveillance Algorithms, by Richard E. Blahut). The famous 1960 technical report by Wilcox (The Synthesis Problem for Radar Ambiguity Functions) was featured prominently in the program and is also published here for the first time. A great effort was made by the lecturers to insure that the participants covered two or all three short courses in detail: mathematicians needed to spend more time in the engineering and physics components, and a corresponding distribution of effort was encouraged for engineers and physicists. One of the main goals was to ensure that people with different backgrounds would help each other, and learn in the process a bit about each others language and approach to problems in Radar and Sonar. We believe that the effort was a great success and offer these notes for the benefit of the wider mathematical sciences community.

We also take this opportunity to thank those agencies whose financial support made the summer program possible: the Air Force Office of Scientific Research, the Army Research Office, the Defense Advanced Research Projects Agency, the National Science Foundation and the Office of Naval Research.

Avner Friedman

Willard Miller, Jr.

THEORY OF REMOTE SURVEILLANCE ALGORITHMS

RICHARD E. BLAHUT

Abstract. Algorithms for remote surveillance imaging have been developed independently in many fields including radar, sonar, medical imaging, and radio astronomy. Recently, it has become apparent that an underlying theory of remote surveillance could be developed. This emerging unified theory may suggest new directions for future developments.

This short course will develop, from the engineer's point of view, the two-dimensional Fourier transform, the ambiguity function, the Radon transform, and the projection-slice theorem. The course will explain how these topics deal with radar and sonar, including the task of imaging, by discussing how they relate to doppler frequency shifts, synthetic aperture radar, beamforming, and the tomographic reconstruction of images. The course will be organized in a way to integrate the vocabulary and methods of the fields of radar, sonar, tomography, radio astronomy, and related fields.

I. INTRODUCTION

We may perceive our environment as a complex tapestry of information-bearing signals of many kinds, some man-made and some not, reaching us from many directions. Some of the signals, such as optical signals and acoustic signals are immediately compatible with our senses. To perceive other of these signals, however, we require a complex apparatus or sensor. A great variety of man-made sensors now exist to collect signals and artificially form some kind of image of an object or a scene of objects. We refer to these as sensors for remote surveillance. There are many kinds of sensors collected under this heading, differing in the size of the observed scene as from microscopes to radio telescopes, in complexity as from the simple lens to synthetic aperture radars, and in the current state of development as from photography to holography and tomography. All of these devices collect raw sensor data and process that data to extract useful information that may be deeply buried within the data. The description of the processing may require a complex mathematical formulation.

Systems for remote surveillance have developed over the years independently and driven by the kind of hardware used to sense the raw data or to do the processing. The principles of operation then are described in a way completely intertwined with a description of the hardware. Recently, there has been interest in abstracting the common signal processing and information-theoretic methods that are used by these sensors to extract the useful information from the raw data. This unification is one way in which to make new advances because then the underlying principles can be more clearly understood and ideas that have already been developed in one area may prove to be useful in others. An integrated formulation is also a valuable way to teach the many topics as one subject.

This short course is intended as one step toward a unified treatment of the mathematical methods that underlie the various systems for remote surveillance. Many of these methods are built around the ubiquitous two-dimensional Fourier transform and the closely related Radon transform and ambiguity function.

1

History

It is interesting to notice that the history of remote surveillance algorithms touches on many Nobel prizes. In each case the Nobel prize went to a body of work that included processing algorithms. We may even find, in our retrospective, some overlap in the ideas - probably more evident now than it was then. We can even whimsically define the subject of remote surveillance algorithms as the intersection of a certain subset of Nobel prize winning work. This thought alone is enough justification to formalize the underlying computational ideas.

The principle that a radar could be used for high-resolution imaging has been attributed to Wiley of Goodyear Aircraft, although he did not publish his ideas nor construct such a radar. Wiley observed (in 1951) that whereas the azimuth resolution of a conventional radar is limited by the width of the radar beam, each reflecting element within the radar beam from a moving aircraft has a doppler frequency shift that depends on the angle between the aircraft velocity vector and the direction to the reflecting element. Thus, he concluded that a precise frequency analysis of the radar reflections would provide finer along-track resolution than the azimuth resolution defined by the antenna beamwidth. The following year, a group at the University of Illinois came to the same idea independently, based upon frequency analysis of experimental radar returns. During the summer of 1953, these ideas were reviewed by the members of a summer study (Project Wolverine) at the University of Michigan which laid plans for the development of an imaging radar, eventually called a synthetic aperture radar from the notion that the physical antenna is moved to sweep out a synthetic aperture. It was recognized that the processing requirements for imaging placed extreme demands on the analog processing technology of the day. Meanwhile, Emmett Leith, at the University of Michigan, turned to the processing ideas of holography and developed optical processing techniques to satisfy the synthetic aperture processing requirements. The first synthetic aperture radar was successfully demonstrated in 1957. These methods were later extended to the imaging of distant rotating objects such as planets and asteroids, as by Paul Green.

The earlier development of search radars for moving target detection is spread more broadly and individual contributions are less spectacular. From the first use of radar, it was recognized that the need to detect moving targets could be satisfied by using the doppler shift of the return. Any doppler-shifted echo corresponds to a moving reflector. However, the magnitude of the doppler shift is only a very small fraction of the transmitted pulse bandwidth. The technology did not exist to filter a faint doppler-shifted signal from a strong background of signals echoed from other stationary emitters, hence, the development of search radars did not depend so much on invention at the conceptual level as it did on the development of technology to support widely understood requirements. By the end of World War II, radars were developed that used doppler filters to suppress the clutter signal reflected from the stationary background. These early radars used simple delay lines to cancel the stationary return from one pulse with the nearly identical return from the previous pulse. In this way, large rapidly-moving objects could be detected from stationary radar stations. Later, the requirements for search radars shifted to moving airborne radar stations for observing small, slowly moving target objects at large range. It then became necessary to employ much more delicate techniques for finding a signal return within a large clutter background. These techniques employ coherent processing with the aid of large special-purpose digital computers.

Meanwhile, the astronomers had come to realize that a large amount of astronomical information reaches the earth in the microwave bands. Astronomers are well grounded in optical theory where beamwidths as small as one arc second are obtained. In the microwave band, comparable beamwidths require receiving antennas that are many miles in diameter. Such antennas must be mechanically rigid under wind, ice, and temperature gradients to a small fraction of an inch. Clearly, such antennas are impractical. Around 1952, Martin Ryle, at the University of Cambridge, began to study methods for synthesizing such an aperture by physically moving antenna elements, or by allowing the earth's rotation to sweep an array of fixed antennas through space. In retrospect, this development may be viewed as a passive counterpart to the development of synthetic aperture radar. The aperture is synthesized by recording the radio signal received at two or more antenna elements and later processing these coherently within a digital processor. The first such radio telescope was the Cambridge One-Mile telescope completed 1964. Many other synthetic aperture radio telescopes are now in operation throughout the world. For his development of synthetic aperture radio telescopes, Ryle was awarded in 1974 Nobel prize in Physics (jointly with Hewish who discovered pulsars with the radio telescope).

Holography was described by Dennis Gabor in a series of papers in 1948, 1949, and 1951 which earned him the 1971 Nobel prize in physics. Gabor realized that whereas conventional photography first processes the optical information to form an image which is then recorded on film, it is also possible to record the raw optical data on the photographic film directly and place the processing in the future with the viewer. His method for recording the raw optical data is called a *hologram.* Because the raw optical data contains more information than a final photographic image, in principle the hologram can be used to create an image superior to a photograph. Most striking in this regard is the creation of three-dimensional images from a two-dimensional hologram. Holography is technically much more difficult than photography because recording of the raw optical data requires precision on the order of optical wavelengths. For this reason holography waited for the invention of the laser to make it more attractive.

Closely related to holography are other kinds of optical processing, many of them using diffraction phenomena that are describable in terms of the two-dimensional Fourier transform. Diffraction of X-rays by crystals was established by Laue as a proof of the wave properties of X-rays. Bragg immediately inverted the point of view to turn X-ray diffraction into a way of probing crystals, which has since evolved into a sophisticated imaging technique. Laue in 1914 and Bragg, with his son, in 1915 won Nobel prizes in Physics for these foundations of crystallography. Because it is not possible to measure the phase of a scattered X-ray wave, only the magnitude of the crystallographic diffraction pattern is available. Herbert Hauptman and Jerome Karle (1953), developed methods to invert the data using magnitude only, which earned them the 1985 Nobel prize in Chemistry. Aaron Klug (1964) developed methods for the imaging of viruses using the diffraction of laser light by electron microscope images to remove the large depth of focus for which he won the 1982 Nobel prize in Chemistry.

The development of tomography in the modern sense of computerized image reconstruction for medical applications began in Great Britain. The key algorithmic feature is the reconstruction of images from their projections; similar ideas can be found in other fields. The methods of algorithmic reconstruction of images from X-ray projections were first developed by Cormack (1963) and reduced to practice by Hounsfield. The 1979 Nobel prize in physiology and medicine was awarded to Hounsfield and Cormack for the development of computerized tomography. Subsequently, emission tomography was proposed by Kuhl and Edwards (1963). Nuclear magnetic resonance imaging was proposed by Damadian (1971, 1972) and demonstrated by Lauterbur (1973). The ideas of tomography are closely related to methods of radio astronomy, especially the formulation of reconstruction algorithms by Bracewell (1956).

II. SIGNALS IN ONE AND TWO DIMENSIONS

Most surveillance systems employ finite-energy functions of time as one-dimensional waveforms that probe the environment. In turn, the environment is usually modeled as a two-dimensional or a three-dimensional spatial region on which is defined a function of interest. This function is the function to be estimated. Loosely then, we can think of the typical computational task as that of estimating a two-dimensional or three-dimensional target function when given a collection of measurements consisting of one-dimensional functions derived from the target function. This kind of problem is sometimes called an inverse problem. Thus we must study one-dimensional functions $s(t)$, two or three dimensional functions $s(x, y)$ or $s(x, y, z)$, and their Fourier transforms. The principle connections relating one-dimensional functions and two-dimensional functions are the *ambiguity function* and the *projection-slice theorem*.

Signals in One Dimension

Every real-valued or complex-valued waveform (or pulse) s(t) of finite energy, E_p, given by

$$E_p = \int\limits_{-\infty}^{\infty} |s(t)|^2 dt < \infty$$

is associated with a complex-valued function, known as its Fourier transform, and defined as

$$S(f) = \int\limits_{-\infty}^{\infty} s(t)e^{-j2\pi ft} dt$$

We may abbreviate this equation by the functional notation $S(f) = F[s(t)]$ and

$s(t) = F^{-1}[S(f)]$. A Fourier transform pair is also denoted by the notation.

$$s(t) \longleftrightarrow S(f)$$

For example, if

$$\text{rect } t = \begin{cases} 1 & \text{if } |t| \leq 1/2 \\ 0 & \text{if } |t| > 1/2 \end{cases}$$

and

$$\text{sinc } t = \frac{\sin \pi t}{\pi t}$$

then

$$\text{rect } \frac{t}{T} \longleftrightarrow T \text{ sinc } (fT)$$

Another important example is the gaussian pulse. If $s(t) = e^{-\pi t^2}$, then $S(f) = e^{-\pi f^2}$.

The standard properties of the Fourier transform include:

- Inverse Fourier transform

$$s(t) = \int_{-\infty}^{\infty} S(f)e^{j2\pi ft}df$$

- Convolution theorem

$$s(t) * r(t) = \int_{-\infty}^{\infty} s(t - \tau)r(\tau)d\tau \longleftrightarrow S(f)R(f)$$

- Parseval's formula

$$\int_{-\infty}^{\infty} s_1(t)s_2^*(t)dt = \int_{-\infty}^{\infty} S_1(f)S_2^*(f)df$$

- Complex modulation theorem

$$s(t)e^{j2\pi f_0 t} \longleftrightarrow S(f - f_0)$$

Signals with finite support in the frequency domain for which $S(f)$ equals zero on some interval containing the origin are called *passband signals*; and signals for which there is no such interval containing the origin on which $S(f)$ equals zero are called *baseband signals*. Every real-valued passband signal can be decomposed as

$$s(t) = s_R(t) \cos 2\pi f_0 t - s_I(t) \sin 2\pi f_0 t$$

where $s_R(t)$ and $s_I(t)$ are real baseband signals called the in-phase and quadrature modulation components respectively. The real passband signal may be represented by the *complex baseband signal*

$$s(t) = s_R(t) + js_I(t)$$

Sometimes the passband signal will be written as $Re[s(t)e^{j2\pi f_0 t}]$ or more simply, in the complex passband form $s(t)e^{j2\pi f_0 t}$ where, here, $s(t)$ is understood to be complex. These several representations are not identical but are similar enough that it is easy to pass between them according to the needs of a discussion.

It is convenient to call both the passband signal and its complex representation by the same name $s(t)$ although they are really different functions. Usually, it is easier to deal with the complex representation in theoretical discussions and in computations. The passband representation is constructed for propagation as the modulation of a carrier electromagnetic or acoustic wave.

Complicated waveforms can be formed by replicating simple waveforms. Let

$$p(t) = \sum_{l=-n}^{n} s(t - lT_r)$$

The waveform $p(t)$ is called a *pulse train* formed from $N = 2n + 1$ uniformly spaced copies of the *pulse* $s(t)$. The Fourier transform of the pulse train is

$$P(f) = S(f)\frac{\sin N\pi f T_r}{\sin \pi f T_r}$$

This is the product of two factors, one due to the simple pulse $s(t)$ and one due to the array. Let

$$\text{dirc}_N(x) = \frac{\sin \pi N x}{\sin \pi x}$$

This function is called a *dirichlet function*. For small x,

$$\text{dirc}_N(x) \approx N \text{ sinc } (Nx)$$

The dirichlet function has its extrema, of magnitude N, at integer values of x. These extrema, called *grating lobes*, alternate in sign if N is even, and have the same sign if N is odd. The grating lobes should be thought of as very narrow because the dirichlet function equals zero at $x = k \pm \frac{1}{N}$ for any integer k.

A pulse $s(t)$ has a *Gabor timewidth* given by

$$T_G = \sqrt{\overline{t^2} - \overline{t}^2}$$

and a *Gabor bandwidth* given by

$$B_G = \sqrt{\overline{f^2} - \overline{f}^2}$$

where the time moments are

$$\overline{t^i} = \int\limits_{-\infty}^{\infty} t^i \frac{|s(t)|^2}{E_p} dt$$

and the frequency moments are

$$\overline{f^i} = \int\limits_{-\infty}^{\infty} f^i \frac{|S(f)|^2}{E_p} df$$

The Gabor timewidth or Gabor bandwidth may be infinite, even for reasonable $s(t)$. There is also a *skew parameter* ρ defined as

$$\rho = \frac{1}{T_G B_G} Re[\overline{tf} - \overline{t}\,\overline{f}]$$

where

$$\overline{tf} = \frac{j}{2\pi} \int\limits_{-\infty}^{\infty} t \frac{s'(t)s^*(t)}{E_p} dt = \frac{j}{2\pi} \int\limits_{-\infty}^{\infty} f \frac{S(f)S'^*(f)}{E_p} df$$

The moments $\overline{t}, \overline{t^2}, \overline{f}$, and $\overline{f^2}$ are all real, but $\overline{tf}$ may be complex. If the pulse $s(t)$ is real, then $\overline{tf}$ is purely imaginary.

The *uncertainty principle* for pulses is

$$T_G B_G \geq \frac{1}{4\pi}$$

This is a consequence of the Schwarz inequality and is satisfied with equality only if $s(t)$ is a gaussian pulse, possibly time-shifted or frequency-shifted. Despite its name there is nothing uncertain about the uncertainty principle.

Signals in Two Dimensions

Every function of two variables $s(x, y)$ (called a two-dimensional function) with finite energy E_p

$$E_p = \int\limits_{-\infty}^{\infty} \int\limits_{-\infty}^{\infty} |s(x, y)|^2 dx dy < \infty$$

has a two-dimensional Fourier transform defined by

$$S(f_x, f_y) = \int\limits_{-\infty}^{\infty} \int\limits_{-\infty}^{\infty} s(x, y) e^{-j2\pi(f_x x + f_y y)} dx dy.$$

The transform relationship between the pair of functions $s(x, y)$ and $S(f_x, f_y)$ is denoted by

$$s(x, y) \Leftrightarrow S(f_x, f_y).$$

Notice the doubly shafted arrow $\Leftrightarrow$ is used instead of the singly shafted arrow used with one-dimensional Fourier transforms.

Standard properties of the two-dimensional Fourier transform include:

- Inverse Fourier transform

$$s(x,y) = \int_{-\infty}^{\infty} \int_{-\infty}^{\infty} S(f_x, f_y) e^{j2\pi(f_x x + f_y y)} df_x df_y$$

- Convolution theorem

$$g(x,y) * *h(x,y) \Leftrightarrow G(f_x, f_y) H(f_x, f_y)$$

where $**$ denotes two-dimensional convolution given by

$$g(x,y) * *h(x,y) = \int_{-\infty}^{\infty} \int_{-\infty}^{\infty} g(\xi, \eta) h(x - \xi, y - \eta) d\xi d\eta$$

- Parseval's formula

$$\int_{-\infty}^{\infty} \int_{-\infty}^{\infty} g(x,y) h^*(x,y) dx dy = \int_{-\infty}^{\infty} \int_{-\infty}^{\infty} G(f_x, f_y) H^*(f_x, f_y) df_x df_y$$

- Modulation: For any real constants a and b.

$$s(x,y) e^{j2\pi(ax+by)} \Leftrightarrow S(f_x - a, f_y - b)$$

- Energy relation:

$$\int_{-\infty}^{\infty} \int_{-\infty}^{\infty} |s(x,y)|^2 dx dy = \int_{-\infty}^{\infty} \int_{-\infty}^{\infty} |S(f_x, f_y)|^2 df_x df_y$$

- Coordinate transformation:

$$s(a_1 x + b_1 y, a_2 x + b_2 y) \Leftrightarrow \frac{1}{|a_1 b_2 - a_2 b_1|} S(A_1 f_x + A_2 f_y, B_1 f_x + B_2 f_y)$$

where

$$\begin{bmatrix} A_1 & B_1 \\ A_2 & B_2 \end{bmatrix} = \begin{bmatrix} a_1 & b_1 \\ a_2 & b_2 \end{bmatrix}^{-1}$$

As an example of a two-dimensional Fourier transform, let

$$\text{rect}\,(x,y) = \text{rect}\,(x)\,\text{rect}\,(y)$$

Then
$$\text{rect}\left(\frac{x}{a}, \frac{y}{b}\right) \Leftrightarrow ab \text{ sinc } (af_x) \text{ sinc } (bf_y)$$

If b is very small and a is very large, then $S(f_x, f_y)$ is narrow in f_x and wide in f_y.

Another important example is the unit circle function defined by

$$\text{circ } (x, y) = \begin{cases} 1 & \text{if } \sqrt{x^2 + y^2} \le 1 \\ 0 & \text{otherwise} \end{cases}$$
$$= \text{rect}\left(2\sqrt{x^2 + y^2}\right)$$

Define the *jinc function* by

$$\text{jinc } (x) = \frac{J_1(2\pi x)}{x}$$

where $J_1(x)$ is the first-order Bessel function of the first kind. The general appearance of the jinc function is very similar to the appearance of the sinc function. However, the zeros of the jinc function are not spaced periodically and the magnitude of the sidelobes fall off more quickly than do those of the sinc function. The two-dimensional Fourier transform pair then is

$$\text{circ } (x, y) \Leftrightarrow \text{jinc } \left(\sqrt{f_x^2 + f_y^2}\right)$$

The Projection-Slice Theorem

Let $s(x, y)$ be a two-dimensional signal of finite energy. The projection of $s(x, y)$ onto the x axis is

$$p(x) = \int_{-\infty}^{\infty} s(x, y) dy$$

More generally, the *projection* of $s(x, y)$ at angle ϕ is defined as

$$p_\phi(t) = \int_{-\infty}^{\infty} s(t \cos \phi - r \sin \phi, t \sin \phi + r \cos \phi) dr$$

The angle ϕ specifies a rotation relating variables t, r to variables x, y. The term projection comes from the sense that $s(x, y)$ is "projected" into $p_\phi(t)$ by integrating along lines at an angle of ϕ from the y axis.

Let $S(f_x, f_y)$ be a two-dimensional signal of finite energy. The *slice* of $S(f_x, f_y)$ along the f_x axis is $S(f, 0)$ and the *slice* of $S(f_x, f_y)$ at angle ϕ is $S(f \cos \phi, f \sin \phi)$.

The projection-slice theorem is a basic theorem of Fourier analysis relating the two-dimensional Fourier transform of a two-dimensional function and the one-dimensional Fourier transform of its one-dimensional projection. The theorem says that a projection of $s(x, y)$ transforms into a slice of $S(f_x, f_y)$.

THEOREM (PROJECTION-SLICE THEOREM). *If $p_\phi(t)$ is the projection of $s(x,y)$ at angle ϕ, and $S(f_x, f_y)$ is the Fourier transform of $s(x,y)$, then*

$$P_\phi(f) = S(f \cos \phi, f \sin \phi)$$

is the Fourier transform of $p_\phi(t)$.

Proof.

$$P_\phi(f) = \int_{-\infty}^{\infty} p_\phi(t) e^{-j2\pi ft} dt$$

$$= \int_{-\infty}^{\infty} \int_{-\infty}^{\infty} s(t \cos \phi - r \sin \phi, t \sin \phi + r \cos \phi) e^{-j2\pi ft} dr dt$$

Now make the change in variable

$$x = t \cos \phi - r \sin \phi$$
$$y = t \sin \phi + r \cos \phi$$

which leads in a straightforward way to the expression of the theorem. □

DEFINITION. Given a two-dimensional signal $s(x,y)$ of finite energy, the *Radon transform* of $s(x,y)$ is the two-dimensional function

$$p(t, \phi) = \int_{-\infty}^{\infty} s(t \cos \phi - r \sin \phi, t \sin \phi + r \cos \phi) dr$$

In two-dimensional space, the Radon transform is also called the *shadow transform* or the *X-ray transform*. In n-dimensional space, the Radon transform and the shadow transform are not the same. A Radon transform, in general, maps an n-dimensional function onto its integral on $(n-1)$-dimensional hyperplanes and the shadow transform maps an n-dimensional function onto its integral on (one-dimensional) lines.

The projection-slice theorem implies that the Radon transform can be inverted to recover $s(x,y)$ from $p(t, \phi)$. Specifically, if $P(f, \phi)$ denotes the one-dimensional Fourier transform of $p(t, \phi)$ in the variable t, then

$$P(f, \phi) = S(f \cos \phi, f \sin \phi)$$

Thus, $P(f, \phi)$ is simply the two-dimensional Fourier transform of $s(x,y)$ expressed in polar coordinates.

We will give several simple examples of the Radon transform. Let

$$s(x,y) = \text{circ}(x,y)$$

Because this is circularly symmetric, the Radon transform is easily seen to be

$$p(t, \phi) = \begin{cases} 2\sqrt{1 - t^2} & t < 1 \\ 0 & t \geq 1 \end{cases}$$

which is independent of ϕ.

The two-dimensional gaussian pulse

$$s(x, y) = e^{-\pi(x^2 + y^2)}$$

is also circularly symmetric. The Radon transform is

$$p(t, \phi) = \int_{-\infty}^{\infty} e^{-\pi(t^2 + r^2)} dr$$

$$= e^{-\pi t^2}$$

III. THE AMBIGUITY FUNCTION

An ambiguity function is a two-dimensional function defined as a functional of a one-dimensional waveform $s(t)$. Every finite-energy waveform is associated with an ambiguity function which provides a surprising amount of insight into the performance of the waveform in radar and sonar applications. The ambiguity function is a two-dimensional point-spread function for the two-dimensional scene observed by the radar. In sonar applications, the wideband ambiguity function, which we do not consider, may replace the use of the (narrowband) ambiguity function.

DEFINITION. The *ambiguity function* of a finite energy pulse $s(t)$ is a function of $s(t)$ given by the complex function of two variables:

$$\chi(\tau, \nu) = \int_{-\infty}^{\infty} s(t + \tau/2) s^*(t - \tau/2) e^{-j2\pi\nu t} dt$$

The magnitude of the ambiguity function $|\chi(\tau, \nu)|$ is called the *ambiguity surface*.

An equivalent form

$$\chi(\tau, \nu) = \int_{-\infty}^{\infty} \left[s(t + \tau/2) e^{-j\pi\nu t} \right] \left[s(t - \tau/2) e^{j\pi\nu t} \right]^* dt$$

shows that $\chi(\tau, \nu)$ can be thought of as the correlation coefficient between the waveform $s(t)$ shifted in time and in frequency with itself shifted in the opposite direction in time and in frequency.

A slightly different form which is asymmetric but essentially equivalent is given by the definition

$$\chi'(\tau, \nu) = \int_{-\infty}^{\infty} s(t) s^*(t - \tau) e^{-j2\pi\nu t} dt$$

While the asymmetrical form appears simpler at the onset, the symmetrical form is chosen for theoretical studies because it simplifies the appearance of many later results. In engineering practice, either form may be used without comment depending on convenience. The two forms are related by a linear phase term. This can be seen by a change in variables to give

$$\chi(\tau, \nu) = e^{j\pi\tau\nu}\chi'(\tau, \nu)$$

By setting $\nu = 0$ in $\chi(\tau, \nu)$, we have the autocorrelation function of $s(t)$.

$$\chi(\tau, 0) = \phi(\tau)$$

By setting $\tau = 0$ in $\chi(\tau, \nu)$, we have the Fourier transform of the square of the pulse.

$$\chi(0, \nu) = \int_{-\infty}^{\infty} |s(t)|^2 e^{-j2\pi\nu t} dt$$

By setting both $\tau = 0$ and $\nu = 0$, we have

$$\chi(0, 0) = \int_{-\infty}^{\infty} |s(t)|^2 dt = E_p$$

Thus, when evaluated at the origin, the value of the ambiguity function is equal to the energy in the pulse.

THEOREM. *The ambiguity function can be written*

$$\chi(\tau, \nu) = \int_{-\infty}^{\infty} S(f + \nu/2)S^*(f - \nu/2)e^{j2\pi f\tau} df$$

Proof.

$$\chi(\tau, \nu) = \int_{-\infty}^{\infty} \left[s(t + \tau/2)e^{-j\pi\nu t}\right] \left[s(t - \tau/2)e^{j\pi\nu t}\right]^* dt$$

The first term has transform $S(f + \nu/2)e^{j\pi(f+\nu/2)\tau}$ and the second term has transform $S(f - \nu/2)e^{-j\pi(f-\nu/2)\tau}$. Then, by Parseval's formula,

$$\chi(\tau, \nu) = \int_{-\infty}^{\infty} \left[S(f + \nu/2)e^{j\pi(f+\nu/2)\tau}\right] \left[S(f - \nu/2)e^{-j\pi(f-\nu/2)\tau}\right]^* df$$

$$= \int_{-\infty}^{\infty} \left[S(f + \nu/2)e^{j\pi f\tau}\right] \left[S(f - \nu/2)e^{-j\pi f\tau}\right]^* df$$

as was to proved. $\square$

COROLLARY.

$$\int_{-\infty}^{\infty} \chi(\tau,\nu)e^{j2\pi\nu t}\,d\nu = s(t+\tau/2)s^*(t-\tau/2)$$

$$\int_{-\infty}^{\infty} \chi(\tau,\nu)e^{-j2\pi f\tau}\,d\tau = S(f+\nu/2)S^*(f-\nu/2)$$

Proof. The first expression is the inverse Fourier transform of the defining equation for $\chi(\tau,\nu)$ with τ fixed. The second expression is just a Fourier transform of the expression in the theorem with ν fixed. $\square$

From the first line of the corollary, simply by replacing t by $\tau/2$. we can immediately write down

$$s(\tau) = \frac{1}{s^*(0)} \int_{-\infty}^{\infty} \chi(\tau,\nu)e^{j\pi\nu\tau}\,d\nu$$

and $|s(0)|^2 = \int_{-\infty}^{\infty} \chi(0,\nu)d\nu$. Thus, $s(t)$ can be recovered from $\chi(\tau,\nu)$.

We can express the inverse two-dimensional Fourier transform of $\chi(\tau,\nu)$ as

$$\Xi(t,f) = \int_{-\infty}^{\infty}\int_{-\infty}^{\infty} \chi(\tau,\nu)e^{j2\pi\nu t}e^{j2\pi\tau f}\,d\nu d\tau$$

$$= \int_{-\infty}^{\infty} s(t+\tau/2)s^*(t-\tau/2)e^{j2\pi\tau f}\,d\tau$$

This superficially resembles the equation defining $\chi(\tau,\nu)$ but is actually quite different because the integration is in the variable τ. The function $\Xi(t,f)$ is called the *Wigner distribution* of the pulse $s(t)$.

There are a few simple pulses whose ambiguity function is convenient to compute in closed form. The ambiguity function of the simple rectangular pulse rect (t/T) is easy to compute. The term $s(t+\tau/2)s^*(t-\tau/2)$ equals 1 only for $|t| \le (T-|\tau|)/2$. Therefore,

$$\chi(\tau,\nu) = \begin{cases} (T-|\tau|)\,\text{sinc}\,\nu(T-|\tau|) & \text{if } |\tau| \le T \\ 0 & \text{if } |\tau| > T \end{cases}$$

It is instructive to notice that for $\nu = 0$, $\chi(\tau,0)$ takes the form of a triangular pulse in τ

$$\chi(\tau,0) = \begin{cases} T-|\tau| & |\tau| \le T \\ 0 & |\tau| > T \end{cases}$$

and for $\tau = 0$, $\chi(0,\nu)$ takes the form of a sinc pulse in ν

$$\chi(0,\nu) = T\,\text{sinc}\,\nu T$$

Also notice that $\chi(\tau,\nu) = 0$ along the curve $\nu(T - |\tau|) = k$, where k is any nonzero integer. Finally notice that $B_G = \infty$ for the rectangular pulse.

The ambiguity function of a gaussian pulse is also easy to compute. Let

$$s(t) = e^{-\pi t^2}$$

Then

$$\chi(\tau,\nu) = \sqrt{\frac{1}{2}} e^{-\pi(\tau^2 + \nu^2)/2}$$

Thus the ambiguity function of a gaussian pulse is a two-dimensional gaussian function.

Another important pulse is one known as a chirp pulse. Any pulse whose instantaneous frequency varies linearly across the duration of the pulse is called a linear FM pulse. If the amplitude of the pulse is constant during its duration, then the linear FM pulse is also called a linear chirp pulse, or more simply, a *chirp pulse*. The chirp pulse has strong properties and is useful for a variety of purposes.

The chirp pulse can be written in the passband form,

$$s(t) = \begin{cases} \cos(2\pi f_0 t + \pi \alpha t^2) & -T/2 \le t \le T/2 \\ 0 & \text{otherwise} \end{cases}$$

The "instantaneous frequency" of the chirp pulse is

$$f(t) = f_0 + \alpha t.$$

Usually, αT is small compared to f_0. The chirp pulse also can be written in the complex baseband form

$$s(t) = \begin{cases} e^{j\pi\alpha t^2} & -T/2 \le t \le T/2 \\ 0 & \text{otherwise} \end{cases}$$

The ambiguity function for the chirp pulse can be obtained using the quadratic phase property of ambiguity functions to be proved in the next section as Property 6. This property says that if

$$s(t) \to \chi(\tau,\nu)$$

then

$$s(t)e^{j\pi\alpha t^2} \to \chi(\tau, \nu - \alpha\tau)$$

Therefore, for the chirp pulse

$$\chi(\tau,\nu) = \begin{cases} (T - |\tau|) \operatorname{sinc}\left[(\nu - \alpha\tau)(T - |\tau|)\right] & |\tau| \le T \\ 0 & |\tau| > T \end{cases}$$

The correlation function of the chirp pulse is now easily written down by setting $\nu = 0$.

$$\begin{aligned} \phi(\tau) &= \chi(\tau, 0) \\ &= (T - |\tau|) \operatorname{sinc}\left[\alpha\tau(T - |\tau|)\right] \quad |\tau| \le T \end{aligned}$$

Sometimes it is necessary to study the similarity of two pulses $s_1(t)$ and $s_2(t)$ after a time delay and frequency shift of one of them. For this purpose, the cross-ambiguity function is defined.

DEFINITION. Let $s_1(t)$ and $s_2(t)$ be finite energy waveforms. The *cross-ambiguity function* of $s_1(t)$ with $s_2(t)$ is

$$\chi_{12}(\tau,\nu) = \int\limits_{-\infty}^{\infty} s_1(t+\tau/2)s_2^*(t-\tau/2)e^{-j2\pi\nu t}\,dt$$

Many properties of the ambiguity function can be carried over to the cross-ambiguity function in an obvious way. Only slight changes are needed to restate and reprove these properties. It is easy to verify that the cross-ambiguity function can be written as

$$\chi_{12}(\tau,\nu) = \int\limits_{-\infty}^{\infty} S_1(f+\nu/2)S_2^*(f-\nu/2)e^{j2\pi f\tau}\,df$$

The cross-ambiguity function can also be defined in the asymmetrical form

$$\chi_{12}(\tau,\nu) = \int\limits_{-\infty}^{\infty} s_1(t)s_2^*(t-\tau)e^{-j2\pi\nu t}\,dt$$

The asymmetrical form of the cross-ambiguity function has the same mathematical form as the modification of the Fourier transform

$$S_\tau(f) = \int\limits_{-\infty}^{\infty} s(t)g(t-\tau)e^{-j2\pi ft}\,dt$$

which goes by the name of the *short-time* Fourier transform. Here the function $g(t)$ is thought of as a weighting "window" on the pulse $s(t)$.

Often, of the two functions entering the cross-ambiguity function, one is known a priori and the second is a measured signal and is thought of as a realization of a random process. Then the cross-ambiguity function is a two-dimensional random process.

$$\chi_c(\tau,\nu) = \int\limits_{-\infty}^{\infty} s(t)v^*(t-\tau)e^{-j2\pi\nu t}\,dt$$

where $v(t)$ denotes a random process. When $v(t)$ is thought of as a sample function or realization of the random process, $\chi_c(\tau,\nu)$ is called a *sample cross-ambiguity function* . This terminology reminds us that $\chi_c(\tau,\nu)$ is then a single realization from an ensemble of ambiguity functions sharing some common probabilistic interpretation.

For example, we often have the situation where

$$v(t) = s(t - \tau_o)e^{-j2\pi\nu_o t} + n(t)$$

where τ_o and ν_o are constants and $n(t)$ is additive gaussian noise. Because $v(t)$ is a random process, $\chi_c(\tau, \nu)$ is a random two-dimensional process.

Properties of the Ambiguity Function

The ambiguity function has a tidy set of interlocking properties. It is unusual for an arbitrary two-dimensional function to satisfy these properties, and so, in a set of possible two-dimensional functions, ambiguity functions are very rare. Usually in radar design problems, one has a pretty good idea of the desired ambiguity surface, and wishes to work backwards to find a corresponding waveform. This, however, is an unsolved problem: no techniques are known for finding a waveform corresponding to a desired ambiguity surface, nor are a set of rules known for ensuring that a desired ambiguity surface is in fact an ambiguity surface; a waveform $s(t)$ that gives rise to this desired ambiguity surface might not exist.

We develop the main properties of the ambiguity function. Some of these are close analogues of properties of the Fourier transform. Proofs that follow easily from the definition are omitted.

Property 1: Let $s'(t) = s(t - \Delta)$. Then $\chi_{s'}(\tau, \nu) = e^{-j2\pi\nu\Delta}\chi_s(\tau, \nu)$.

Property 2: Let $s'(t) = s(t)e^{j2\pi f t}$. Then $\chi_{s'}(\tau, \nu) = e^{-j2\pi f\tau}\chi_s(\tau, \nu)$.

Property 3: (Symmetry) $\chi(\tau, \nu) = \chi^*(-\tau, -\nu)$

Property 4: (Maximum) The largest value of the ambiguity function always is at the origin.

$$|\chi(\tau, \nu)| \leq |\chi(0, 0)| = E_p$$

with strict inequality if $(\tau, \nu) \neq (0, 0)$.

Proof: A straightforward application of the Schwarz inequality gives

$$|\chi(\tau, \nu)|^2 = \left| \int\limits_{-\infty}^{\infty} [s(t + \tau/2)e^{-j\pi\nu t}][s(t - \tau/2)e^{j\pi\nu t}]^* dt \right|^2$$

$$\leq \int\limits_{-\infty}^{\infty} |s(t + \tau/2)e^{-j\pi\nu t}|^2 dt \int\limits_{-\infty}^{\infty} |s(t - \tau/2)e^{j\pi\nu t}|^2 dt$$

$$= \int\limits_{-\infty}^{\infty} |s(t)|^2 dt \int\limits_{-\infty}^{\infty} |s(t)|^2 dt$$

Property 5: (Scaling) Let $s'(t) = s(at)$. Then

$$\chi_{s'}(\tau, \nu) = \frac{1}{|a|}\chi_s(a\tau, \nu/a).$$

Property 6: (Quadratic Phase Property) Let $s'(t) = s(t)e^{j\pi\alpha t^2}$.

Then

$$\chi_{s'}(\tau, \nu) = \chi_s(\tau, \nu - \alpha\tau)$$

Proof:

$$\chi_{s'}(\tau, \nu) = \int_{-\infty}^{\infty} s'(t + \tau/2)s'^*(t - \tau/2)e^{-j2\pi\nu t}\,dt$$

$$= \int_{-\infty}^{\infty} s(t + \tau/2)s^*(t - \tau/2)e^{j2\pi\alpha\tau t}e^{-j2\pi\nu t}\,dt$$

$$= \chi_s(\tau, \nu - \alpha\tau)$$

Property 7: (Volume Property) $\int_{-\infty}^{\infty}\int_{-\infty}^{\infty} |\chi(\tau, \nu)|^2\,d\tau\,d\nu = |\chi(0,0)|^2 = E_p^2.$

Proof: Recall that $\chi(\tau, \nu)$ can be expressed in the following two ways:

$$\chi(\tau, \nu) = e^{j\pi\tau\nu}\int_{-\infty}^{\infty} s(t)s^*(t - \tau)e^{-j2\pi\nu t}\,dt$$

$$\chi(\tau, \nu) = e^{j\pi\tau\nu}\int_{-\infty}^{\infty} S(f + \nu)S^*(f)e^{j2\pi f\tau}\,df$$

Therefore, we have the four-fold integral

$$\int_{-\infty}^{\infty}\int_{-\infty}^{\infty} |\chi(\tau, \nu)|^2\,d\tau\,d\nu = \iiiint s(t)s^*(t - \tau)S^*(f + \tau)S(f)e^{-j2\pi(\nu t + f\tau)}\,dt\,df\,d\tau\,d\nu$$

Now interchange the order of integration and identify terms with the Fourier transforms:

$$\int_{-\infty}^{\infty} s^*(t - \tau)e^{-j2\pi f\tau}\,d\tau = S^*(f)e^{-j2\pi ft}$$

and

$$\int_{-\infty}^{\infty} S^*(f + \nu)e^{-j2\pi\nu t}\,d\nu = s^*(t)e^{j2\pi ft}$$

This gives

$$\int_{-\infty}^{\infty}\int_{-\infty}^{\infty} |\chi(\tau, \nu)|^2\,d\tau\,d\nu = \int_{-\infty}^{\infty}\int_{-\infty}^{\infty} |s(t)|^2|S(f)|^2\,dt\,df = [\chi(0,0)]^2$$

as was to be proved.

The maximum property and the volume property strongly constrain the set of two-dimensional functions that can be ambiguity surfaces. The volume under the surface must equal the square of the maximum. Any attempt to push down the function in one place makes it pop up somewhere else.

Shape and Resolution Parameters

The shape near the origin of the ambiguity surface of a waveform determines the accuracy of range and frequency measurements in a radar using that waveform, and also determines the ability of the radar to resolve two closely spaced targets. In this section, we shall relate the shape of the main lobe of $\chi(\tau, \nu)$ to properties of the waveform $s(t)$.

The ambiguity function of $s(t)$ is

$$\chi(\tau, \nu) = \int_{-\infty}^{\infty} s(t)s^*(t - \tau)e^{-j2\pi\nu t}dt$$

The shape parameters of $\chi(\tau, \nu)$ describe the shape of the main lobe of $\chi(\tau, \nu)$. They are coefficients of a Taylor series expansion whenever the derivatives of $\chi(\tau, \nu)$ exist. Expand $\chi(\tau, \nu)$ in a Taylor series as follows.

$$\chi(\tau, \nu) = \chi(0,0) + \tau\frac{\partial\chi}{\partial\tau} + \nu\frac{\partial\chi}{\partial\nu} + \frac{1}{2}\tau^2\frac{\partial^2\chi}{\partial\tau^2} + \tau\nu\frac{\partial^2\chi}{\partial\tau\partial\nu} + \frac{1}{2}\nu^2\frac{\partial^2\chi}{\partial\nu^2} + \dots$$

This expansion exists whenever the partial derivatives exist. Notice however, that the ambiguity function of a rectangular pulse fails to have a first partial derivative with respect to τ.

We want to relate the partial derivatives of $\chi(\tau, \nu)$ to properties of the pulse $s(t)$. To do so, write

$$s(t - \tau) = s(t) - \tau\dot{s}(t) + \frac{1}{2}\tau^2\ddot{s}(t) + \dots$$

and

$$e^{j2\pi\nu t} = 1 + j2\pi\nu t - 2\pi^2\nu^2 t^2 + \dots$$

Then, up to terms of second order,

$$\chi(\tau, \nu) = \int_{-\infty}^{\infty} s(t)s^*(t - \tau)e^{j2\pi\nu t}dt$$

$$= \int_{-\infty}^{\infty} \left[|s(t)|^2 - \tau s(t)\dot{s}^*(t) + \frac{\tau^2}{2}s(t)\ddot{s}^*(t)\right]\left[1 + j2\pi\nu t - 2\pi^2\nu^2 t^2\right]dt$$

$$= \chi(0,0)\left[1 + j2\pi\nu\overline{t} - 2\pi^2\nu^2\overline{t^2} + j\tau 2\pi\overline{f} - 2\pi^2\tau^2\overline{f^2} - 4\pi^2\tau\nu\overline{tf}\right]$$

where the time and frequency moments are as were defined earlier.

The shape of the ambiguity surface $|\chi(\tau, \nu)|$ is closely related to the shape of $\chi(\tau, \nu)$. The shape of the squared ambiguity surface near the origin in given by

$$|\chi(\tau, \nu)|^2 = \chi(0,0)^2|1 + j2\pi\nu\overline{t} - 2\pi^2\nu^2\overline{t^2} + j2\pi\tau\overline{f} - 2\pi^2\tau\overline{f^2} - 4\pi^2\tau\nu\overline{tf}|^2.$$

Up to terms of second order, this becomes

$$|\chi(\tau, \nu)| = \chi(0,0)\left[1 - 2\pi^2\nu^2 T_G^2 - 4\pi^2\tau\nu T_G B_G \rho - 2\pi^2\tau^2 B_G^2\right]$$

The Gabor bandwidth and the Gabor timewidth are external descriptors of the waveform $s(t)$ in the sense that they can be measured from $|\chi(\tau, \nu)|$ without knowledge of the details of $s(t)$.

The reciprocal of the Gabor timewidth of the pulse measures the width of the main lobe of the ambiguity surface in the ν direction and the reciprocal of the Gabor bandwidth measures the width of the main lobe of the ambiguity surface in the τ direction. The shape of the main lobe can be described more completely by describing the intersection of the main lobe with a horizontal plane. If the plane is high enough it slices through the main lobe just below the peak and the shape of the cut describes the shape of the main lobe near the peak. Within the quadratic approximation to $\chi(\tau, \nu)$ this intersection is described by the equation of an ellipse in τ and ν

$$\tau^2 B_G^2 + 2\tau\nu T_G B_G \rho + \nu^2 T_G^2 = C$$

which is known as the *uncertainty ellipse*. The uncertainty ellipse provides a summary description of the main lobe of the ambiguity function. Because we are only interested in the shape of the uncertainty ellipse, but not its size, the constant C has no special importance.

Suppose we are given a pulse $s(t)$ whose uncertainty ellipse is given by

$$B_G^2 \tau^2 + T_G^2 \nu^2 = C$$

Then the chirped pulse $s(t)e^{j\pi\alpha t^2}$ has uncertainty ellipse

$$B_G^2 \tau^2 + T_G^2(\nu - \alpha\tau)^2 = C$$

which can be rewritten as

$$\left(B_G^2 + \alpha^2 T_G^2\right)\tau^2 - 2\alpha T_G^2 \nu\tau + T_G^2 \nu^2 = C$$

Consequently, if $s(t)$ has Gabor bandwidth B_G, then $s(t)e^{j\pi\alpha t^2}$ has Gabor bandwidth $\sqrt{B_G^2 + \alpha^2 T_G^2}$. The uncertainty ellipses for the pulses $s(t)$ and $s(t)e^{j\pi\alpha t^2}$ are shown in Figure 1.

The Gabor bandwidth, the Gabor timewidth, and the eccentricity parameter ρ describe the shape and width of the main lobe of the ambiguity function near the maximum. They could also be used as a rough measure of resolution for decomposing the sum

$$\chi(\tau, \nu) + \chi(\tau + \Delta\tau, \nu + \Delta\nu).$$

However, the width of the main lobe may not be accurately described by the curvature at the peak. Better measures of resolution are used to measure the width of the main lobe in a more appropriate way.

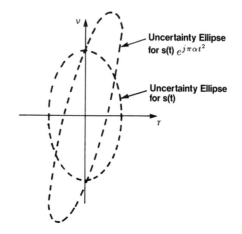

Figure 1. Some Uncertainty Ellipses.

Ambiguity Functions of Pulse Trains

Long duration signals can be generated by periodically repeating a suitable short-duration waveform called a pulse. A pulse train, then, is a waveform consisting of a finite or infinite number of nonoverlapping pulses. Usually, we will consider uniformly spaced pulse trains. Let $s(t)$ be the pulse. Then, with the first of N pulses centered at time zero, the pulse train is

$$p(t) = \sum_{l=0}^{N-1} s(t - lT_r)$$

or, with the pulse train itself centered at time zero

$$p(t) = \sum_{l=0}^{N-1} s\left(t - lT_r + \frac{1}{2}(N-1)T_r\right)$$

where T_r is the pulse repetition interval (PRI) and $s(t)$ has a width smaller than T_r. We shall relate $\chi_p(\tau, \nu)$, the ambiguity function of $p(t)$, to $\chi(\tau, \nu)$, the ambiguity function of $s(t)$.

THEOREM. Let

$$p(t) = \sum_{l=0}^{N-1} s\left(t - lT_r + \frac{1}{2}(N-1)T_r\right)$$

Then

$$\chi_p(\tau, \nu) = \sum_{n=-(N-1)}^{N-1} \chi(\tau - nT_r, \nu) \frac{\sin \pi \nu T_r(N - |n|)}{\sin \pi \nu T_r}.$$

Proof. We will work with the pulse train in the form

$$p(t) = \sum_{l=0}^{N-1} s(t - lT_r)$$

and later slide it left. By definition

$$\chi_p(\tau, \nu) = \sum_{n=0}^{N-1} \sum_{m=0}^{N-1} \int_{-\infty}^{\infty} s\left(t + \frac{\tau}{2} - nT_r\right) s^*\left(t - \frac{\tau}{2} - mT_r\right) e^{-j2\pi\nu t} dt$$

Replace t by $t + \frac{m+n}{2}T_r$

$$\chi_p(\tau, \nu) = \sum_{n=0}^{N-1} \sum_{m=0}^{N-1} e^{-j2\pi\frac{m+n}{2}\nu T_r} \int_{-\infty}^{\infty} s\left(t + \frac{\tau}{2} + \frac{m-n}{2}T_r\right) s^*\left(t - \frac{\tau}{2} - \frac{m-n}{2}T_r\right) e^{-j2\pi\nu t} dt$$

$$= \sum_{n=0}^{N-1} \sum_{m=0}^{N-1} e^{-j2\pi\frac{m+n}{2}\nu T_r} \chi(\tau + (m-n)T_r, \nu)$$

The sum over the N by N grid is indexed along rows and columns by n and m. We can instead sum the N^2 elements of any N by N array by summing first down the main diagonal, then summing all subdiagonals above the main diagonal and all subdiagonals below the main diagonal. This leads to the general identity.

$$\sum_{n=0}^{N-1} \sum_{m=0}^{N-1} A_{nm} = \sum_{k=0}^{N-1} A_{kk} + \sum_{n=1}^{N-1} \sum_{k=0}^{N-1-n} A_{k(k+n)} + \sum_{n=1}^{N-1} \sum_{k=0}^{N-1-n} A_{(k+n)k}$$

Then,

$$\chi_p(\tau, \nu) = \sum_{n=0}^{N-1} \sum_{k=0}^{N-1-n} e^{-j2\pi\frac{2k+n}{2}\nu T_r} \chi(\tau + nT_r, \nu) + \sum_{n=1}^{N-1} \sum_{k=0}^{N-1-n} e^{-j2\pi\frac{2k+n}{2}\nu T_r} \chi(\tau - nT_r, \nu)$$

Now use the relationship

$$\sum_{k=0}^{N-1} e^{j2Ak} = e^{j(N-1)A} \frac{\sin NA}{\sin A}$$

to obtain

$$\chi_p(\tau, \nu) = e^{-j\pi\nu T_r(N-1)} \left[\sum_{n=0}^{N-1} \chi(\tau + nT_r, \nu) \frac{\sin[\pi\nu T_r(N-n)]}{\sin \pi\nu T_r} \right.$$

$$\left. + \sum_{n=1}^{N-1} \chi(\tau - nT_r, \nu) \frac{\sin[\pi\nu T_r(N-n)]}{\sin \pi\nu T_r} \right]$$

The two sums can be combined by changing the sign of n in the first term and replacing n by $|n|$ in the argument of the sine. Then

$$\chi_p(\tau, \nu) = e^{-j\pi\nu T_r(N-1)} \sum_{n=-(N-1)}^{N-1} \chi(\tau - nT_r, \nu) \frac{\sin[\pi\nu T_r(N - |n|)]}{\sin \pi\nu T_r}$$

The phase term drops out when the pulse train is centered by redefining the time origin, and so the proof is complete. □

The summands in the theorem are a product of two terms. For each n, the first term is the ambiguity function of a single pulse delayed by n PRI multiples. The second term is a dirichlet function due to the pulse train. For fixed n, the dirichlet function has grating lobes at those values of ν that are integer multiples of $1/T_r$.

The derivation for the ambiguity function of a pulse train does not presume any simple structure for the pulse $s(t)$. The pulse may itself be of complex structure. For example, $s(t)$ may itself be a train of pulses with its own PRI. The formula for the pulse train ambiguity function may then be embedded into itself to describe the pulse train of pulse trains.

The main lobe of the ambiguity function is described by the uncertainty ellipse. The structure of the ambiguity function of the pulse train $p(t)$ can be portrayed by combining this uncertainty ellipse with the grating lobes of the dirichlet function for each value of the delay offset nT_r. One then obtains the depiction of $\chi_p(\tau, \nu)$ as is shown in Figure 2.

The uncertainty ellipse of $s(t)$ repeats in $\chi_p(\tau, \nu)$ whenever τ is a multiple of T_r, and so T_r is called the range (or delay) ambiguity. Similarly, the dirichlet function

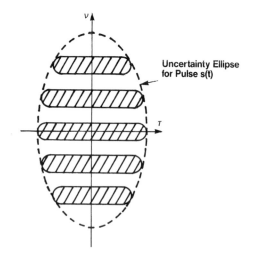

Figure 2. Illustrating the Formation of Doppler Grating Lobes
- Simple Pulse Shape.

has grating lobes in the ν direction that repeat for ν a multiple of $1/T_r$. The ν separation is called the doppler ambiguity. Generally one wishes that the range ambiguity and the doppler ambiguity are both large, but a compromise is always necessary with a uniform pulse train.

Next, consider the uncertainty ellipse of $s(t)e^{j\pi\alpha t^2}$ where $s(t)$ is a real pulse. The uncertainty ellipse of the pulse has poor resolution along the line of the major axis. The ambiguity function of the pulse train, however, is relatively narrow in both the τ and the ν directions. Thus, as illustrated in Figure 3, a pulse train of chirp pulses can be used to give good resolution in both τ and ν even though the pulse width is very wide in comparison to the width of the ambiguity function in the ν direction.

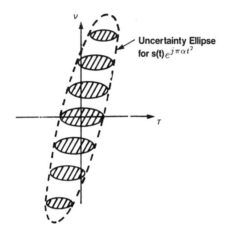

Figure 3. Illustrating the Formation of Doppler Grating Lobes
- Chirped Pulse.

Whenever ambiguities in the τ and the ν directions are undesirable, one must design a waveform with the periodicity suppressed. There are many ways to do this while still maintaining the basic structure of a pulse train. One can use an irregular spacing of the pulses.

$$p(t) = \sum_{l=0}^{N-1} s\left(t - \sum_{i=0}^{l} T_i\right)$$

where T_i is the ith pulse spacing, or one can use an irregular pattern of phase shifts of the pulses

$$p(t) = \sum_{l=0}^{N-1} s(t - lT_r)e^{-j\theta_l}$$

where θ_l is the phase angle of the lth pulse, or one can use an irregular pattern of frequency shifts of the pulses

$$p(t) = \sum_{l=0}^{N-1} s(t - lT_r)e^{-j2\pi\Omega_l t}.$$

where Ω_l is the frequency of the lth pulse.

One can develop these waveforms in a variety of ways in hopes of getting a satisfactory ambiguity surface. We shall study only one such method in the remainder of this section, the method of using an irregular pattern of frequencies.

Let

$$p(t) = \sum_{l=0}^{N-1} s(t - lT_r)e^{-j2\pi\Omega_l t}$$

where

$$\Omega_l = \frac{\theta_l}{T}$$

and the θ_l are a permutation of the integers

$$\{\theta_0, \theta_1, \ldots, \theta_{N-1}\} = \{1, 2, 3, \ldots, N\}$$

By definition

$$\chi_p(\tau, \nu) = \sum_{n=0}^{N-1} \sum_{m=0}^{N-1} \int_{-\infty}^{\infty} s\left(t + \frac{\tau}{2} - nT_r\right)e^{-j2\pi\Omega_n(t+\tau/2)} s^*\left(t - \frac{\tau}{2} - mT_r\right)e^{j2\pi\Omega_m(t-\tau/2)}e^{-j2\pi\nu t}\,dt$$

Replace t by $t + \frac{m+n}{2}T_r$ and identify $\chi(\tau, \nu)$ as the ambiguity function of the single pulse $s(t)$ to write

$$\chi_p(\tau, \nu) = \sum_{n=0}^{N-1} \sum_{m=0}^{N-1} e^{-j2\pi\frac{m+n}{2}\nu T_r} e^{-j2\pi(\Omega_n+\Omega_m)\tau/2} \chi(\tau + (m-n)T_r, \nu + \Omega_n - \Omega_m)$$

There are two kinds of terms in the double sum. Those with $m = n$ and those with $m \neq n$. Thus,

$$\chi_p(\tau, \nu) = \sum_{n=0}^{N-1} e^{-j2\pi n\nu T_r} e^{-j2\pi\Omega_n \tau}\chi(\tau, \nu)$$

$$+ \sum_{n=0}^{N-1} \sum_{m \neq n} e^{-j2\pi\frac{m+n}{2}\nu T_r} e^{-j2\pi(\Omega_n+\Omega_m)\tau/2}\chi(\tau + (m-n)T_r, \nu + \Omega_n - \Omega_m)$$

Our strategy will be to treat the first term as a desired term and the second term as an undesired "self-noise" term which will be made small by the choice of permutation.

We abbreviate this as

$$\chi_p(\tau, \nu) = \chi_p^{(1)}(\tau, \nu) + \chi_p^{(2)}(\tau, \nu)$$

We shall see that the first term will produce the main lobe of the ambiguity function, the second term will produce the sidelobes. Our goal is to make the self-noise term small by choice of the "firing sequence" $\{\theta_0, \theta_1, \ldots, \theta_{N-1}\}$.

Although we cannot evaluate the sum in the first term analytically for arbitrary (τ, ν), we can evaluate it along the τ and ν axes. First set $\tau = 0$ and $T_r = T$. Then

$$|\chi_p^{(1)}(0, \nu)| = |NT \text{ sinc } \nu NT|$$

which is the same as a pulse of duration NT. Now compute the function along the τ axis. Set $\nu = 0$. Then aside from the phase term, which can be removed by redefining the time origin,

$$\chi_p^{(1)}(\tau, 0) = N(T - |\tau|)\frac{\sin N\pi\tau/T}{\sin \pi\tau/T}$$

Thus in the ν direction, $\chi_p^{(1)}(\tau, \nu)$ behaves as if there were a single pulse of duration NT while roughly by the width in the τ direction, it behaves as if there were a single pulse of duration T/N.

Now we turn to the "self-noise" term $\chi_p^{(2)}(\tau, \nu)$. With the phase terms abbreviated as ϕ_{nm}, we can write this as

$$\chi_p^{(2)}(\tau, \nu) = \sum_{n=0}^{N-1} \sum_{m \neq n} e^{-j\phi_{nm}} \chi(\tau + (m - n)T, \nu + \Omega_m - \Omega_n)$$

and

$$\left|\chi_p^{(2)}(\tau, \nu)\right| \leq \sum_{n=0}^{N-1} \sum_{m \neq n} \left|\chi\left(\tau + (m - n)T, \nu + (\theta_m - \theta_n)T^{-1}\right)\right|$$

The term in the summation with indices m, n has a peak at $\tau = (n - m)T$ and $\nu = (\theta_n - \theta_m)/T$. To keep the self-noise small everywhere, we want to choose the "firing sequence" $\theta_0, \theta_1, \ldots, \theta_{N-1}$ so that each summand takes its peak at a different place in the (τ, ν) plane. This means that for a given pair of integers (r, s), of all pairs (m, n) such that $n - m = s$, at most one such pair can satisfy $\Omega_n - \Omega_m = r$. This requirement motivates the following definition.

DEFINITION. A *Costas array* $\mathbf{A}$ is an N-by-N array of 0's and 1's such that

$$\sum_{i=0}^{N-1} A_{ij} = \sum_{j=0}^{N-1} A_{ij} = 1$$

$$\sum_{i=0}^{N-1} \sum_{j=0}^{N-1} A_{ij} A_{i+r,j+s} \leq 1 \qquad \begin{array}{l} \text{if } r \neq 0 \\ s \neq 0 \end{array}$$

(with the understanding that $A_{i+r,j+s} = 0$ if either $i + r$ or $j + s$ is not in the set of indices $\{0, \ldots, N - 1\}$.)

An example of a Costas array is the 4-by-4 array

$$\mathbf{A} = \begin{bmatrix} 0 & 0 & 0 & 1 \\ 1 & 0 & 0 & 0 \\ 0 & 0 & 1 & 0 \\ 0 & 1 & 0 & 0 \end{bmatrix}$$

This array has firing sequence $2, 4, 3, 1$. Figure 4 shows how this Costas array is used to form a frequency-hopping pulse with four "chips." One can form much larger Costas pulses, say with hundreds of chips, that have very sharp ambiguity functions, but this requires the construction of large Costas arrays.

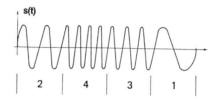

Figure 4. Waveform Based on n=4 Costas Array.

General rules for constructing or classifying Costas arrays are not known, but several constructions are known for special cases. A simple construction is available when $N + 1$ is a prime. Let p be a prime and $N = p - 1$, and let π be a primitive element of the integer arithmetic system modulo p. An N-by-N *Welch-Costas array* is an N-by-N array of 0's and 1's such that $A_{ij} = 1$ if and only if $j = \pi^i$. For example, with $p = 7$ and the primitive element $\pi = 3$, we have the firing sequence $3, 2, 6, 4, 5, 1$.

Computing the Cross-Ambiguity Function

Computation of the sample cross-ambiguity function

$$\chi_c(\tau, \nu) = \int_{-\infty}^{\infty} s(t) v^*(t + \tau) e^{-j2\pi\nu t} dt$$

from the received signal $v(t)$ and the transmitted signal $s(t)$ is a central task of radar and sonar signal processing. Depending on the circumstances of the application, this might be a massive computation or might be a trivial computation, sometimes employing very gross approximations.

The processing might be performed by analog circuits, either baseband or passband, by optical processors, or by digital processors. To keep the computations manageable one might adopt either of two strategies. One strategy is to develop fast computational algorithms that compute $\chi_c(\tau, \nu)$ for an arbitrary $s(t)$. A second strategy is to choose $s(t)$ cleverly to make the computation easier.

It is only a slight exaggeration to say that the main concern of the designers of the early or simple radars is to choose $s(t)$ so as to make the computation trivial. We may choose $s(t)$ to be a short pulse of duration T such that $\nu T \ll 1$ for all ν of interest. Then $\chi_c(\tau, \nu)$ does not depend on ν for the range of interest and it is enough to do the computations for $\nu = 0$. Then the computation is

$$\chi_c(\tau, 0) = \int\limits_{-T/2}^{T/2} s(t)v^*(t + \tau)dt$$

which can be computed by passing $v(t)$ through a filter with impulse response $s^*(t)$. A radar with such a waveform is called a *pulse radar*.

Alternatively, we may choose $s(t)$ to be a long pulse of narrow bandwidth B so that $B\tau \ll 1$ for all τ of interest. Then $\chi_c(\tau, \nu)$ does not depend on τ over the range of interest and it is enough to do the computation for $\tau = 0$. Then, the computation is

$$\chi_c(0, \nu) = \int\limits_{-\infty}^{\infty} s(t)v^*(t)e^{-j2\pi\nu t}dt$$

Moreover, if $s(t)$ is a rectangular pulse

$$\chi_c(0, \nu) = \int\limits_{-T/2}^{T/2} v^*(t)e^{-j2\pi\nu t}dt$$

This kind of radar is called a *doppler radar*. In this case, $\chi_c(0, \nu)$ can be computed on a discrete grid by passing $v(t)$ through a bank of passband filters as can be described by a Fourier transform

$$\chi_c(0, \nu_k) = \int\limits_{-T/2}^{T/2} v^*(t)e^{-j2\pi\nu_k t}dt$$

for some discrete set of ν_k.

A waveform that combines these approximations to get good resolution in both τ and ν is the pulse train

$$p(t) = \sum_{l=0}^{N-1} s(t - lT_r)$$

where $s(t) = 0$ for $|t| \geq T/2$. Then

$$\chi_c(\tau, \nu) = \int\limits_{-\infty}^{\infty} \sum_{l=0}^{N-1} s(t - lT_r)v^*(t + \tau)e^{-j2\pi\nu t}dt$$

$$= \sum_{l=0}^{N-1} e^{-j2\pi\nu l T_r} \int\limits_{-T/2}^{T/2} s(t)v^*(t + lT_r + \tau)e^{-j2\pi\nu t}dt$$

There are now N integrals, each from $-T/2$ to $T/2$, and if T is small in comparison to the pulse repetition interval T_r there can be a considerable computational savings because of using a pulse train. Moreover, if T is chosen so that $\nu T \ll 1$ for all ν of interest, this becomes

$$\chi_c(\tau, \nu) = \sum_{l=0}^{N-1} e^{-j2\pi\nu l T_r} \int_{-T/2}^{T/2} s(t) v^*(t + lT_r + \tau) dt$$

What this amounts to is that each returned pulse is filtered to provide an output, say $u_l(\tau)$. Then the sample cross-ambiguity function takes the form, for each value of τ, of a one-dimensional Fourier transform

$$\chi_c(\tau, \nu) = \sum_{l=0}^{N-1} u_l(\tau) e^{-j2\pi\nu l T_r}$$

A popular pulse train is a train of chirped pulses. One reason for this popularity is that the computation can be made to take on the form of a two-dimensional Fourier transform. Replace $s(t)$ by $e^{j\pi\alpha t^2}$ and write

$$\chi_c(\tau, \nu) = \sum_{l=0}^{N-1} e^{-j2\pi\nu l T_r} \int_{-T/2}^{T/2} e^{j\pi\alpha t^2} v^*(t + lT_r + \tau) dt$$

$$= \sum_{l=0}^{N-1} e^{-j2\pi\nu l T_r} u_l(\tau)$$

Now use the identity

$$e^{j\pi\alpha t^2} = e^{j\pi\alpha(t+\beta)^2} e^{-j2\pi\alpha t\beta} e^{-j\pi\alpha\beta^2}$$

to write

$$u_l(\tau) = e^{-j\pi\alpha\beta^2} \int_{-T/2}^{T/2} e^{-j2\pi\alpha t\beta} \left[e^{j\pi\alpha(t+\beta)^2} v^*(t + lT_r + \tau) \right] dt$$

Next let $t' = t + \beta$ and suppress the exponential in front as uninteresting.

$$u_l(\tau) = \int_{-T/2+\beta}^{T/2+\beta} e^{-j2\pi\alpha t'\beta} \left[e^{j\pi\alpha t'^2} v^*(t' + lT_r + \tau - \beta) \right] dt'$$

The next step is to choose $\beta = \tau - \bar{\tau}$ where $\bar{\tau}$ is a convenient reference value in the range of τ, such as $\bar{\tau} = (\tau_{\max} + \tau_{\min})/2$. Define the lth dechirped pulse as

$$v_l(t) = e^{-j\pi\alpha t^2} v(t + lT_r + \bar{\tau})$$

Then

$$u_l(\tau) = \int_{-T/2+\beta}^{T/2+\beta} e^{-j2\pi\alpha t(\tau-\bar{\tau})} v_l(t) dt$$

$$\approx \int_{-T/2}^{T/2} e^{-j2\pi\alpha t(\tau-\bar{\tau})} v_l(t) dt$$

The approximation assumes that $\tau - \bar{\tau}$ is small compared to $T/2$. It means that part of the signal returned from reflectors near the edge of the scene will be wasted, so the image will be attenuated near the edge of the scene in proportion to $2(\tau - \bar{\tau})/T$.

Finally,

$$\chi(\tau, \nu) = \sum_{l=0}^{N-1} e^{-j2\pi\nu l T_r} \int_{-T/2}^{T/2} e^{-j2\pi\alpha t(\tau-\bar{\tau})} v_l(t) dt$$

This computation is now in the form of a two-dimensional Fourier transform of the dechirped signal. One axis is a discrete Fourier transform, which could be approximated by an integral, and one axis is a continuous Fourier transform, which could be approximated by a discrete sum. Depending on the processing technology to be used, the computation may be put either in the form of a continuous two-dimensional Fourier transform or a discrete two-dimensional Fourier transform.

Notes

The ambiguity function was first introduced to the study of radar waveforms by Woodward (1953) having been defined earlier by Ville (1948). The basic theorems were developed by Siebert (1956), and by Lerner (1958), Wilcox (1960), Sussman (1962), and by Price and Hofstetter (1965). The inverse problem of finding a pulse s(t) that corresponds to a given ambiguity surface is very difficult. Algebraic properties of the mapping that maps a function s(t) into its ambiguity function were studied by Auslander and Tolimieri (1985).

The uncertainty principle of waveforms is discussed by Gabor (1946) and by Kay and Silverman (1957). Additional properties of the ambiguity function are described by Helstrom (1960), Rihaczek (1965), and Grünbaum (1984). Costas (1984) introduced the Costas array as one way of designing a frequency hopping pulse with a good ambiguity function. Efficient digital computation of the ambiguity function of an arbitrary waveform has been studied by Tolimieri and Winograd (1985).

IV. ANTENNA SYSTEMS

An antenna is a linear device that forms the interface between free propagation of electromagnetic signals and guided propagation of electromagnetic signals. An antenna can be used either to transmit an electromagnetic signal or to receive an electromagnetic signal. During transmission, the function of the antenna is to concentrate the signal into a beam that points in the desired spatial direction.

During reception, the function of the antenna is to collect the incident signal and deliver it to the receiver. An important theorem of antenna theory, known as the *reciprocity theorem*, allows us to study the antenna either as a transmitting device or as a receiving device depending on which is more convenient for a particular discussion.

Antenna Aperture and Pattern

A physical antenna is the interface between an electronic system and free space. A planar antenna can be described as a device for setting up an *illumination function* $i(x, y)$, possibly complex, on a region of space called the aperture; the integration of signal radiated by this aperture into various directions then constitutes the radiation pattern. The province of the antenna designer is to configure a set of conducting objects and their electrical feeds so as to generate the illumination function $i(x, y)$ within the aperture. We will not study the design of antennas; we study only the relationship between the illumination function $i(x, y)$ and the radiation pattern.

It is not important for the mathematical analysis to specify a particular kind of radiation, only that the radiation be waves of frequency f_0 and velocity c. At the point (x, y) of the aperture is launched a spherical wave of amplitude $i(x, y)e^{j2\pi f_0 t} dx dy$. The magnitude and phase of $i(x, y)$ describe the magnitude and phase of the infinitesimal wave launched from this point (x, y). One may wish to visualize the aperture as a screen with a hole in it that is illuminated from behind by a plane wave of frequency f_0. By visualizing the hole filled with a semitransparent medium with attenuation and phase described by $i(x, y)$, we have a visualization of the aperture.

As far as its effect on the signal in the far field is concerned, an antenna is described completely by its antenna radiation pattern. The *antenna radiation pattern* is a function $A(\phi, \psi)$ of the spherical coordinates (ϕ, ψ). (The squared magnitude of $A(\phi, \psi)$ is also called the antenna radiation pattern.) The function $A(\phi, \psi)$ is a complex function which gives the magnitude and phase of the signal radiated in direction ϕ, ψ. If $A(\phi, \psi)$ is a constant, the antenna is called *omnidirectional*.

A one-dimensional antenna radiation pattern can be related to a one-dimensional illumination function by integrating all signals propagating in a given direction; a signal arriving from an infinitesimal element of the aperture is weighted by the illumination function and delayed by an amount depending on the path difference. The differential delay τ of an infinitesimal element at x with respect to the delay from the infinitesimal element at the origin is obtained by reference to Figure 5 as

$$\tau = \frac{x}{c} \sin \phi$$

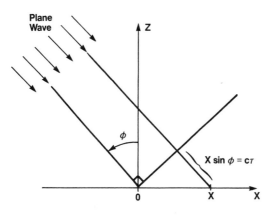

Figure 5. Incident Wave in Two Dimensions.

Therefore, the waveform $c(t)$ at a carrier frequency of f_0 gives rise to a spatially distributed signal in the aperture given by $c(t)e^{j2\pi f_0 t}i(x)$. This leads to a transmitted signal at angle ϕ given by

$$s(t, \phi) = \int_{-\infty}^{\infty} i(x)c\left(t - \frac{x}{c}\sin\phi\right)e^{j2\pi f_0\left(t - \frac{x}{c}\sin\phi\right)}dx$$

The narrowband assumption, that the bandwidth of $c(t)$ is very small compared to f_0, allows the approximation that $c\left(t - \frac{x}{c}\sin\phi\right)$ is equal to $c(t)$. Then

$$s(t, \phi) = c(t)e^{j2\pi f_0 t}\int_{-\infty}^{\infty} i(x)e^{-j2\pi\sin\phi\frac{x}{\lambda}}dx$$

It is conventional to express this in terms of a normalized illumination function so that the physical dimensions scale with wavelength. Let $u = x/\lambda$ be the normalized coordinate. Let $p(u) = i(\lambda u)$. Then

$$s(t, \phi) = c(t)e^{j2\pi f_0 t}\int_{-\infty}^{\infty} \lambda p(u)e^{-j2\pi\sin\phi u}du$$

$$= c(t)e^{j2\pi f_0 t}\lambda P(\sin\phi)$$

where $P(f_x)$ is the Fourier transform of $p(x)$. The function $A(\phi)$ defined as

$$A(\phi) = \lambda P(\sin\phi)$$

is called the *antenna radiation pattern*.

The transform of a rectangle is a sinc pulse, so that if $i(x) = \text{rect}\,(x/L)$ the antenna pattern is

$$A(\phi) = L \, \text{sinc} \left(\frac{L}{\lambda} \sin \phi \right)$$

If $\frac{L}{\lambda}$ is large, we can make the approximation

$$A(\phi) \approx L \, \text{sinc} \left(\frac{L}{\lambda} \phi \right)$$

A two-dimensional antenna radiation pattern is related to a two-dimensional illumination function in the same way that a one-dimensional antenna radiation pattern is related to a one-dimensional illumination function. The delay in the signal received at point (x, y) is $\frac{1}{c}(x \cos \psi + y \sin \psi) \sin \phi$. To see this more clearly, temporarily rotate the coordinate system by angle ψ about the z axis into a new (x', y', z) coordinate system such that the far-field point is in the (x', z) plane. The delay then is $\frac{1}{c}(x' \sin \phi)$. Replacing x' by $x \cos \psi + y \sin \psi$ gives the delay. Consequently, as before

$$s(t, \phi, \psi) = c(t)e^{j2\pi f_0 t} \int\limits_{-\infty}^{\infty} \int\limits_{-\infty}^{\infty} i(x, y)e^{-j2\pi\left(\sin \phi \cos \psi \frac{x}{\lambda} + \sin \phi \sin \psi \frac{y}{\lambda}\right)} dx\,dy$$

$$= c(t)e^{j2\pi f_0 t}\lambda P(\sin \phi \cos \psi, \sin \phi \sin \psi)$$

where $P(f_x, f_y)$ is the two-dimensional Fourier transform of $p(u, v)$ and $p(u, v) = i(\lambda u, \lambda v)$ is the normalized illumination function. Thus,

$$A(\phi, \psi) = P(\sin \phi \cos \psi, \sin \phi \sin \psi)$$

Figure 6 shows a two-dimensional antenna pattern for an illumination function that is a uniformly illuminated circle of radius r,

$$i(x, y) = \text{circ} \left(\frac{x}{r}, \frac{y}{r} \right)$$

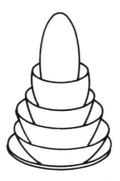

Figure 6. A Circularly Symmetric Antenna Pattern.

Then

$$A(\phi, \psi) = 2\pi \text{ jinc} \left(2\pi \frac{r}{\lambda} \sin \phi\right)$$

In this case, the sidelobes shown in Figure 6 are the sidelobes of the jinc function as a function of $\sin \phi$. Because $\sin \phi \leq 1$, there are only a finite number of sidelobes depending on the ratio of r/λ. If r is such that

$$2\pi \frac{r}{\lambda} \leq 1.22$$

then the first zero of the jinc function will not be seen for any value of ϕ.

DEFINITION. The *gain* of an antenna in direction ϕ, ψ is defined as

$$G(\phi, \psi) = \frac{|A(\phi, \psi)|^2}{\frac{1}{4\pi} \iint |i(x, y)|^2} dx dy$$

The *peak-of-beam gain* is defined as $G = \max_{\phi, \psi} G(\phi, \psi)$.

THEOREM. *The antenna gain G of an aperture of area A satisfies*

$$G \leq 4\pi \frac{A}{\lambda^2}$$

with equality if and only if the aperture is uniformly illuminated in magnitude and phase is linear.

Proof. The proof uses the Schwarz inequality. The function $i(x, y)$ is nonzero only for x, y in the region A. Hence

$$|\lambda A(\phi, \psi)|^2 = \iint_{A} i(x, y) e^{-j2\pi(x \sin \phi \cos \psi + y \sin \phi \sin \psi)} dx dy$$

$$\leq \iint_{A} |i(x, y)|^2 dx dy \iint_{A} dx dy$$

from which the inequality of the theorem follows.

The *effective area* A_e of an antenna aperture is that value satisfying

$$G = 4\pi \frac{A_e}{\lambda^2}$$

The effective area is the area of a uniformly illuminated aperture that has the same gain. The effective area is never larger than the actual area.

Antenna Arrays

Large antennas can be formed by combining the elements of an array of simpler antennas just as a pulse train can be formed by combining many copies of a single pulse. In this context, the simple antenna is called an antenna element, and the large antenna is called an antenna array. Ideally, each element radiates with a pattern just as when it is alone, and the total radiation pattern is the sum of the individual patterns.

An antenna array may be one-dimensional or two-dimensional. For ease of development, we shall study a uniformly spaced one-dimensional array. Denote the radiation pattern of the element by $A_e(\phi)$, and that of the array by $A(\phi)$. The array radiation pattern is

$$A(\phi) = A_e(\phi) \sum_{k=0}^{K-1} e^{j[2\pi(d/\lambda)\sin\phi]k}$$

The sum has been encountered previously in connection with finding the Fourier transform of a pulse train. The sum is

$$A(\phi) = A_e(\phi)e^{j(K-1)\pi(d/\lambda)\sin\phi} \frac{\sin[K\pi(d/\lambda)\sin\phi]}{\sin[\pi(d/\lambda)\sin\phi]}$$

and

$$|A(\phi)|^2 = |A_e(\phi)|^2 \frac{\sin^2[K\pi(d/\lambda)\sin\phi]}{\sin^2[\pi(d/\lambda)\sin\phi]}$$

The last term is the square of a dirichlet function in $\sin\phi$ that determines the main beam of the array. This beam is broadside to the array and has a first null at

$$K(d/\lambda)\sin\phi = 1$$

or

$$\phi = \sin^{-1}[\lambda/(Kd)]$$

The dirichlet function has its first grating lobe at

$$(d/\lambda)\sin\phi = 1$$

or

$$\phi = \sin^{-1}\left(\frac{\lambda}{d}\right)$$

We can regard a two-dimensional K by N rectangular array of antennas as a one-dimensional array along the y axis of antennas each of which is an array of antenna elements along the x axis. Consequently, with the origin at the center of the array so that the main beam is real, the pattern for the array is

$$A(\phi, \psi) = A_e(\phi, \psi) \frac{\sin[N\pi(d_1/\lambda)\sin\phi\cos\psi]}{\sin[\pi(d_1/\lambda)\sin\phi\cos\psi]} \frac{\sin[K\pi(d_2/\lambda)\sin\phi\sin\psi]}{\sin[\pi(d_2/\lambda)\sin\phi\sin\psi]}$$

Thus far, we have simply summed the outputs of the antenna elements to produce the array pattern. This places the main lobe of the array pattern along the z axis. Instead, the main beam of an antenna array can be steered by an appropriate phase shift on each element signal prior to summation. In this case, the antenna array is called a *phased array*. Suppose that we wish to steer the main beam of a one-dimensional phased array to the angle ϕ_o. The steering is described by the equation

$$g(t) = \sum_{k=0}^{K-1} g_k(t) e^{-j[2\pi(d/\lambda)\sin\phi_o]k}$$

This leads to the new radiation pattern, denoted $A_{\phi_o}(\phi)$, for the steered array given by

$$A_{\phi_o}(\phi) = A_e(\phi) \sum_{k=0}^{K-1} e^{j[2\pi(d/\lambda)(\sin\phi - \sin\phi_o)]k}$$

$$= A_e(\phi) e^{j(K-1)\pi(d/\lambda)(\sin\phi - \sin\phi_o)} \frac{\sin[K\pi(d/\lambda)(\sin\phi - \sin\phi_o)]}{\sin[\pi(d/\lambda)(\sin\phi - \sin\phi_o)]}$$

and

$$|A_{\phi_o}(\phi)|^2 = |A_e(\phi)|^2 \frac{\sin^2[K\pi(d/\lambda)(\sin\phi - \sin\phi_o)]}{\sin^2[\pi(d/\lambda)(\sin\phi - \sin\phi_o)]}$$

The main lobe of the dirichlet function now occurs when $\phi = \phi_o$. The first null occurs at

$$\phi = \sin^{-1}[\lambda/(Kd) + \sin\phi_o]$$

Interferometry

There are in use a broad range of devices that depend on the relationship between the angle of arrival of a wavefront with respect to several antenna elements; the wavelength of the wavefront; and the relative phase angle of the signal at the antenna elements. These devices can be used to measure direction-of-arrival, wavelength, or frequency. They use the antenna aperture in a special way distinct from its use as a beamformer.

An antenna aperture used as a beamformer integrates into a simple time function all of the signal incident on the aperture, but weighted by an illumination function. Alternatively, the distribution of the phase of the signal across the aperture can be processed in other ways to estimate the direction of arrival. An *interferometer* samples the aperture by segmenting it into several subapertures and receives a signal in each subaperture. The several integrated signals within the several subapertures are then compared in phase to estimate some parameter of interest. The phase comparison is a nonlinear operation. Consequently, an interferometer is distinctly different from a beamforming antenna system.

The simplest interferometer breaks the aperture into two halves with centers separated by d. From a radiation source that is far from the aperture in comparison to d, the range difference is $d \sin \phi$. Hence the phase difference in the signal as received at the two antennas is

$$\Delta\theta = 2\pi \frac{d}{\lambda} \sin \phi \qquad \text{(radians)}$$

The equation for $\Delta\theta$ can be inverted to give

$$\phi = \sin^{-1} \left[\frac{\lambda}{d} \frac{\Delta\theta}{2\pi} \right]$$

from which the direction of arrival ϕ can be computed from the measured phase difference $\Delta\theta$. However, whenever d/λ is larger than one, there will be ambiguities because there will be more than one value of ϕ solving the equation. Ambiguities in an interferometric angle-of- arrival measurement are quite routine and are routinely resolved by using a second interferometer pair with a different value of the separation d.

Crystallography

Crystallography may be viewed as a mathematical extension of the theory of array antennas to three-dimensional arrays, and so we discuss it here. A crystal is a three-dimensional array of identical molecular structures. Elementary crystals may consist of an array of simple molecules, and these are the more familiar examples, but some very large molecules such as proteins can also be used to form crystals. By forming such a crystal and using it as a three-dimensional diffraction grating, one has a method of imaging the molecule. The wavelength of the incident radiation must be comparable to the crystal spacing, so X-ray frequencies are the natural frequencies to use for most crystallography. From a mathematical point of view, however, the formation of crystallographic diffraction patterns is an exercise in Fourier transform theory, in this case the theory of three-dimensional Fourier transforms.

Let $s(x, y, z)$ be a finite energy function in three variables. The function $s(x, y, z)$ need not be centered in any way at the origin. Indeed, the origin is only a point of reference though we would normally choose the origin so that $s(x, y, z)$ is concentrated near the origin.

Let the 3-by-3 matrix $\mathbf{M}$ specify a three-dimensional lattice. The lattice is the set of points given by

$$\mathbf{v} = \mathbf{Mi}$$

where $\mathbf{i}$ is any three-dimensional vector of integers. For example, a common lattice is a rectangular lattice. In this case, $\mathbf{M}$ is a diagonal matrix.

An infinitely large crystal is defined as a three-dimensional convolution of $s(x, y, z)$ with a three-dimensional array of impulses on the lattice points.

$$c(x, y, z) = s(x, y, z) * * * \sum_{i_1} \sum_{i_2} \sum_{i_3} \delta(\mathbf{x} - \mathbf{Mi})$$

$$= \sum_{i_1} \sum_{i_2} \sum_{i_3} s(\mathbf{x} - \mathbf{Mi})$$

where $s(\mathbf{x} - \mathbf{Mi})$ designates the translate of $s(x, y, z)$ to the lattice point $\mathbf{Mi}$. A finite crystal is obtained if the lattice points are restricted to a finite region of space.

Thus, a finite rectangular crystal can be written

$$c(x, y, z) = \sum_{i_1=0}^{N_1-1} \sum_{i_2=0}^{N_2-1} \sum_{i_3=0}^{N_3-1} s(x - d_1 i_1, y - d_2 i_2, z - d_3 i_3)$$

When infinitesimal volume at x, y, z is illuminated with a plane wave, it reradiates, in phase, in all directions a spherical wavelet with amplitude $c(x, y, z)dxdydz$. As an incident wave sweeps across the crystal, each infinitesimal volume reradiates some of its energy; we shall assume that the fraction reradiated is negligible, and the incident wave maintains its amplitude as it sweeps through the crystal. This assumption allows us to model the radiation from each element of the crystal as equal except for phase.

Consider a single scattering site $s(x, y, z)$. We must find the difference in phase at a point in the far field between a signal originating at the point $(0,0)$ and a signal originating at the point (x, z). Rotate the coordinate system to a new system (x', z') with

$$z' = x \sin \phi + z \cos \phi$$

and the difference in distance is z'. Hence the phase difference due to the path difference after scattering is

$$\Delta \theta = \frac{1}{c}[x \sin \phi + z \cos \phi]$$

There will also be a contribution due to the path difference in the incident wave prior to scattering. This contribution is

$$\Delta \theta = \frac{1}{c}[-x \sin \phi_o - z \cos \phi_o]$$

and the contribution to the signal in the far field is

$$dv(x, y, z) = s(x, y, z)e^{-j2\pi(f_0/c)[x \sin \phi + z \cos \phi - x \sin \phi_o - z \cos \phi_o)]}dxdydz$$

The total signal in the far field then is the superposition of all these contributions

$$v(x, y, z) = \int_{-\infty}^{\infty} \int_{-\infty}^{\infty} \int_{-\infty}^{\infty} s(x, y, z)e^{-j(2\pi/\lambda)[x(\sin \phi - \sin \phi_o) + z(\cos \phi - \cos \phi_o)]}dxdydz$$

which is the equation of a Fourier transform. More generally,

$$v(x,y,z) = S\left(\frac{\cos\psi\sin\phi - \cos\psi_0\sin\phi_0}{\lambda}, \frac{\sin\psi\sin\phi - \sin\psi_o\sin\phi_o}{\lambda}, \frac{\cos\phi - \cos\phi_o}{\lambda}\right)$$

To write this more concisely, let $\mathbf{i}, \mathbf{j}$ and $\mathbf{k}$ be unit vectors along the x, y and z axes, and let $\mathbf{s}_i, \mathbf{s}_o$ be unit vectors along the incident direction and the scattered direction respectively. Then

$$v(x,y,z) = S(f_x, f_y, f_z)$$

where

$$f_x = \frac{1}{\lambda}\mathbf{i}\cdot(\mathbf{s}_i - \mathbf{s}_o)$$
$$f_y = \frac{1}{\lambda}\mathbf{j}\cdot(\mathbf{s}_i - \mathbf{s}_o)$$
$$f_z = \frac{1}{\lambda}\mathbf{k}\cdot(\mathbf{s}_i - \mathbf{s}_o)$$

For the complete crystal,

$$v(x,y,z) = S(f_x, f_y, f_z)\,\text{dirc}_{N_1}(d_1 f_x)\,\text{dirc}_{N_2}(d_2 f_y)\,\text{dirc}_{N_3}(d_3 f_z)$$

This signal $v(x,y,z)$ will be large only when all three dirichlet functions are large. This is so when all have a grating lobe. This requires that the following three equations, known as the Bragg-Laue equations, are satisfied

$$(\mathbf{s}_i - \mathbf{s}_o)\cdot\mathbf{i} = l_x\lambda/d_1$$
$$(\mathbf{s}_i - \mathbf{s}_o)\cdot\mathbf{j} = l_y\lambda/d_2$$
$$(\mathbf{s}_i - \mathbf{s}_o)\cdot\mathbf{k} = l_z\lambda/d_3$$

where $l_x, l_y,$ and l_z are integers. The left side of each of these equations cannot be larger than two which gives a largest value for l_x, l_y and l_z at which a grating lobe occurs. At other values of x, y, z, the scattered signal $v(x,y,z)$ will be negligibly small.

The grating lobes can be observed for

$$l_x \leq 2d_1/\lambda$$
$$l_y \leq 2d_2/\lambda$$
$$l_z \leq 2d_3/\lambda$$

For this reason λ must be chosen much smaller than the crystal spacing. Measuring $v(x,y,z)$ along the grating lobes gives $S(l_x/d_1, l_y/d_2, l_z/d_3)$, the Fourier coefficients of $s(x,y,z)$. In principle, the Fourier coefficients can be measured and inverted to find $s(x,y,z)$. In practice, however, only the magnitude of the Fourier coefficients, not the phase, is measured. This means that the inversion must use some a-priori knowledge.

V. RADAR IMAGING SYSTEMS

A radar consists of a transmitter that illuminates a region of interest, a receiver that collects the signal reflected by objects in that region, and a processor that extracts information of interest from the received signal. A radar processor consists of a preprocessor, a detection and estimation function, and a postprocessor. The preprocessor is where the signal is extracted from the noise, and the entire signal reflected from the same reflector is integrated into a single statistic. The detection and estimation function is where individual target elements are recognized, and parameters associated with these target elements are estimated. The postprocessor refines postdetection data, and establishes track histories on detected targets.

A radar can be described as a device for forming a two-dimensional convolution of the reflectivity density function of an illuminated scene with the ambiguity function of the radar waveform; the two-dimensional convolution describes the output of the preprocessor. This unifying description, which we will develop, is very powerful because it submerges all of the details of the processing output into a simple two-dimensional filter in which the radar ambiguity function plays the role of a point-spread function. As long as the time-varying geometry of the situation in adaquately approximated by constant range rates, this approximation is quite adequate.

The Received Signal

Electromagnetic signals travel through space at a finite speed, in free space at the free space speed of light c (approximately 3×10^8 meters/second). Electromagnetic waves consist of propagating vector fields. Because the electric field is a vector, it has an orientation called the polarization of the signal that is orthogonal to the direction of propagation, but otherwise the orientation is arbitrary. For the most part, polarization has little bearing on the signal processing.

If a signal $s(t)$ is emitted at one point, then the signal, $v(t)$, received at a distant point at distance R is

$$v(t) = as(t - R/c)$$

The signal $v(t)$ differs from $s(t)$ by virtue of an amplitude attenuation a and a time delay R/c.

Similarly, if a signal $s(t)$ is transmitted, travels a distance R_1 to a reflector, then travels a distance R_2 to a receiver, the received signal is

$$v(t) = a\rho s(t - (R_1 + R_2)/c)$$

Now the time delay is $(R_1 + R_2)/c$ and the amplitude attenuation is $a\rho$ where the constant ρ represents the fraction of the signal incident on the reflector that is actually reflected towards the receiver, and a represents the signal attenuation from all other causes. This simple model ignores the possibility of certain bizarre artifacts (such as multiple reflections) that may occur in a scattered radar signal.

The parameter ρ is called the *reflectivity*. In general, it is a complex number because there can be a phase shift during the process of reflection. The *radar cross-section* of the reflector is defined as $\sigma = |\rho|^2$.

For a more complicated object, the scattering is usually described adequately by a linear model. The scene is broken into infinitesimal cells of area $dx\,dy$ and the reflectivity from a cell at x, y is denoted $\rho(x, y)dx\,dy$. Then $\rho(x, y)$ is called the *reflectivity density* at x, y. It is a complex function of x and y.

For our purposes, the environment is defined as the reflectivity density $\rho(x, y)$ (or perhaps $\rho(x, y, z)$). The relationship between $\rho(x, y)$ and the properties of a configuration of physical objects is left to the study of *scattering theory*. In particular we ignore the deep question of estimating more fundamental properties of the scatterer from the radar signal directly or from the estimated reflectivity density.

The *radar cross section density* or *scattering function* at x, y is a real positive function given by

$$\sigma(x, y) = |\rho(x, y)|^2$$

The integral of $\sigma(x, y)$ over an object or a portion of an object is the *radar cross section* of that object or portion of that object.

The reflectivity density and the radar cross section density depend on the angle of incidence of the incident radiation as measured by the spherical angles (ϕ, ψ).

One may also be interested in applications in which the transmitter and the receiver are located in different places. Then one is interested in the signal from direction (ϕ, ψ) that is reflected into direction (ϕ', ψ'). In this way, one can also define the bistatic (or bidirectional) reflectivity density $\rho(x, y, \phi, \psi, \phi', \psi')$ and the bistatic radar cross section density $\sigma(x, y, \phi, \psi, \phi', \psi')$.

For sufficiently smooth and perfectly conducting bodies, in the limit of small wavelength, the bistatic cross section is equal to the monostatic cross section at the bisector of the bistatic angle between the direction to the transmitter and receiver. This implies that the cross section is unchanged if the positions of the transmitter and receiver are interchanged.

The reflectivity depends on the polarization of the incident signal. One may consider the incident signal to be vertically or horizontally polarized, and the reflected signal to be vertically or horizontally polarized. Consequently, there are really four reflectivity density functions. These can be arranged in a matrix of functions

$$\boldsymbol{\rho} = \begin{bmatrix} \rho_{11}(x, y) & \rho_{12}(x, y) \\ \rho_{21}(x, y) & \rho_{22}(x, y) \end{bmatrix}$$

Similarly, for the radar cross section density

$$\boldsymbol{\sigma} = \begin{bmatrix} \sigma_{11}(x, y) & \sigma_{12}(x, y) \\ \sigma_{21}(x, y) & \sigma_{22}(x, y) \end{bmatrix}$$

This matrix is known as the *scattering matrix*. Each element of the scattering matrix may be written as a function also of ϕ, ψ, ϕ', and ψ'.

Suppose that the transmitter at time t is at spatial coordinates $(x'(t), y'(t), z'(t))$ and the receiver at time t is at location $(x(t), y(t), z(t))$. Then

$$R(t) = \sqrt{(x'(t) - x(t))^2 + (y'(t) - y(t))^2 + (z'(t) - z(t))^2}$$

The received signal is the retarded signal

$$g(t) = s(t - R(t)/c)$$

A radar system makes use of echoes, so we need the range between transmitter and reflector and the range between reflector and receiver. In a bistatic radar system the transmitter, receiver, and reflector are all at different points. Suppose that the transmitter is at location $(x'(t), y'(t), z'(t))$, the echo is from an object at location $(x(t), y(t), z(t))$, and the receiver is at location $(x''(t), y''(t), z''(t))$. Then we have two range expressions

$$R_1(t) = \sqrt{(x'(t) - x(t))^2 + (y'(t) - y(t))^2 + (z'(t) - z(t))^2}$$
$$R_2(t) = \sqrt{(x''(t) - x(t))^2 + (y''(t) - y(t))^2 + (z''(t) - z(t))^2}$$

and

$$g(t) = s\left(t - \frac{R_1(t)}{c} - \frac{R_2(t)}{c}\right)$$

In a monostatic radar system, the transmitter and receiver are at the same place and

$$g(t) = s(t - 2R(t)/c)$$

The various cases are subsumed by one equation

$$g(t) = s(t - \tau(t))$$

where

$$\tau(t) = (R_1(t) + R_2(t))/c$$

or

$$\tau(t) = 2R(t)/c$$

or even

$$\tau(t) = R(t)/c$$

for an emitting target.

Whenever there is motion in the system, $\dot{R}(t)$ is nonzero, and the retarded signal becomes quite involved. In the general case $R(t)$ is not a straight line. It may be written as a Taylor series expansion

$$R(t) = R_o + \dot{R}_o t + \frac{1}{2}\ddot{R}_o t^2 + \ldots$$

Usually, $R(t)$ can be adequately approximated by the first several terms of the series. One should not forget that the acceleration term, and higher order terms, are in the kinematics. They are sometimes important. However, in many applications, the elementary narrowband "delay/doppler" approximation, which we will develop, is entirely adequate. In other applications one may make simple compensation corrections to the received signal and then proceed using the delay/doppler approximation.

In the simplest case, $R(t)$ is adequately approximated by

$$R(t) = R + vt$$

In the case of a single range delay, a received baseband pulse would be

$$g(t) = s\left(\left(1 - \frac{v}{c}\right)t - \frac{R}{c}\right)$$

This is a compression or expansion of the time axis (depending on the sign of v) plus a range delay. In sonar applications, the dilation of the time axis may be important. In radar applications, the velocity v is very small compared to c (by a factor of a million) so that most properties of $s(t)$ are scarcely changed. Notice, however, that a shortened pulse has less energy. Even though this effect may be miniscule in radar applications, it does violate the law of conservation of energy. (This small, hardly noticeable flaw in the theory requires the development of the special theory of relatively to correct. For our purposes it can be ignored.)

A passband signal is changed by the motion in a very important way; its frequency is changed. The transmitted complex passband signal $s(t)e^{j2\pi f_0 t}$ after a single range delay, is received as

$$g(t) = s\left(\left(1 - \frac{v}{c}\right)t - \frac{R}{c}\right)e^{j2\pi f_0\left(1 - \frac{v}{c}\right)t}e^{-j2\pi f_0 R/c}$$

The dilation term $\left(1 - \frac{v}{c}\right)$ in the pulse $s(t)$ can be ignored for narrowband signals. The received signal will be expressed compactly as the complex passband signal

$$g(t) = s(t - \tau_o)e^{-j2\pi\nu_o t}e^{-j2\pi\theta}e^{j2\pi f_0 t}$$

or the complex baseband signal

$$g(t) = s(t - \tau_o)e^{-j2\pi\nu_o t}e^{-j2\pi\theta}$$

where $\tau_o = R/c$ is a modulation delay, $\nu_o = f_0 v/c$ is a doppler shift, $\theta = f_0\tau_o$ is a phase offset. The same equation also can be used for the bistatic radar case by setting $\tau_o = (R_1 + R_2)/c$ and $\nu_o = f_0(\dot{R}_1 + \dot{R}_2)/c$, or for the monostatic radar case by setting $\tau_o = 2R/c$ and $\nu_o = 2f_0 v/c$.

Although the dilation of the time axis might be unnoticeable in the modulation term $s(t)$, in the carrier term it appears as a frequency shift that is quite noticeable. This frequency offset $\nu_o = (v/c)f_0$ is known as the doppler shift (after Christian Doppler, 1803-1853, Austrian physicist). Typically v/c is in the range of 10^{-6} to 10^{-7}. Then for f_0 in the VHF band (above 30×10^6 Hertz), the doppler shift is tens of Hertz or greater. This is easily measured with a properly designed waveform and suitable equipment.

The Imaging Kernel

A radar computes the sample cross-ambiguity function of the received waveform $v(t)$ and the transmitted waveform $s(t)$. In this section we show that the output can be given a mathematical description in terms of two-dimensional filter theory. We shall see that, as shown in Figure 7, the expected output is related to the reflectivity density $\rho(\tau, \nu)$ by a two-dimensional convolution with the ambiguity function of the imaging waveform, $\chi(\tau, \nu)$, which now plays the role of a point-spread function. Based on this formulation, one can devise methods for the manipulation and processing of radar images by using two-dimensional Fourier transform techniques.

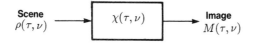

Figure 7. A Filtering View of Imaging.

The approximation that we need in this section is that the range to each point of the scene is changing linearly with time. Thus the approximation $R(t) = R_0 + vt$ is assumed to be adequate across the scene. In a later section we show how to include the quadratic term αt^2 in a simple way. Later still, we discuss the possibility of switching to polar coordinates when the quadratic approximation breaks down.

If $s(t)e^{j2\pi f_0 t}$ is a transmitted signal, then the complex baseband representation of the signal returned by a unit reflector at delay τ' and doppler ν' is

$$v(t) = s(t - \tau')e^{-j\theta'}e^{-j2\pi\nu' t} + n(t)$$

and $\theta' = 2\pi f_0 \tau'$ is a phase delay due to the time of propagation.

The cross-ambiguity function is

$$\chi_c(\tau, \nu) = \int_{-\infty}^{\infty} s(t)v^*(t + \tau)e^{-j2\pi\nu t}\,dt$$

and the expected value is

$$E[\chi_c(\tau, \nu)] = \int_{-\infty}^{\infty} s(t)E[v^*(t + \tau)]e^{-j2\pi\nu t}\,dt$$

$$= \int_{-\infty}^{\infty} s(t)s^*(t + \tau - \tau')e^{j2\pi\theta'}e^{j2\pi\nu'(t+\tau)}e^{-j2\pi\nu t}\,dt$$

In most applications $\nu\tau$ is very small compared to a radian and can be ignored. Thus, neglecting the annoying term $e^{-j\pi\nu'\tau}$, we have

$$M(\tau,\nu) = E\left[\chi_c(\tau,\nu)\right] = e^{j\theta}\chi(\tau - \tau', \nu - \nu')$$

To form $M(\tau,\nu)$, the ambiguity function of $s(t)$ is displaced by τ' in the τ direction and by ν' in the ν direction and is multiplied by an exponential phase term.

The sample cross-ambiguity function is linear in the received signal. Suppose that $v(t)$ is composed of echoes from two reflectors:

$$v(t) = s(t - \tau')e^{-j\theta'}e^{-j2\pi\nu't} + s(t - \tau'')e^{-j\theta''}e^{-j2\pi\nu''t}$$

Then, clearly

$$M(\tau,\nu) = e^{j\theta'}\chi(\tau - \tau', \nu - \nu') + e^{j\theta''}\chi(\tau - \tau'', \nu - \nu'')$$

The principle of superposition continues to be valid in this way for any number of reflectors, and even for a continuum of reflectors. Suppose that an extended target has complex reflectivity density of $\rho(\tau,\nu)$. We may redefine the complex reflectivity in such a way that the range-dependent phase shift θ is lumped into the phase of the complex reflectivity. Then the image is

$$M(\tau,\nu) = \int\limits_{-\infty}^{\infty}\int\limits_{-\infty}^{\infty} \rho(\tau',\nu')\chi(\tau - \tau', \nu - \nu')d\tau'd\nu'$$

This is a two-dimensional convolution. We may think of the reflectivity density as a two-dimensional signal which is the input into a two-dimensional filter, and of the image as the two-dimensional output. The ambiguity function $\chi(\tau,\nu)$ of the waveform $s(t)$ is the point-spread function of the two-dimensional filter.

The approximations needed to state the imaging equations as a two-dimensional convolution are the approximations that $\ddot{R}(t)$ can be neglected (or compensated), that $\rho(\tau,\nu)$ remains essentially constant during the duration of the illumination, and that the antenna pattern does not move significantly with respect to the scene during the duration of the illumination.

If it were possible, one would choose $\chi(\tau,\nu)$ equal to a two-dimensional impulse.

$$\chi(\tau,\nu) = E_p\delta(\tau,\nu)$$

Then the image would be equal to the reflectivity density function of the scene. In practice the imaging filter $\chi(\tau,\nu)$ has a finite width in τ and ν and resolution in the scene can be no better than the resolution of the filter. The resolution in range is determined by the τ width of $\chi(\tau,\nu)$; the resolution in crossrange is determined by the ν width of $\chi(\tau,\nu)$.

The development of this section makes it quite clear that any waveform $s(t)$ can be used for imaging. The quality of the image is determined by properties of the ambiguity function $\chi(\tau,\nu)$ of that waveform. However, a uniform pulse train is a commonly used waveform. In this case, the imaging radar is called synthetic aperture radar.

Synthetic Aperture Imaging

A real antenna array consists of an arrangement of identical antenna elements. A pulse is simultaneously transmitted through all elements in the array and the return is simultaneously received by all elements of the array. The same result can be obtained if an identical sequence of pulses is transmitted, one pulse from each antenna in turn but every echo received at all antennas. Then the sequence of returns at all antennas is recorded and later processed to emulate the case where all antennas are illuminated together.

An imaging radar of the kind that uses doppler of the reflected waveform to improve image resolution is called a synthetic aperture radar. The term "synthetic aperture" is intended to capture the idea of a single antenna element that is moved from position to position over a series of pulses as shown in Figure 8. At each position the antenna is pulsed and the return is recorded. The recorded sequence of returns later is coherently added together to mimic a real array. The notion of a synthetic antenna aperture is appealing and there is a long history of describing an imaging radar in this way. However, the analogy is flawed in several ways and can lead to false conclusions.

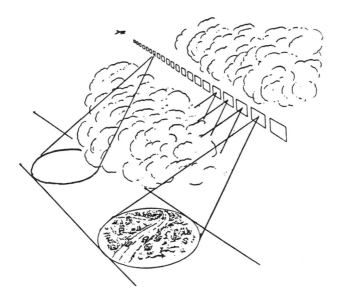

Figure 8. Swath Mode Synthetic Aperture Radar.

A synthetic aperture is superficially similar to a real array, but there are a number of significant differences. A synthetic array can only receive a return at the same antenna element from which the pulse was transmitted. The signal that might have been received at the other antenna elements is lost because those antenna elements don't exist. Hence, the synthetic array processes less information. The array element is moving while transmitting and receiving. Hence the returned pulse is doppler shifted. Because of these differences, one should be wary of pushing the analogies between a real array and a synthetic array too far.

The simplest and most common arrangement for a synthetic aperture imaging radar has the transmitter and receiver located on the same moving platform. This is called a *monostatic* imaging system. A less common arrangement has the transmitter and receiver at different places. This is called a *bistatic* imaging system.

The transmitter or receiver of a synthetic aperture radar system must be in motion with respect to the target scene. The two most popular arrangements are the swath-mapping SAR, shown in Figure 8, and the spotlight-mapping SAR, shown in Figure 9. A swath-mapping SAR ideally has a straight-line constant-velocity trajectory. It illuminates and images a continuous strip parallel to the trajectory. The resolution is limited by the amount of time that a reflecting element is illuminated which in turn is limited by the beamwidth of the real antenna. A spotlight mode SAR enhances resolution by keeping the antenna beam on a target area for a longer time by rotating the antenna beam. A much different geometry for a monostatic radar, which arises in radar astronomy, is shown in Figure 10. Lines of constant τ and ν are shown on a sphere in the far field.

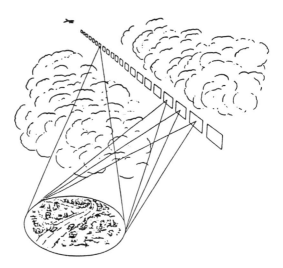

Figure 9. Spotlight Mode Synthetic Aperture Radar.

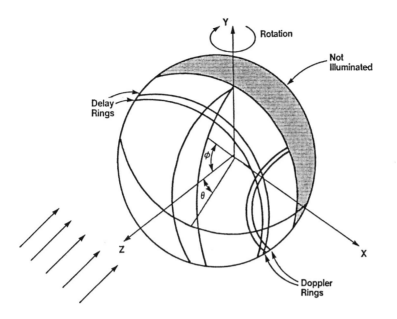

Figure 10. Delay-Doppler Coordinates on a Distant Sphere.

The image of a synthetic aperture radar is created in τ, ν coordinates. This leads to two considerations. First, is the requirement of converting the image from τ, ν coordinates into a more practical system of rectangular coordinates. Second, the effect of quadratic phase terms must be compensated to form the image.

Imaging Performance

The image of the point reflector $\delta(x, y)$ at the center of the scene is

$$|M(x,y)| = |\chi(\tau(x,y), \nu(x,y))|$$

and the arguments of χ are related to x and y by

$$\tau(x,y) = \frac{2}{c} R(x, y, x', y')$$
$$\nu(x,y) = 2\frac{f_0}{c} \dot{R}(x, y, x', y')$$

For a sidelooking configuration, forming an image of a small scene centered at range R lying in a direction perpendicular to the aircraft velocity vector of magnitude V, we have the approximations

$$\tau(x,y) = \frac{2}{c} x$$
$$\nu(x,y) = \frac{2}{c} f_0 \frac{V}{R} y$$

Therefore, with a simple scaling, τ and ν become range x and crossrange y variables.

Choose any measure of resolution. The resolution in τ and ν are related to resolution in R and $\dot{R}$ by

$$\Delta\tau = \frac{2}{c}\Delta R$$

$$\Delta\nu = 2\frac{f_0}{c}\Delta\dot{R}$$

For a sidelooking configuration with the velocity vector along the y coordinate axis and with the x coordinate axis in the range direction

$$\Delta R = \Delta x$$

$$\Delta\dot{R} = \Delta y\frac{V}{R} = V\sin\phi\Delta\phi$$

Combining these gives

$$\Delta x = \frac{c}{2}\Delta\tau$$

$$\Delta y = \frac{R}{V}\frac{c}{2f_0}\Delta\nu$$

For the resolution in range, we use the τ resolution of

$$\Delta\tau = \frac{1}{B}$$

where B is the bandwidth of the transmitted waveform. Therefore the range resolution is

$$\Delta x = \frac{c}{2B}$$

The ambiguity function of the uniform pulse train has its first null at $1/(NT_r)$. Therefore

$$\Delta\nu = \frac{1}{NT_r}$$

and the y resolution is

$$\Delta y = \frac{R}{2}\frac{c}{V}\frac{1}{NT_r f_0}$$
$$= \frac{\lambda R}{2L}$$

where $L = VNT_r$ is the distance that the antenna moves during the duration of the pulse $s(t)$. Because of this interpretation, L is called the "synthetic aperture length."

It is interesting to compare the Δy resolution to the resolution of an uniformly illuminated real array of length L. The Δy resolution of a real aperture of length L is given by

$$\Delta y = R\Delta\phi$$
$$= \frac{\lambda R}{D}$$

because for a real aperture of length D that is uniformly illuminated

$$\Delta\phi = \frac{\lambda}{D}$$

The crossrange resolution of a real aperture will be equal to the crossrange resolution of the synthetic aperture if $D = 2L$. A real aperture must be twice as long as a synthetic aperture to obtain the same crossrange resolution.

An implicit assumption of our analysis of the imaging radar is that the antenna beam is directed at the scene throughout the duration of the waveform $s(t)$. For a swath mode synthetic aperture radar, this limits VT. Figure 11 shows an idealized antenna beam illuminating the scene. The width of the beam at range R is $R\Delta\phi$. If VT is larger than $R\Delta\phi$, then the beam can move off of a target element during the waveform. The integration time T is then set by

$$VT = R\Delta\phi$$

Therefore, the crossrange resolution of a swath-mode imaging radar is

$$\Delta y = \frac{\lambda R}{2VT} = \frac{\lambda R}{2R\Delta\phi} = \frac{\lambda R}{2R(\lambda/D)} = \frac{D}{2}$$

The crossrange resolution is limited to one half the physical size of the real antenna unless the real antenna is rotated to keep the reflecting element illuminated.

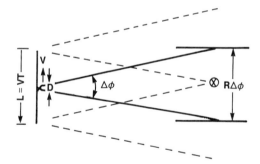

Figure 11. Swath Mode Limit on Illumination Time.

Focusing and Motion Compensation

Because of changing geometry, objects will not normally have persistent τ, ν coordinates. To get good ν resolution, long waveforms are needed, and during the course of this waveform ν itself will change. In general, objects will be out of focus or smeared because their τ and ν coordinates are changing during the exposure. This imposes a requirement of stabilizing the coordinates of the reflecting elements. We can think of this as focusing the image.

The purpose of focusing and motion compensation is to stabilize the signal from each reflecting element so that it has constant coordinates in τ, ν space. Because the received signal actually is the composite of many echoes from many points, the compensation is different for every point of the scene. However, if the scene is not too large, the compensation is approximately the same for every point in the scene.

The delay in the received signal depends on the location of the reflector and on time. The standard processing equations use the approximation

$$\tau(x, y, t) = \tau_o(x, y) + \dot{\tau}_o(x, y)t$$

Neglecting the higher order terms is called a focusing error. It causes blurring and loss of resolution.

We may now look more carefully at the approximation with the intention of bringing in higher order terms as needed, but in the simplest possible way. Specifically the next more complicated approximation is

$$\tau(x, y, t) \approx \tau_o(x, y) + \dot{\tau}_o(x, y)t + \frac{1}{2}\ddot{\tau}_o(0, 0)t^2$$

which may be needed in the carrier phase. The additional term is a "focusing" compensation. The significance of choosing a term that is independent of x and y is that the focusing compensation need not be applied separately for each point in the scene. It can be incorporated into the equations simply by multiplying $v(t)$ by a phase term determined by this focusing term.

A better approximation of the same form is

$$\tau(x, y, t) = \tau_o(x, y) + \dot{\tau}_o(x, y)t + \frac{1}{2}\ddot{\tau}_o(0, 0)t^2 + \delta\tau(t)$$

The last term will account for the deviations in the motion of the radar from a straight line. That term is referred to as the motion compensation.

Imaging Interferometry

The image of a synthetic aperture radar is two-dimensional, presenting the image as if it were viewed from directly overhead. If it is important to know the elevation of the individual reflectors, then one can incorporate an interferometer for direction finding into the synthetic aperture radar. Figure 12 shows the concept. An image is formed for each of two antennas separated by distance d. If a delay/doppler cell has only a single dominant reflecting element, then the relative phase between $\chi^{(1)}(\tau, \nu)$ and $\chi^{(2)}(\tau, \nu)$ is simply related to the difference in slant range to that reflector. An interfermetric calculation gives an angle to that reflector.

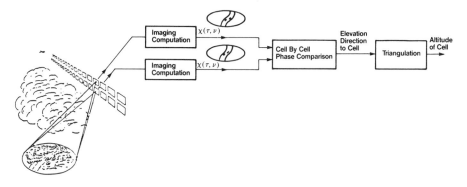

Figure 12. Imaging Interferometry.

VI. RADAR SEARCH SYSTEMS

A radar processor consists of a preprocessor, a detection and estimation function, and a postprocessor. A received radar signal on entering the receiver first encounters the preprocessor, which can almost always be viewed as the computation of the sample cross-ambiguity function though sometimes in a crudely approximated form. A search radar is one that examines isolated peaks in the sample cross-ambiguity function so as to detect objects in the environment and to estimate parameters associated with those objects.

A reflecting object may be made up of many individual reflecting elements, such as corners, edges and so on. When the resolution is coarse compared with the size of the individual reflecting elements, then the reflecting object appears as a single peak. The search radar detects the peak and estimates the delay and doppler coordinates of the reflector considered as a single object.

The Radar Range Equation

The performance of a search radar depends in large measure on the amount of energy E_{pr} in the pulse that reaches the receiver in comparison with the internal thermal noise inevitably generated within the receiver. The equation that expresses E_{pr} in terms of the transmitted pulse energy E_{pt} is known as the radar range equation. One form of the radar range equation is given by

$$E_{pr} = \left(\frac{\lambda^2}{4\pi}G_r\right)\left(\frac{1}{4\pi R_r^2}\right)\sigma\left(\frac{1}{4\pi R_t^2}\right)G_t E_{pt}$$

where G_t and G_r are the gains of the transmitting and receiving antennas in the direction of propagation, σ is the radar cross section of the reflector, R_t and R_r are the ranges from transmitter to reflector and from reflector to receiver, and λ is the wavelength of the radiation. The equation may also be expressed in terms of power rather than of energy by replacing E_{pr} and E_{pt} by P_r and P_t.

The range equation has been introduced with the terms arranged from right to left to tell the story of the energy bookkeeping. Transmitted energy E_{pt} appears at the output of the antenna as the effective radiated energy $G_t E_{pt}$. The energy spreads in spherical waves. The term $4\pi R_t^2$ is the area of a sphere of radius R_t. By dividing $G_t E_{pt}$ by this term, we have the energy per unit area that passes through a surface at a distance of R_t from the transmitter. A reflector reradiates a fraction of this energy given by the scattering cross section σ. Of this reflected energy a fraction is captured by the receiving antenna. By dividing the reflected energy by $4\pi R_r^2$, we have the energy per unit area that passes through a spherical surface at a distance of R_r from the reflector. The energy captured by the receiving antenna is determined by effective area A_e of the receiving antenna. It is related to antenna gain by antenna theory as

$$A_e = G_r \frac{\lambda^2}{4\pi}$$

The product of all the factors leads to the radar range equation. When $R_t = R_r = R$, the equation can be multiplied out and written in the form

$$E_{pr} = G_r G_t \frac{\lambda^2 \sigma}{(4\pi)^3 R^4} E_{pt}$$

The radar range equation is written in many alternative forms by grouping the terms differently or by renaming groups of terms. Another form of the equation is

$$E_{pr} = G_r \left(\frac{\lambda^2}{4\pi R_r^2}\right) \left(\frac{\sigma}{4\pi \lambda^2}\right) \left(\frac{\lambda^2}{4\pi R_t^2}\right) G_t E_{pt}$$

The energy reflected from an object determines the magnitude of the peak in the sample cross-ambiguity function corresponding to that object. Each resolvable object will have a peak and the magnitude of that peak is the energy reflected from that object.

Figure 13 illustrates how two objects are detected in the sample cross-ambiguity surface by testing at which points the surface is larger than some fixed threshold. Also, shown is a false alarm at a point where noise in the surface breaks threshold. The value of the threshold is set high enough so that the probability of a false alarm is small. Setting the threshold too high however will mean that weak objects will have a small probability of detection.

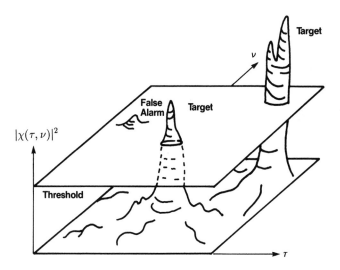

Figure 13. Thresholding the Sample Ambiguity Surface.

Clutter

The cumulative signal at the radar receiver arising as radar reflections from background objects other than those of interest is called *clutter*. Usually the term clutter carries the connotation that the clutter-causing objects are a dense set of small, unresolvable elements, such as vegetation or perhaps moving elements such as raindrops, birds, swaying tree branches, or ocean waves. The effect of clutter on the radar ambiguity function is similar to that of thermal noise in that it creates a background intensity above which the main lobes of sample ambiguity functions of targets must rise if they are to be detected. Clutter is different from thermal noise, however, in that it is dependent on the transmitted signal and has statistical properties that may be exploited to improve performance.

The quality of a radar in clutter is judged by a number called the *subclutter visibility*. This is a term of imprecise definition. It is the number of decibels by which the maximum allowable clutter-to-signal ratio at the receiver input exceeds the maximum allowable noise-to-signal ratio at the receiver input.

Moving Target Detection

A radar waveform that is designed to have an ambiguity function with good resolution in the doppler direction can be used to detect moving targets even though there may be stationary or slowly moving reflectors that are many orders of magnitude larger. A *moving target detection* (MTD) radar is one that is designed to detect the presence of a moving target by using a waveform with an appropriate ambiguity function.

The ambiguity function can be used to study the effect of clutter on a linear stationary receiver. The radar return from moving targets is cluttered by echoes from stationary or slowly moving reflectors in the antenna main beam or sidelobes. We will examine the effect of clutter on the sample cross-ambiguity function. Figure

14 shows how the clutter density $\rho(\tau, \nu)$ is two-dimensionally convolved with the ambiguity function of the radar waveform. The clutter is supported only on a strip of the τ, ν plane and will contaminate certain regions of the sample ambiguity function as determined by the doppler ambiguities, but other regions will be nearly clutter-free. Moving objects that lie in the clutter-free region will be readily detected if they are sufficiently strong. The waveform is unsuitable for the clutter model if the doppler ambiguities are comparable to the width of the clutter support because the ambiguity function then leads to a clutter-free region that is too small.

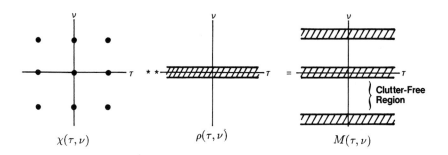

Figure 14. Plan View for Narrowband Clutter.

VII. IMAGE RECONSTRUCTION

Suppose that we have one or more partial images of a multidimensional object, but that each of these images is distorted or limited in some way. For example, the images may be projections of a multidimensional object onto a lower dimensional space, or may be images that are blurred or poorly resolved in various directions. Several such images of a common object can be combined to estimate a single enhanced image.

Images may be reconstructed from the phase of the Fourier transform or the magnitude of the Fourier transform. Images may be constructed by converting phase information in the signal into intensity information. Images may be reconstructed from a collection of projections.

Techniques for combining several poor images into a single improved image are known collectively as tomography (Greek tomo: a cut + graphy). This is different from the practice of enhancing a single image by processing techniques, although of course, the two problems are closely related.

Image processing refers to the computation of a higher quality image from a single lower quality image. Tomography refers to the generation of an image from a collection of its projections. Intermediate between these two is the computation of a higher quality image from a set of lower quality images.

Tomography

The reconstruction of images from projections is called tomography. Familiar examples of tomography are in the field of medical imaging. Within a region of the body, an unknown function $s(x, y, z)$ is to be estimated based on an observation of its projections. The projection data may be generated by absorption, emission, or scattering. Absorption refers to the attenuation of a signal such as an X-ray while passing through a region at (x, y, z). Emission refers to the generation of a signal, such as radioactive particles, from a region at (x, y, z). Scattering refers to the reflection of a signal into a specified direction from a region at (x, y, z). In each case, $s(x, y, z)$ refers to the magnitude of a phenomena such as the logarithm of the attenuation and we wish to estimate $s(x, y, z)$ by observing projections.

Mathematically, the two-dimensional form of the problem is as follows. Given the projections

$$p_\phi(t) = \int\limits_{-\infty}^{\infty} s(t\cos\phi - r\sin\phi,\ t\sin\phi + r\cos\phi)dr$$

for some values of the parameter ϕ, find $s(x, y)$. The projection-slice theorem says that $P_\phi(f)$, the Fourier transform of $p_\phi(t)$ is given by

$$P_\phi(f) = S(f\cos\phi, f\sin\phi)$$

Therefore an equivalent task is to estimate $S(f_x, f_y)$ when given $P_\phi(f)$ for some values of the parameter ϕ.

A more general form of the a problem is to estimate $s(x, y)$ when given

$$v_\phi(t) = p_\phi(t) + n_\phi(t)$$

for some values of the parameter ϕ, where $n_\phi(t)$ is a noise process. We shall study the simple problem in which there is no noise and the projection is known for all ϕ. The mathematical inversion problem can then be solved exactly.

When $p_\phi(t)$ is known for all ϕ from 0 to π, the set of projections constitutes the Radon transform of $s(x, y)$ denoted by

$$p(t, \phi) = \int\limits_{-\infty}^{\infty} s(t\cos\phi - r\sin\phi, t\sin\phi + r\cos\phi)dr$$

The next theorem gives an inverse for the Radon transform.

THEOREM (BACK-PROJECTION THEOREM).

$$s(x, y) = \int\limits_{0}^{\pi}\int\limits_{-\infty}^{\infty} P(f, \phi)e^{j2\pi f(x\cos\phi + y\sin\phi)}|f|df d\phi$$

Proof.

$$s(x,y) = \int\limits_{-\infty}^{\infty} \int\limits_{-\infty}^{\infty} S(f_x, f_y) e^{j2\pi(f_x x + f_y y)} df_x df_y$$

Change the integration to polar coordinates by setting

$$f_x = f \cos \phi$$
$$f_y = f \sin \phi$$

Then

$$s(x,y) = \int\limits_{0}^{\pi} \int\limits_{-\infty}^{\infty} S(f \cos \phi, f \sin \phi) e^{j2\pi f(x \cos \phi + y \sin \phi)} |f| df d\phi$$

which completes the proof of the theorem. □

COROLLARY.

$$s(r \cos \beta, r \sin \beta) = \int\limits_{0}^{\pi} \int\limits_{-\infty}^{\infty} \left[P_\phi(f) |f| e^{j2\pi f r \cos(\beta - \phi)} df \right] d\phi$$

where $x = r \cos \beta$ and $y = r \sin \beta$.

Proof. Immediate

The structure of the back projection may be easier to see if it is decomposed into two steps

$$g(t, \phi) = \int\limits_{-\infty}^{\infty} |f| P_\phi(f) e^{j2\pi ft} dt$$

$$s(x,y) = \int\limits_{0}^{\pi} g(x \cos \phi + y \sin \phi, \phi) d\phi$$

The first integral is in the form of a Fourier transform of the product $|f| P_\phi(f)$. Consequently, it can be described formally as the convolution of $k(t)$ with $p_\phi(t)$ where $k(t)$ is the inverse Fourier transform of $|f|$. This inverse Fourier transform does not exist as a function but, if desired, one can introduce a generalized function that behaves the right way under formal manipulations. Specifically the second derivative of $|f|$ is a delta function and so $|f|$ formally has inverse Fourier transform $-(2\pi f)^{-2}$.

Alternatively, we can avoid the use of generalized functions by dealing with a filtered signal $\tilde{s}(x,y)$. This will replace $|f|$ by $|f| H(f)$ and will also ensure that the two-dimensional Fourier transform falls off with frequency sufficiently quickly for our purposes. We may wish to consider a filter on either the signal $s(x,y)$ or on the projections $p_\phi(t)$. The following theorem gives a condition for which these two are equivalent.

Let $H(f)$ be a one-dimensional filter, and let

$$H'(f_x, f_y) = H \left(\sqrt{f_x^2 + f_y^2} \right)$$

be a two-dimensional circularly symmetric filter defined in terms of $H(f)$.

THEOREM. *Let $p_\phi(t)$ be the projection at angle ϕ of $s(x,y)$. If $\tilde{p}(t) = h(t) * p(t)$ and $\tilde{s}(x,y) = h'(x,y) ** s(x,y)$, then $\tilde{p}_\phi(t)$ is the projection at angle ϕ of $\tilde{s}(x,y)$.*

Proof. The projection of $\tilde{s}(x,y)$ at angle ϕ will have Fourier transform

$$\tilde{S}(f\cos\phi, f\sin\phi) = H'(f\cos\phi, f\sin\phi)S(f\cos\phi, f\sin\phi)$$
$$= H(f)P_\phi(f)$$

which completes the proof of the theorem. □

With the aid of the previous corollary, we can now write

$$\tilde{s}(x,y) = \int_0^\pi \left[\int_{-\infty}^\infty P_\phi(f)H(f)|f|e^{j2\pi fr\cos(\beta-\phi)}df \right] d\phi$$

The filter $H(f)$ can be chosen so that all frequencies of interest in $s(x,y)$ are passed. In that case the filter will have no effect on $s(x,y)$ but will be of considerable benefit in the computations. The kernal of the reconstruction is defined as

$$k(t) = \int_{-\infty}^\infty |f|H(f)e^{j2\pi ft}df$$

Let

$$g(t,\phi) = k(t) * p_\phi(t)$$

Then

$$s(x,y) = \int_0^\pi g(x\cos\phi + y\sin\phi, \phi)d\phi$$

We now have the back projection computation broken into a two-step process: filter every projection using a filter with impulse response $k(t)$; then evaluate an integral. The integral is not a particularly convenient one for numerical processing because there is an implicit conversion from data in a polar format into data in a rectangular format. On a discrete grid, this requires interpolation.

Tomography and Imaging Radar

The processing for an imaging radar can also be formulated as a tomographic system. This formulation clarifies the behavior of an imaging radar by exploring it from another point of view. It may be a superior point of view when the synthetic aperture is so long that a polar (r, ϕ) coordinate system fits the situation better than a rectangular (x,y) coordinate system.

A heuristic approach to a tomographic formulation is as follows. Suppose that the reflectivity density of a scene $\rho(x,y)$ is observed through a set of I distinct point-spread functions

$$m^{(i)}(x,y) = \chi^{(i)}(x,y) ** \rho(x,y)$$

for $i = 1, \ldots, I$. In the frequency domain this is

$$M^{(i)}(f_x, f_y) = \Xi^{(i)}(f_x, f_y)R(f_x, f_y)$$

for $i = 1, \ldots, I$. The terms $M(f_x, f_y), \Xi(f_x, f_y)$, and $R(f_x, f_y)$ denote, respectively, the transforms of $m(x, y), \chi(x, y)$, and $\rho(x, y)$. If the union of the supports of the I Wigner distributions $\Xi^{(i)}(f_x, f_y)$ covers all (or most) of the support of $R(f_x, f_y)$, then $R(f_x, f_y)$ can be estimated from the I degraded images $M^{(i)}(f_x, f_y)$.

One way to obtain the I point-spread functions $\chi^{(i)}(x, y)$ is to transmit in turn I different pulses. Another way is to always use the same pulse, the ith copy is transmitted and received at its own viewing angle. Each pulse is then processed individually to obtain one aspect angle on the scene. The batch of such projections is processed to form a tomographic image.

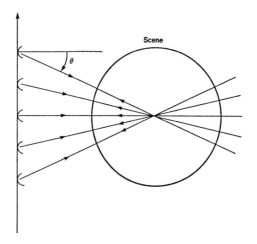

Figure 15. Tomographic Depiction of Synthetic Aperture Radar.

We may fully process each received pulse by computing a sample cross-ambiguity function on a pulse-by-pulse basis. The result for one pulse is

$$m^{(i)}(\tau, \nu) = \chi(\tau, \nu) * *\rho(\tau, \nu)$$

where $\chi(\tau, \nu)$ is the ambiguity function of the single pulse $s(t)$. For each pulse, the τ, ν coordinate system has τ axis from the antenna to the center of the scene. We can regard the τ, ν coordinate system as a polar coordinate system. The reason for processing each pulse individually is that for each pulse a different relationship between the x, y fixed rectangular coordinate systems and the τ, ν coordinate system can be used.

Because an individual pulse is very short, we may choose to ignore the doppler shift within the individual pulse. Thus, although the radar antenna is physically moved from pulse to pulse to get different viewing angles, we may regard the antenna as stationary during each pulse. This slight approximation gives up usable information in order to give a crisper tomographic interpretation to the data.

The pulse is very short, so $\chi^{(p)}(\tau, \nu)$ is very wide in the ν direction. It may be wider than the illuminated scene, so that $\chi(\tau, \nu)$ can be approximated as independent of τ. Thus

$$\chi(\tau, \nu) = \lambda(\tau)$$

The image computed for one pulse then is a function of a single coordinate

$$m^{(i)}(\tau, \nu) = \lambda(\tau) * *\rho(\tau, \nu)$$

$$= \lambda(\tau) * \int\limits_{-\infty}^{\infty} \rho(\tau, \nu) d\nu$$

In the (r, ϕ) polar coordinate systems, centered relative to the center of a scene at distance R, the integrated reflectivity from all reflectors at range r is

$$p_\phi(r) = \int\limits_{-\infty}^{\infty} \rho(r, r') dr'$$

Thus, rewriting the computed image in a polar coordinate system, we see that the pulse transmitted at angle ϕ provides a projection at angle ϕ.

$$m_\phi(r) = \lambda(r) * p_\phi(r)$$

These projections can be used to form an image by the methods of tomography, as by using the projection-slice theorem.

In the special case where the pulse is a chirp pulse the processing takes on a special form. The radar return seen by the pulse at angle ϕ is

$$v_\phi(t) = \int\limits_{-\infty}^{\infty} p_\phi(r) s \left(t - \frac{2}{c}(R + r) \right) dr$$

If the pulse duration T is long in comparison with the dispersion in delay across the scene we can be a little careless with regard to end points of the pulse $s(t)$ without causing significant error, Thus,

$$v_\phi(t) = \int\limits_{-\infty}^{\infty} p_\phi(r) e^{-j\left[2\pi f_0\left(t - \frac{2(R+r)}{c}\right) + \pi\alpha\left(t - \frac{2(R+r)}{c}\right)^2\right]} dr$$

Define the dechirped pulse as

$$c_\phi(t) = v_\phi(t) e^{j\left[2\pi f_0\left(t - \frac{2R}{c}\right) + \pi\alpha\left(t - \frac{2R}{c}\right)^2\right]}$$

$$= \int\limits_{-\infty}^{\infty} p_\phi(r) e^{j\left[2\pi f_0 \frac{2r}{c} + \pi\alpha\left(\frac{4r}{c}t - \frac{8Rr}{c^2} - \frac{4r^2}{c^2}\right)\right]} dr$$

This has the appearance of a Fourier transform except for the phase term that is quadratic in r. We will neglect this term under the assumption that the scene size is such that $4\alpha r^2/c^2$ is a small phase angle (alternatively, one may devise methods of approximate compensation). Then

$$c_\phi(t) = \int\limits_{-\infty}^{\infty} p_\phi(r)e^{j2\pi\frac{2}{c}(f_0+\alpha(t-\tau_o))r}dr$$

where $\tau_o = \frac{2R}{c}$. Consequently, we now recognize $c_\phi(t)$ as a Fourier transform. Specifically,

$$c_\phi(t) = P_\phi\left(\frac{2}{c}(f_0 + \alpha(t - \tau_0))\right)$$

and $c_\phi(t)$ is the Fourier transform of a projection. This means that a pulse transmitted at angle ϕ will generate a slice of the transform at angle ϕ simply by dechirping. A collection of such pulses for various ϕ gives a sequence of projections.

Phase Contrast Imaging

The earliest application of spatial filtering arose in the microscopic imaging of transparent objects (such as organisms) that effect light, not by absorption, but by a spatially dependent delay, which appears in the light signal as a spatially dependent phase variation. Because both film and the eye detect only the intensity of light, but not its phase, objects that are completely transparent are impossible to see. However, they can be seen if a device is placed in the optical path that will convert phase modulation to amplitude modulation. One method of phase-contrast imaging known as the *schlieren method* is popular in applications such as wind tunnel photographs in which the air stream has nonuniform index of refraction as may be caused by pressure differences or temperature differences. This method turns phase modulation into amplitude modulation based on the use of a filter known as a knife-edge filter. Another technique is the method known as the *central dark ground method* in which a small spot in the frequency domain is used to modify the component of the Fourier transform near zero frequency. We shall describe only the dark ground method. The schlieren method differs only in the choice of filter.

Suppose that a transparent object has a complex transmittance consisting of phase variation only

$$t(x,y) = e^{j\theta(x,y)}$$

and that it is coherently illuminated in an image forming system. The image produced by a conventional optical imaging system would have intensity

$$|t(x,y)|^2 = 1$$

which is independent of the signal, and the signal is lost. To preserve the desired signal, $t(x,y)$ is first passed through a *Zernike filter* defined in the frequency domain as

$$H(f_x, f_y) = \begin{cases} ja & \text{if } \sqrt{f_x^2 + f_y^2} \le \epsilon \\ 1 & \text{otherwise} \end{cases}$$

where ϵ is a small number and a is a real constant.

The output of the filter in the frequency domain is $H(f_x, f_y)T(f_x, f_y)$. If $\theta(x,y)$ is less than one radian, then we have the approximation

$$t(x,y) \approx 1 + j\theta(x,y)$$

and

$$T(f_x, f_y) = \delta(f_x, f_y) + j\Theta(f_x, f_y)$$

Because $\Theta(f_x, f_y)$ is negligible at the origin, the filter output is approximately

$$H(f_x, f_y)T(f_x, f_y) = ja\delta(f_x, f_y) + j\Theta(f_x, f_y)$$

Consequently

$$h(x,y) * *t(x,y) \approx j(a + \theta(x,y))$$

which has intensity

$$|(a + \theta(x,y))|^2 = a^2 + 2a\theta(x,y)$$

Whereas the intensity of the unfiltered signal is independent of $\theta(x,y)$, the intensity for the filtered signal does depend on the object.

VIII. PASSIVE SURVEILLANCE

Many surveillance problems are passive, including problems encountered in radio astronomy, seismic data analysis, electromagnetic surveillance, and sonar. Passive surveillance systems collect a propagating signal such as a waveform that the system itself does not generate. The waveform is usually unknown. The incidence waveform is collected at one or more sites in the propagating medium; possibly the receiving sites are moving. From this set of noisy signals, the location of the set of radiation sources is estimated, or their configuration is imaged.

Passive location can be quite difficult when the signals are weak compared to the noise, or when there are a large number of radiation sources. A propagating medium may be teeming with a multitude of minute signals even though it appears superficially to be empty. For example, an acoustic medium such as water may appear quite still and yet may contain tiny pressure waves originating in various submerged objects and reflecting off of other objects. A passive sonar system will measure these waves and extract useful information from the raw data. Likewise a seismographic sensor or an array of such sensors on the surface of the earth can measure minute vibrations and deduce the location of distance earthquakes or of other geophysical disruptions.

Radio Astronomy

The task of radio astronomy is to detect and image astronomical - often extra-galactic - sources of electromagnetic waves in the radio bands. Although the wavelength in these bands is large, implying poor resolution, one can create an array of radio dishes limited in size only by the diameter of the earth, thereby obtaining good resolution. Consequently, in recent years radio astronomy has outperformed classical optical astronomy in many ways.

Consider any pair of antenna elements with a signal $s(t)$ arriving from a point source at infinity at angle ϕ with respect to the perpendicular bisector of the baseline between the antenna elements. With the phase reference defined so that there is no phase delay at the first element, the received signals are

$$v_1(t) = s(t)$$
$$v_2(t) = s(t)e^{-j2\pi(d/\lambda)\sin\phi}$$

We take the transmitted signal $s(t)$ to be a random process, usually gaussian noise. Consequently

$$\int_0^T v_1(t)v_2^*(t)dt = e^{j2\pi(d/\lambda)\sin\phi}\int_0^T |s(t)|^2 dt$$
$$= ce^{j2\pi(d/\lambda)\sin\phi}$$

where c is a constant denoting the energy in the intercepted segment of $s(t)$.

Next suppose there are two point sources at angles ϕ_1 and ϕ_2 that are emitting independent gaussian noise $s_1(t)$ and $s_2(t)$. The received signals are

$$v_1(t) = s_1(t) + s_2(t)$$
$$v_2(t) = s_1(t)e^{-j2\pi(d/\lambda)\sin\phi_1} + s_2(t)e^{-j2\pi(d/\lambda)\sin\phi_2}$$

The sample correlation is

$$\int_0^T v_1(t)v_2^*(t)dt = c(\phi_1)e^{j2\pi(d/\lambda)\sin\phi_1} + c(\phi_2)e^{j2\pi(d/\lambda)\sin\phi_2} + n_e(t)$$

where

$$c(\phi_1) = \int_0^T |s_1(t)|^2 dt$$
$$c(\phi_2) = \int_0^T |s_2(t)|^2 dt$$

and the so-called "self-noise" term is the cross term

$$n_e(t) = e^{j2\pi(d/\lambda)\sin\phi_2}\int_0^T s_1(t)s_2^*(t)dt + e^{j2\pi(d/\lambda)\sin\phi_1}\int_0^T s_2(t)s_1^*(t)dt$$

For large enough T, the self noise term can be neglected in comparison with the other terms because $s_1(t)$ and $s_2(t)$ are uncorrelated.

Now suppose there is a continuum of noise signals in the angle ϕ, denoted $s(t, \phi)$ and satisfying

$$E[s(t, \phi)s^*(t, \phi')] = \bar{c}(\phi)\delta(\phi - \phi')$$

The received signals are

$$v_1(t) = \int_0^{2\pi} s(t, \phi)d\phi$$

$$v_2(t) = \int_0^{2\pi} s(t, \phi)e^{-j2\pi(d/\lambda)\sin\phi}d\phi$$

Then, neglecting all self-noise terms, the sample crosscorrelation is

$$\int_0^T v_1(t)v_2^*(t)dt = \int_0^{2\pi} c(\phi)e^{j2\pi(d/\lambda)\sin\phi}d\phi$$

$$= C\left(\frac{d}{\lambda}\right)$$

under this approximation $\sin\phi = \phi$. Consequently, the sample cross-correlation output is an estimate of one Fourier coefficient of $E[c(\phi)] = \bar{c}(\phi)$.

To get additional Fourier coefficients, simply add more antenna elements. A linear array of three elements with spacings d_1 and d_2 provides three pairwise correlations and hence, three Fourier coefficients $C\left(\frac{d_1}{\lambda}\right), C\left(\frac{d_2}{\lambda}\right), C\left(\frac{d_1+d_2}{\lambda}\right)$. Similarly four planar elements gives six pairwise correlations and six Fourier coefficients.

More generally, a two-dimensional array can be used to obtain a two-dimensional Fourier transform of a two-dimensional radiation source. Every pair of elements gives one sample of the Fourier transform. Moreover, if the array rotates, as by rotation of the earth, then a later observation gives a new set of Fourier samples.

Passive Source Location

When the source of intercepted radiation is not at infinity it is possible to estimate the location of the source in both range and direction. Passive location is a task encountered in electromagnetic surveillance, passive sonar, and seismic data analysis. At each of several widely separated receivers, some of which may be moving, we observe an attenuated and delayed version of a signal $s(t)$ embedded in receiver noise. The source of the signal $s(t)$ is at an unknown location, and that location is to be estimated. A formal statement of the problem for the case of stationary receivers is as follows: Given

$$v_i(t) = s(t - R_i/c) + n_i(t) \qquad i = 1, \ldots, I$$

where $s(t)$ is a passband signal, c is a constant and

$$R_i = \sqrt{(x - x_i)^2 + (y - y_i)^2}$$

form an estimate $(\hat{x}, \hat{y})$ of the source location (x, y).

More generally, a signal transmitted at center frequency f_0 by a radiation source, possibly moving, when received at a moving point, has undergone a time delay and a frequency shift. The ith received complex signal can be written as

$$v_i(t) = s(t - R_i(t)/c)e^{j2\pi(f_0/c)\dot{R}_i(t)t}e^{j\theta_i} + n_i(t)$$

where $s(t)$ is the complex envelope of the transmitted signal; $R_i(t)$ given by

$$R_i(t) = \sqrt{(x(t) - x_i(t))^2 + (y(t) - y_i(t))^2}$$

is the distance from the emitter to the receiver, $\dot{R}_i(t)$ given by

$$\dot{R}_i(t) = \frac{(x(t) - x_i(t))(\dot{x}(t) - \dot{x}_i(t)) - (y(t) - y_i(t))(\dot{y}(t) - \dot{y}_i(t))}{\sqrt{(x(t) - x_i(t))^2 + (y(t) - y_i(t))^2}}$$

is the rate of change of these distances (range rates); and θ_i is the constant phase shift.

We may introduce both differential delay and differential doppler as intermediate variables.

$$\tau_{ij} = (R_i - R_j)/c$$
$$\nu_{ij} = (f_0/c)(\dot{R}_i - \dot{R}_j)$$

from which $x, y, \dot{x}, \dot{y}$ can be computed.

The sample cross-ambiguity surface

$$|\chi_{ij}(\tau, \nu)| = \left| \int\limits_{-\infty}^{\infty} v_i(t)v_j^*(t + \tau)e^{j2\pi\nu t}dt \right|$$

will have a peak, in the absence of noise, at τ_{ij}, ν_{ij}.

REFERENCES

[1] WOODWARD, P. M., Probability and Information Theory, with Applications to Radar, Pergamon Press, Oxford (1953).

[2] VILLE, J., Theory and Application of the Notion of the Complex Signal, Cables et Transmission, Vol. 2 (1948), pp. 67-74.

[3] SIEBERT, W. M., A Radar Detection Philosophy, IRE Transactions on Information Theory, Vol. IT-2 (1956), pp. 204-221.

[4] LERNER, R. M., Signals with Uniform Ambiguity Functions, IRE National Convention Record, Part 4 (1958), pp. 27-33.

[5] WILCOX, C. H., The Synthesis Problem for Radar Ambiguity Functions, Univ. Wisconsin Math Res. Center, Tech. Summary Report 157, April 1960.

[6] SUSSMAN, S. M., Least-Square Synthesis of Radar Ambiguity Functions, IRE Transactions on Information Theory, Vol. IT-8 (1962), pp. 246-254.

[7] PRICE, R., AND E. M. HOFSTETTER, Bounds on the Volume and Height Distribution of the Ambiguity Function, IEEE Transactions on Information Theory, Vol. IT-11 (1965), pp. 207-214.

[8] AUSLANDER, L. AND R. TOLIMIERI, Radar Ambiguity Functions and Group Theory, Siam Journal Math. Analysis, Vol. 16 (1985), pp. 577-601.

[9] GABOR, D., Theory of Communication, Journal of the IEE, Vol. 23 (1946), pp. 429-457.

[10] KAY, I., AND R. A. SILVERMAN, On the Uncertainty Relation for Real Signals, Information and Control, Vol. 1 (1957), pp. 64-75.

[11] HELSTROM, C. W., Statistical Theory of Signal Detection, Pergamon Press, Oxford, 1960.

[12] RIHACZEK, A. W., Radar Signal Design for Target Resolution, Proceedings of the IEEE, Vol. 53 (1965), pp. 116-128.

[13] COSTAS, J. P., A Study of a Class of Detection Waveforms Having Nearly Ideal Range-Doppler Ambiguity Properties, Proceedings of the IEEE, Vol. 72 (1984), pp. 996-1009.

[14] GRÜNBAUM, F. A., A Remark on Radar Ambiguity Functions, IEEE Transactions on Information Theory, Vol. IT-30, (1984), pp. 126-127.

[15] TOLIMIERI, R., AND S. WINOGRAD, Computing the Ambiguity Surface, IEEE Transactions on Acoustics Speech and Signal Processing, Vol. ASSP-33 (1985), pp. 1239-1245.

[16] CUTRONA, L. J., W. E. VIVIAN, E. N. LEITH, AND G. O. HALL, A High-Resolution Radar Combat-Surveillance System, IRE Transactions on Military Electronics, Vol MIL-5 (1961), pp. 127-131.

[17] BROWN, W. M., AND L. J. PORCELLO, An Introduction to Synthetic Aperture Radar, IEEE Spectrum, September, 1969, pp. 52–62.

[18] BROWN, W. M., AND R. FREDERICKS, Range-Doppler Imaging with Motion Through Resolution Cells, IEEE Transactions on Aerospace and Electronic Systems, Vol AES-5 (1969), pp. 98-102.

[19] WALKER, J. L., Range-Doppler Imaging of Rotating Objects, IEEE Transactions on Aerospace and Electronic Systems, Vol AES-16 (1980), pp. 23-52.

[20] JENSEN, H., L. C. GRAHAM, L. J. PORCELLO, E. N. LEITH, Side-Looking Airborne Radar, Scientific American, October 1977, pp. 84-95.

[21] MUNSON, D. C., JR., J. D. O'BRIEN, AND W. K. JENKINS, A Tomographic Formulation of Spotlight-Mode Synthetic Aperture Radar, Proceedings of the IEEE, Vol 71 (1983), pp. 917-925.

[22] GRAHAM, L. C., Synthetic Interferometer Radar for Topograpic Mapping, Proceedings of the IEEE, Vol 62 (1974), pp. 763-768.

[23] MARCUM, J. R., A Statistical Theory of Target Detection of Pulsed Radar, IRE Transactions on Information Theory, Vol IT-6 (1960), pp. 145-267.

[24] SWERLING, P., Probability of Detection for Fluctuating Models, IRE Transactions on Information Theory, Vol IT-6 (1960), pp. 269-308.

[25] SELIN, I., Detection of Coherent Radar Returns of Unknown Doppler Shift, IEEE Transactions on Information Theory, Vol IT-11 (1965), pp. 396-400.

[26] EMERSON, R. C., Some Pulsed Doppler, MTI, and AMTI Techniques, Rand Corp. Rep R-274 (1954).

[27] URKOWITZ, H., Filters for Detection of Small Targets in Clutter, Journal of Applied Physics, vol 24 (1953), pp. 1024-1031.

[28] GREEN, P. E., JR., Radar Astronomy Measurement Techniques, Technical Report No. 282, Lincoln Laboratory, MIT (1962).

TOPICS IN HARMONIC ANALYSIS
WITH
APPLICATIONS TO RADAR AND SONAR

WILLARD MILLER, JR.*

Abstract. This minicourse is an introduction to basic concepts and tools in group represen-tation theory, both commutative and noncommutative, that are fundamental for the analysis of radar and sonar imaging. Several symmetry groups of physical interest will be studied (circle, line, rotation, $ax + b$, Heisenberg, etc.) together with their associated transforms and representation theories (DFT, Fourier transform, expansions in spherical harmonics, wavelets, etc.). Through the unifying concepts of group representation theory, familiar tools for commutative groups, such as the Fourier transform on the line, extend to transforms for the noncommutative groups which arise in radar-sonar.

The insight and results obtained will be related directly to objects of interest in radar-sonar, such as the ambiguity function. The material will be presented with many examples and should be easily comprehensible by engineers and physicists, as well as mathematicians.

TABLE OF CONTENTS

*School of Mathematics and IMA, University of Minnesota. The research contribution of this paper was supported in part by the National Science Foundation under grant DMS 88–23054

66

§1. INTRODUCTION

These notes are intended as an introduction to those basic concepts and tools in group representation theory, both commutative and noncommutative, that are fundamental for the analysis of radar and sonar imaging. Several symmetry groups of physical interest will be studied (circle, line, rotation, Heisenberg, affine, etc.) together with their associated transforms and representation theories (DFT, Fourier transforms, expansions in spherical harmonics, Weyl-Heisenberg frames, wavelets, etc.) Through the unifying concepts of group representation theory, familiar tools for commutative groups, such as the Fourier transform on the line, extend to transforms for the noncommutative groups which arise in radar and sonar.

The insights and results obtained will be related directly to objects of interest in radar-sonar, in particular, the ambiguity and cross-ambiguity functions. (We will not, however, take up the study of tomography, even though this field has group-theoretic roots.) The material is presented with many examples and should be easily comprehensible by engineers and physicists, as well as mathematicians.

The main emphasis in these notes is on the matrix elements of irreducible representations of the Heisenberg and affine groups, i.e., the narrow and wide band ambiguity and cross-ambiguity functions of radar and sonar. In Chapter 2 we introduce the ambiguity functions in connection with the Doppler effect. Chapters 3 and 4 constitute a minicourse in the representation theory of groups. (Much of the material in these chapters is adapted from the author's textbook [M5] .) Chapters 5 and 6 specialize these ideas to the Heisenberg and affine groups. Chapters 7 and 8 are devoted to frames associated with the Heisenberg group. (Weyl-Heisenberg) and with the affine group (wavelets). We conclude with a chapter touching on the Schrödinger group and the metaplectic formula.

It is assumed that the reader is proficient in linear algebra and advanced calculus, and some concepts in functional analysis (including the basic properties of countable Hilbert spaces) are used frequently. (References such as [AG1], [K2], [K8], [NS] and [RN] contain all the necessary background information.) The theory presented here is largely algebraic and (sometimes) formal so as not to obscure the clarity of the ideas and to keep the notes short. However, the needed rigor can be supplied. (The knowledgeable reader can invoke Fubini's theorem when we interchange the order of integration, the Lebesgue dominated convergence theorem when we pass to a limit under the integral sign, etc.)

Finally, the author (who is not an expert on radar or sonar) wishes to thank the experts whose writings form the core of these notes, e.g., [AT1], [AT3], [D4], [G1], [HW], [N3], [S2], [W5].

§2. THE DOPPLER EFFECT

2.1 Wideband and narrow-band echos. We begin by reviewing the Doppler effect as it relates to radar and sonar. Consider a stationary transmitter/detector and a moving (point) target located at a distance $R(t)$ from the transmitter at time t. We assume that the distance from the transmitter to the target changes linearly with time,

$$(2.1) \qquad R(t) = r + vt.$$

Now suppose that the transmitter emits an electromagnetic pulse $s(t)$. The pulse is transmitted at a constant speed c in the ambient medium (e.g., air or water) impinges the target, and is reflected back to the transmitter/receiver where it is detected as the echo $e_{r,v}(t)$. (We assume $c > |v|$.) Let $\Delta_{r,v}(t)$ be the time delay experienced by the pulse which is received as an echo at time t. (Thus this pulse is emitted at time $t - \Delta$ and travels a distance $c\Delta$ before it is received as an echo at time t.) It follows that the pulse impinges the target at time $t - \Delta/2$ and we have the identity

$$c\Delta = 2R(t - \Delta/2) = 2r + 2(t - \Delta/2)v.$$

Thus

$$(2.2) \qquad \Delta_{r,v}(t) = \frac{2v}{c+v} t + \frac{2r}{c+v}.$$

The echo $e_{r,v}(t)$ is proportional to the signal at time $t - \Delta_{r,v}(t)$:

$$(2.3) \qquad e_{r,v}(t) = -\sqrt{\frac{c-v}{c+v}} s\left(t\left[\frac{c-v}{c+v}\right] - \frac{2r}{c+v}\right).$$

Here the -1 factor in the amplitude of the echo is based on the boundary conditions for the normal component of the electric field at the surface of a (conducting) target, [J1]. The factor $\sqrt{\frac{c-v}{c+v}}$ is needed if we require that the energy of the pulse is conserved:

$$\int_{-\infty}^{\infty} |s(t)|^2 dt = \int_{-\infty}^{\infty} |e(t)|^2 dt.$$

Changing notation, we write the echo as

$$(2.4) \qquad e_{x,y}(t) = -\sqrt{y} s(y[t+x])$$

where $y = (c-v)/(c+v)$ and $x = -2r/(c-v)$. Note that the transformation $(x,y) \to (r,v)$ is one-to-one. The result (2.3) is the exact **(wideband)** solution for the given assumptions.

It is very common to approximate the wideband solution under the assumption that $|v| \ll c$ and t is small (of the order of r/c) over the period of observation. (See [CB] and [S7] for careful analyses of the relationship between the wideband solution and the narrow-band approximation.) One writes the signal in the form

$$s(t) = a(t)e^{2\pi i \omega_0 t}$$

where $a(t)$ is assumed to be a slowly varying complex function of t (the **envelope** of the waveform) with respect to the exponential factor.

Then we have

$$e_{x,y}(t) = -\sqrt{y}a(y[t+x])e^{2\pi i(y[t+x])\omega_0}$$

or, since $y \approx 1 - 2\beta$, $yx \approx -(2r/c)(1-\beta)$ where $\beta = v/c$, we have

(2.5)
$$e_{x,y}(t) \approx -a(t - 2r/c)e^{2\pi i\omega_0[t(1-2\beta)-2(r/c)(1-\beta)]}$$
$$\approx -s(t - 2r/c)e^{-4\pi i\beta\omega_0[t-2r/c]}.$$

This is called the **narrow-band approximation** of the Doppler effect. (Generally speaking, the narrow-band approximation is usually adequate for radar applications but is less appropriate for sonar.)

To see the significance of this approximation, consider the Fourier transform of the signal and the echo:

(2.6)
$$S(\omega) = \int_{-\infty}^{\infty} s(t)e^{-2\pi i\omega t}\,dt,$$

where, [DM2],

(2.7)
$$s(t) = \int_{-\infty}^{\infty} S(\omega)e^{2\pi i\omega t}\,d\omega.$$

Then

(2.8)
$$S(\omega) \to E(\omega) = -S(\omega + 2\beta\omega_0)e^{-4\pi i\omega r/c}$$

in the narrow-band approximation, whereas the exact result is

(2.9)
$$S(\omega) \to E(\omega) = -\frac{1}{\sqrt{y}}S\left(\frac{\omega}{y}\right)e^{2\pi i\omega x}.$$

Clearly (2.8) is a good approximation to (2.9) if the support of $S(\omega)$ lies in a narrow band around ω_0. In the narrow-band approximation the signal is delayed by the time interval $2r/c$ and the frequency changes by $2\omega_0 v/c$.

2.2 Ambiguity functions. To determine the position and velocity of a target from a narrow-band echo (2.5) one computes the inner product (cross-correlation) of the echo with a test signal

$$s_{r',v'}(t) = -s\left(t - \frac{2r'}{c}\right)e^{-4\pi i\omega_0 v'\left[t-\frac{2r'}{c}\right]/c},$$

(2.10)
$$\langle e_{r,v}, s_{r',v'}\rangle = \int_{-\infty}^{\infty} e_{r,v}(t)\overline{s}_{r',v'}(t)\,dt$$
$$= e^{\frac{4\pi i\omega_0}{c}\left[\frac{2r'v'}{c}+\frac{2r}{c}v\right]}\int_{-\infty}^{\infty} s(t-2r/c)\overline{s}\left(t-\frac{2r'}{c}\right)e^{\frac{4\pi i\omega_0}{c}[(v'-v)t]}\,dt$$

Considering $|\langle e_{r,v}, s_{r',v'}\rangle|$ as a function of r', v', one tries to maximize this function for a signal $s(t) \in L_2(R)$, the square integrable functions on the real line. Assume that s is normalized to have unit energy: $||s||^2 = \langle s, s, \rangle = 1$. Then by the Schwarz inequality we have

$$|\langle e_{r,v}, s_{r',v'}\rangle| \le ||e_{r,v}|| \cdot ||s_{r',v'}|| = 1.$$

Furthermore the maximum is assumed for $e_{r,v} \equiv e^{i\alpha} s_{r',v'}$ where α is a real constant. Hence, the maximum is assumed if and only if $r' = r, v' = v$.

We can simplify the notation by remarking that

$$(2.11) \qquad \begin{aligned} I(u, w) &= |\langle e_{r,v}, s_{r',v'}\rangle|^2 \\ &= |A_s(u_0 - u, w_0 - w)|^2 \end{aligned}$$

where

$$(2.12) \qquad \begin{aligned} A_s(u, w) &= \int_{-\infty}^{\infty} s\left(t - \frac{u}{2}\right) \bar{s}\left(t + \frac{u}{2}\right) e^{4\pi i t w} dt \\ &= \int_{-\infty}^{\infty} S(\omega - w)\overline{S}(\omega + w) e^{-2\pi i \omega u} d\omega, \end{aligned}$$

$$u_0 = \frac{2r}{c}, u = \frac{2r'}{c}, w_0 = \frac{\omega_0 v}{c}, w = \frac{\omega_0 v'}{c}.$$

Here, $A_s(u, w)$ is known as the **radar (self-) ambiguity function**.

To determine the velocity from the ambiguity function one typically chooses a single-frequency signal of the form $s(t) = \chi_T(t)e^{2\pi i \omega_0 t}$ where $T \gg \omega_0^{-1}$ and

$$\chi_T(t) = \begin{cases} 1 \text{ if } -T \le t \le T \\ 0 \text{ otherwise}. \end{cases}$$

It can be shown in this case that as a function of v' the ambiguity function has a sharp peak about $v' = v$, whereas it is relatively insensitive to changes in r'. To estimate the range one typically chooses $s(t) = G_\sigma(t)e^{2\pi i \omega_0 t}$ where $G_\sigma(t)$ is a very peaked Gaussian wave function, centered at $t = 0$ and with standard deviation $G \ll 1$. This gives an ambiguity function with a sharp peak about $r' = r$, but which is relatively insensitive to changes in v'.

Now we consider the wideband case and generalize to allow a distribution of moving targets with density function $D(x, y)$, where the support of this function is contained in the set $\{(x, y) : y > 0\}$. Then from (2.4) the echo from the signal $s(t)$ is

$$(2.13) \qquad e(t) = \int_0^{\infty} \int_{-\infty}^{\infty} \sqrt{y} s(y[t + x]) D(x, y) dx dy.$$

Correlating the echo with a "test" echo $e'(t)$ generated from the signal $s(t)$ and the "test" density function $D'(x, y)$ we obtain

$$(2.14) \qquad \langle e, e' \rangle = \int_{-\infty}^{\infty} \int_0^{\infty} \int_{-\infty}^{\infty} \int_0^{\infty} A(x, y, \tilde{x}, \tilde{y}) D(x, y)\overline{D'(\tilde{x}, \tilde{y})} dx dy d\tilde{x} d\tilde{y}$$

where

(2.15)
$$A(x, y, \tilde{x}, \tilde{y}) = \int_{-\infty}^{\infty} \sqrt{y\tilde{y}} s(y[t + x]) \bar{s}(\tilde{y}[t + \tilde{x}]) dt$$

is the **radar (self-) ambiguity function** in the wideband case. (A very similar construction of a moving target distribution can be carried out in the narrow-band case.) Note that if $D'(\tilde{x}, \tilde{y}) = \delta(\tilde{x} - x_1, \tilde{y} - y_1)$ and $D(x, y) = \delta(x - x_0, y - y_0)$ then

(2.16)
$$\langle e, e' \rangle (x_0, y_0, x_1, y_1) = A(x_0, y_0, x_1, y_1).$$

As with the narrow-band ambiguity function, if we normalize the signal s to have unit energy, then by the Schwarz inequality

(2.17)
$$|A(x, y, \tilde{x}, \tilde{y})| \leq ||s|| \cdot ||s|| = 1$$

and the maximum is assumed for $y = \tilde{y}, x = \tilde{x}$. Note also that

$$A(x, y, \tilde{x}, \tilde{y}) = A\left((x - \tilde{x})\tilde{y}, \frac{y}{\tilde{y}}, 0, 1\right).$$

Thus to study the ambiguity function one can restrict to the case

(2.18)
$$A(x, y) = A(x, y, 0, 1) = \sqrt{y} \int_{-\infty}^{\infty} s(y[t + x]) \bar{s}(t) dt.$$

Suppose $\{s_n\}$ is a basis for $L_2(\mathbb{R})$, the standard space of square integrable functions on the real line. Then we can consider the cross-correlation of s_m with the echo e_n from s_n:

(2.19)
$$A_{nm}(x, y) = \langle e_n, s_m \rangle = \sqrt{y} \int_{-\infty}^{\infty} s_n(y[t + x]) \bar{s}_m(t) dt.$$

We call A_{nm} the **cross-ambiguity function** of s_n and s_m. Similarly the **(narrow-band)** cross-ambiguity function is

$$\psi_{nm}(u, w) = \int_{-\infty}^{\infty} s_n\left(t - \frac{u}{2}\right) \bar{s}_m\left(t + \frac{u}{2}\right) e^{4\pi itw} dt.$$

The ambiguity and cross-ambiguity functions are of fundamental importance in the theory of radar/sonar. The use of the ambiguity function to estimate the range and velocity of point targets and, more generally, of the cross-ambiguity function to estimate target distribution functions in both the wideband and narrow band cases is basic to the theory.

In these lectures we shall summarize and elucidate this theory by exploiting the intimate relationship between the cross-ambiguity functions defined above and the theory of group representations. In particular the wide-band cross-ambiguity functions can be interpreted as matrix elements of unitary irreducible representations of the two-dimensional affine group; the narrow-band cross-ambiguity functions are matrix elements of unitary irreducible representations of the three-dimensional Heisenberg group. The basic properties of these functions thus emerge as consequences of analysis on affine and Heisenberg groups.

2.3 Exercises.

2.1 Consider the Gaussian pulse

$$s(t) = (\frac{2}{\pi T^2})^{\frac{1}{4}} e^{-t^2/T^2 + 2\pi i \omega_0 t},$$

normalized to have unit energy. Verify that the ambiguity function is given by

$$A_s(u, w) = e^{-\frac{1}{2}\left(\frac{u^2}{T^2} + 4\pi^2 w^2 T^2\right)} e^{-2\pi i \omega_0 u}.$$

Describe the level curves $|A_s(u, w)| = k$ in the $u - w$ plane. Discuss the effect of varying the pulse length T on the problem of estimating the range and velocity of the target.

2.2 Show that the area enclosed by a level curve $|A_s(u, w)| = k$ in Exercise 2.1 is independent of T.

2.3 Consider the normalized frequency modulated pulse

$$s(t) = (\frac{2}{\pi T^2})^{\frac{1}{4}} e^{-t^2/T^2 + 4\pi i (\omega_0 t + \gamma t^2)}.$$

The ambiguity function is given by

$$A_s(u, w) = e^{-\frac{1}{2}\left[(1 + 16\pi^2\gamma^2 T^4)\frac{u^2}{T^2} - 8\pi^2\gamma T^2 wu + 4\pi^2 w^2 T^2\right]} e^{-2\pi i \omega_0 u}.$$

Describe the level curves $|A_s(u, w)| = k$ in the $u - w$ plane. Discuss the effect of varying the pulse length T and the "compression ratio" $m = \sqrt{1 + 16\pi^2\gamma^2 T^4}$ on the problem of estimating the range and velocity of the target.

2.4 Consider the rectangular pulse with unit energy

$$s(t) = \frac{1}{\sqrt{2T}} \chi_T(t) e^{2\pi i \omega_0 t}$$

where

$$\chi_T(t) = \begin{cases} 1 \text{ if } -T \le t \le T \\ 0 \text{ otherwise .} \end{cases}$$

Show that the ambiguity function is

$$A_s(u, w) = \begin{cases} \sin[(1 - \frac{|u|}{2T})(4\pi wT)]/4\pi wT \text{ if } |u| \le 2T \\ 0 \text{ if } |u| > 2T. \end{cases}$$

(This isn't easy.) Describe the level curves $|A_s(u, w)| = k$ in the $u - w$ plane. Show that for $k = 1 - c^2$ with c very close to zero, the level curves can be approximated by $\frac{|u|}{2T} + \frac{8}{3}\pi^2 w^2 T^2 = c^2$.

§3. A GROUP THEORY PRIMER

3.1 Definitions and examples. A group is an abstract mathematical entity which expresses the intuitive concept of symmetry.

DEFINITION. A **group** G is a set of objects $\{g, h, k, \cdots\}$ (not necessarily countable) together with a binary operation which associates to any ordered pair of elements g, h in G a third element gh. the binary operation (called **group multiplication**) is subject to the following requirements:

(1) There exists an element e in G called the **identity element** such that $ge = eg = g$ for all $g \in G$.

(2) For every $g \in G$ there exists in G an **inverse element** g^{-1} such that $gg^{-1} = g^{-1}g = e$.

(3) Associative law. The identity $(gh)k = g(hk)$ is satisfied for all $g, h, k \in G$.

Any set together with a binary operation which satisfies conditions (1) - (3) is called a group. If $gh = hg$ we say that the elements g and h commute. If all elements of G commute then G is a **commutative** or **abelian** group. If G has a finite number $n(G)$ of elements it has **finite order**; otherwise G has **infinite order**.

A **subgroup** H of G is a subset which is itself a group under the group multiplication defined in G. The subgroups G and $\{e\}$ are called **improper** subgroups of G; all other subgroups are **proper**. It can be shown that the identity element e is unique. Also, every element g of G has a unique inverse g^{-1}.

Examples of groups.

(1) The real numbers R with addition as the group product. The product of $x_1, x_2 \in R$ is their sum $x_1 + x_2$. The identity is 0 and the inverse of x is $-x$. Here, R is an infinite abelian group. Among the proper subgroups of R are the integers and the even integers.

(2) The nonzero real numbers in R with multiplication of real numbers as the (commutative) group product. The identity is 1 and the inverse of x is $1/x$. The positive real numbers form a proper subgroup.

(3) The set of matrices

$$R' = \left\{ \begin{pmatrix} 1 & x \\ 0 & 1 \end{pmatrix} : x \in R \right\}$$

with matrix multiplication as the group product. The identity element is the identity matrix. Here,

$$\begin{pmatrix} 1 & x_1 \\ 0 & 1 \end{pmatrix} \begin{pmatrix} 1 & x_2 \\ 0 & 1 \end{pmatrix} = \begin{pmatrix} 1 & x_1 + x_2 \\ 0 & 1 \end{pmatrix}$$

so the one-to-one mapping

$$x \longleftrightarrow r(x) = \begin{pmatrix} 1 & x \\ 0 & 1 \end{pmatrix} \in R'$$

relating the group element x in R to r in R' takes products to products. (A one-to-one mapping from a group G onto a group G' which takes products to products is called a **group isomorphism**. We can identify the two groups in the sense that they have the same multiplication table.) Thus R and R' are isomorphic groups.

More generally we define a **homomorphism** $\mu : G \to G'$ as a mapping from the group G into a group G' which transforms products into products. Thus to every $g \in G$ there is associated $\mu(g) \in G'$ such that $\mu(g_1 g_2) = \mu(g_1)\mu(g_2)$ for all $g_1, g_2 \in G$. (It follows that $\mu(e) = e'$ where e' is the identity element in G', and $\mu(g^{-1}) = \mu(g)^{-1}$.) If μ is one-to-one and onto (i.e. if $\mu(G) = G'$) then μ is an isomorphism of G and G'.

(4) The **symmetric group** S_n. Let n be a positive integer. A **permutation** s of n objects (say the set $X = \{1, 2, \cdots, n\}$) is a one-to-one mapping of X onto itself. We can write such a permutation as

(3.1)
$$s = \begin{pmatrix} 1 & 2 & \cdots & n \\ p_1 & p_2 & \cdots & p_n \end{pmatrix}$$

where 1 is mapped into p_1, 2 into p_2, $\cdots$, n into p_n. The numbers $p_1, \cdots, p_n$ are a reordering of $1, 2, \cdots, n$, and no two of the p_j are the same. The order in which the columns of (3.1) are written is unimportant. The **inverse** permutation s^{-1} is given by

$$s^{-1} = \begin{pmatrix} p_1 & p_2 & \cdots & p_n \\ 1 & 2 & \cdots & n \end{pmatrix}$$

and the **product** of two permutations s and

$$t = \begin{pmatrix} q_1 & q_2 & \cdots & q_n \\ 1 & 2 & \cdots & n \end{pmatrix}$$

is the permutation

$$st = \begin{pmatrix} q_1 & q_2 & \cdots & q_n \\ p_1 & p_2 & \cdots & p_n \end{pmatrix}.$$

(Here we read the product from right to left: t maps q_i to i and s maps i to p_i, so st maps q_i to p_i.) The **identity** permutation is

$$e = \begin{pmatrix} 1 & 2 & \cdots & n \\ 1 & 2 & \cdots & n \end{pmatrix}.$$

It is straightforward to show that the permutations of n objects form a group S_n of finite order $n(S_n) = n!$. For $n > 2$, S_n is not commutative.

(5) The **real general linear group** $GL(n, R)$. The group elements A are nonsingular $n \times n$ matrices with real coefficients:

$$GL(n, R) = \{A = (A_{ij}), 1 \leq i, j \leq n : A_{ij} \in R \text{ and } \det A \neq 0\}.$$

Group multiplication is ordinary matrix multiplication. The identity element is the identity matrix $E = (\delta_{ij})$ where δ_{ij} is the Kronecker delta. The inverse of an element A is its matrix inverse. This group is infinite and for $n \geq 2$ it is non-abelian. Among the subgroups of $GL(n, R)$ are the **real special linear group**

$$SL(n, R) = \{A \in GL(n, R) : \det A = 1\}$$

and the **orthogonal group**

$$O(n) = \{A \in GL(n, R) : AA^t = E\}$$

where $A^t = (A_{ji})$ is the **transpose** of the $n \times n$ matrix $A = (A_{ij})$.

Similarly the **complex general linear group**

$$GL(n, \mathcal{C}) = \{A = (A_{ij}), 1 \leq i, j \leq n : A_{ij} \in \mathcal{C} \text{ and } \det A \neq 0\},$$

where $\mathcal{C}$ is the field of complex numbers, is a group under matrix multiplication. Among its subgroups are the **complex special linear group**

$$SL(n, \mathcal{C}) = \{A \in GL(n, \mathcal{C}) : \det A = 1\},$$

and the **unitary group**

$$U(n) = \{A \in GL(n, \mathcal{C}) : A\bar{A}^t = E\}$$

where $\bar{A} = \{\bar{A}_{ij}\}$ is the **complex** conjugate of $A = \{A_{ij}\}$. For $n = 1$, $U(1) = \{z : |z| = 1\}$ is the **circle group**.

It is not difficult to show that every finite group is isomorphic to a subgroup of S_n. Furthermore, every finite group is isomorphic to a (finite) subgroup of $GL(n, R)$.

(6) The group Z_n. This is the finite abelian group whose elements are the integers $0, 1, 2, \cdots, n-1$ for fixed $n \geq 1$, and group multiplication is addition mod n. Thus 0 is the identity element and $n - k$ is the inverse of $k = 1, \ldots, n - 1$. This group is isomorphic to the multiplicative group of 1×1 matrices

$$C_n = \{\exp(2\pi i k/n) : k = 0, 1, \cdots, n - 1\},$$

where

$$k \longmapsto \exp(2\pi i k/n)$$

is the isomorphism.

(7) The affine or $ax + b$ group. The affine group G_A is the subgroup of $GL(n, R)$ consisting of matrices

$$(a, b) = \begin{pmatrix} a & b \\ 0 & 1 \end{pmatrix}, \quad a, b \in R, \ a > 0.$$

Clearly the group product is

$$(a_1, b_1) \cdot (a_2, b_2) = \begin{pmatrix} a_1 a_2 & a_1 b_2 + b_1 \\ 0 & 1 \end{pmatrix}$$

$$= (a_1 a_2, a_1 b_2 + b_1),$$

the identity element is $(1, 0)$, and the inverse of (a, b) is $(1/a, -b/a)$.

(8) The (three-dimensional real) **Heisenberg group**

$$H_R = \left\{ (x_1, x_2, x_3) = \begin{pmatrix} 1 & x_1 & x_3 \\ 0 & 1 & x_2 \\ 0 & 0 & 1 \end{pmatrix} : x_i \in R \right\}.$$

This is a subgroup of $GL(3, R)$ with group product

$$(x_1, x_2, x_3) \cdot (y_1, y_2, y_3) = (x_1 + y_1, x_2 + y_2, x_3 + y_3 + x_1 y_2).$$

The identity element is the identity matrix $(0, 0, 0)$ and $(x_1, x_2, x_3)^{-1} = (-x_1, -x_2, x_1 x_2 - x_3)$. Note that H_R is non abelian. The **center** of H_R, i.e., the subgroup C_R of all elements which commute with every element of H_R,

$$C_R = \{h \in H_R : hg = gh, \forall g \in H_R\}$$

consists of the elements $(0, 0, x_3), x_3 \in R$. A finite analog of H_R is

$$H_n = \left\{ \begin{pmatrix} 1 & a_1 & a_3 \\ 0 & 1 & a_2 \\ 0 & 0 & 1 \end{pmatrix} : a_i \in Z_n \right\},$$

a finite group under matrix multiplication where addition and multiplication of the number a_i is carried out mod n.

(9) Let V be a finite dimensional vector space, real or complex, and denote by $GL(V)$ the set of all invertible linear transformations of $\mathbf{T}$ of V onto V. Then $GL(V)$ is a group with the product of $\mathbf{T}_1, \mathbf{T}_2 \in GL(V)$ given by $\mathbf{T}_1 \mathbf{T}_2$ where $(\mathbf{T}_1 \mathbf{T}_2) \mathbf{v} = \mathbf{T}_1 (\mathbf{T}_2 \mathbf{v})$ for each $\mathbf{v} \in V$, i.e. , $\mathbf{T}_1 \mathbf{T}_2$ is the usual product of linear transformations. The identity element is the operator $\mathbf{E}$ such that $\mathbf{E} \mathbf{v} = \mathbf{v}$ for all $\mathbf{v} \in G$. The inverse of $\mathbf{T} \in GL(V)$ is the inverse linear operator $\mathbf{T}^{-1}$ where $\mathbf{T} \mathbf{v} = \mathbf{w}$ for $\mathbf{v}, \mathbf{w} \in V$ if and only if $\mathbf{T}^{-1} \mathbf{w} = \mathbf{v}$.

Suppose $\dim V = n$ and let $\mathbf{v}_1, \cdots, \mathbf{v}_n$ be a basis for V. The matrix of the operator $\mathbf{T}$ with respect to the basis is $T = (T_{ij})$ where $\mathbf{T} \mathbf{v}_j = \sum_{i=1}^{n} T_{ij} \mathbf{v}_i, 1 \leq j \leq n$. It is easy to see that this correspondence defines an isomorphism between the group $GL(V)$ and $GL(n)$.

Many of the most important applications of groups to the sciences are expressed in terms of the group representation, to which we now turn.

3.2 Group representations.

DEFINITION. A **representation (rep)** of a group G with **representation space** V is a homomorphism $\mathbf{T} : g \rightarrow \mathbf{T}(g)$ of G into $GL(V)$. The **dimension** of the representation is the dimension of V.

It follows from this definition that

$$
(3.2) \qquad
\begin{aligned}
\mathbf{T}(g_1)\mathbf{T}(g_2) &= \mathbf{T}(g_1 g_2), \quad \mathbf{T}(g)^{-1} = \mathbf{T}(g^{-1}), \\
\mathbf{T}(e) &= \mathbf{E}, \quad g_1, g_2, g \in G,
\end{aligned}
$$

(Initially we shall consider only finite-dimensional reps of groups on complex rep spaces. Later we will lift these finiteness restrictions.)

DEFINITION. An **n-dimensional matrix rep** of G is a homomorphism $T : g \rightarrow T(g)$ of G into $GL(n, \mathcal{C})$.

The $n \times n$ matrices $T(g), g \in G$, satisfy multiplication properties analogous to (3.2). Any group rep $\mathbf{T}$ of G with rep space V defines many matrix reps, since if $\{\mathbf{v}_1, \cdots, \mathbf{v}_n\}$ is a basis of V, the matrices $T(g) = (T(g)_{kj})$ defined by

$$
(3.3) \qquad \mathbf{T}(g)\mathbf{v}_k = \sum_{j=1}^{n} T(g)_{jk}\mathbf{v}_j, \quad 1 \leq k \leq n
$$

form an n-dimensional matrix rep of G. Every choice of basis for V yields a new matrix rep of G defined by $\mathbf{T}$. However, any two such matrix reps T, T' are equivalent in the sense that there exists a matrix $S \in GL(n, \mathcal{C})$ such that

$$
(3.4) \qquad T'(g) = ST(g)S^{-1}
$$

for all $g \in G$. Indeed, if T, T' correspond to the bases $\{\mathbf{v}_i\}, \{\mathbf{v}_i'\}$ respectively, then for S we can take the matrix (S_{ji}) defined by

$$
(3.5) \qquad \mathbf{v}_i = \sum_{j=1}^{n} S_{ji}\mathbf{v}_j', \quad i = 1, \cdots, n.
$$

DEFINITION. Two complex n-dimensional matrix reps T and T' are **equivalent** $(T \approx T')$ if there exists an $S \in GL(n, \mathcal{C})$ such that (3.4) holds.

Thus equivalent matrix reps can be viewed as arising from the same operator rep. Conversely, given an n-dimensional matrix rep $T(g)$ we can define many n-dimensional operator reps of G. If V is an n-dimensional vector space with basis $\{\mathbf{v}_i\}$ we can define the group rep $\mathbf{T}$ by (3.3), i.e., we define the operator $\mathbf{T}(g)$ by the right-hand side of (3.3). Every choice of a vector space V and a basis $\{\mathbf{v}_i\}$ for V yields a new operator rep defined by T. However, if V, V' are two such n-dimensional vector spaces with bases $\{\mathbf{v}_i\}, \{\mathbf{v}_i'\}$ respectively, then the reps $\mathbf{T}$ and $\mathbf{T}'$ are related by

$$
(3.6) \qquad \mathbf{T}'(g) = \mathbf{S}\mathbf{T}(g)\mathbf{S}^{-1},
$$

where **S** is an invertible operator from V onto V' defined by

$$\mathbf{S}\mathbf{v}_i = \mathbf{v}'_i \quad 1 \le i \le n.$$

DEFINITION. Two n-dimensional group reps $\mathbf{T}, \mathbf{T}'$ of G on the spaces V, V' are **equivalent** ($\mathbf{T} \cong \mathbf{T}'$) if there exists an invertible linear transformation $\mathbf{S}$ of V onto V' such that (3.6) holds.

Clearly, there is a one-to-one correspondence between classes of equivalent operator reps and classes of equivalent matrix reps. In order to determine all possible reps of a group G it is enough to find one rep $\mathbf{T}$ in each equivalence class. It is a matter of choice whether we study operator reps or matrix reps.

The following are examples of group reps:

(1) The matrix groups $GL(n, \mathscr{C}), SL(n, \mathscr{C})$ are n-dimensional matrix reps of themselves.

(2) Let G be a finite group of order n. We formally define an n-dimensional vector space R_G consisting of all elements of the form

$$\sum_{g \in G} x(g) \cdot g, \quad x(g) \in \mathscr{C}.$$

Two vectors $\sum x(g) \cdot g$ and $\sum y(g) \cdot g$ are equal if and only if $x(g) = y(g)$ for all $g \in G$. The sum of two vectors and the scalar multiple of a vector are defined by

$$(3.7) \qquad \begin{aligned} \sum x(g) \cdot g + \sum y(g) \cdot g &= \sum [x(g) + y(g)] \cdot g, \\ \alpha \sum x(g) \cdot g &= \sum \alpha x(g) \cdot g, \quad \alpha \in \mathscr{C}. \end{aligned}$$

The zero vector of R_G is $\theta = \sum 0 \cdot g$. The vectors $1 \cdot g_0$, $g_0 \in G$, form a natural basis for R_G. (From now on, we make the identification $1 \cdot g = g \in R_G$.) We define the **product** of two elements $x = \sum x(g) \cdot g$, $y = \sum y(h) \cdot h$ in a natural manner:

$$(3.8) \qquad \begin{aligned} xy = \left(\sum x(g) \cdot g\right)\left(\sum y(h) \cdot h\right) &= \sum_{g,h \in G} x(g)y(h) \cdot gh \\ &= \sum_{k \in G} xy(k) \cdot k, \end{aligned}$$

where

$$(3.9) \qquad xy(k) = \sum_{h \in G} x(h)y(h^{-1}k).$$

(Here $xy(g)$ is called the **convolution** of the functions $x(g), y(g)$.) It is easy to verify the following relations:

$$(3.10) \qquad \begin{aligned} (x + y)z &= xz + yz, \quad x(y + z) = xy + xz, \quad x, y, z \in R_G, \\ (xy)z &= x(yz), \quad \alpha(xy) = (\alpha x)y = x(\alpha y), \\ ex &= xe = x, \quad \alpha \in \mathscr{C}, \end{aligned}$$

where e is the identity element of G. Thus R_G is an algebra, called the **group ring** of G.

The mapping $\mathbf{L}$ of G into $GL(R_G)$ given by

(3.11) $$\mathbf{L}(g)x = gx, \quad x \in R_G$$

defines an n-dimensional rep of G, the **(left) regular** rep. Indeed,

$$\mathbf{L}(g_1 g_2)x = g_1 g_2 x = \mathbf{L}(g_1)g_2 x = \mathbf{L}(g_1)\mathbf{L}(g_2)x$$
$$\mathbf{L}(e)x = ex = x$$

and the $\mathbf{L}(g)$ are linear operators. Similarly, the **(right) regular** rep of G is defined by

(3.12) $$\mathbf{R}(g)x = xg^{-1}, \quad x \in R_G, \ g \in G.$$

Let $\mathbf{T}$ be a rep of the finite group G on a finite-dimensional inner product space V. The rep $\mathbf{T}$ is **unitary** if for all $g \in G$

(3.13) $$\langle \mathbf{T}(g)\mathbf{v}, \mathbf{T}(g)\mathbf{w} \rangle = \langle \mathbf{v}, \mathbf{w} \rangle, \quad \mathbf{v}, \mathbf{w} \in V,$$

i.e., if the operators $\mathbf{T}(g)$ are unitary. Recall that an **orthonormal (ON)** basis for the n-dimensional space V is a basis $\{\mathbf{v}_1, \cdots, \mathbf{v}_n\}$ such that $\langle \mathbf{v}_i, \mathbf{v}_j \rangle = \delta_{ij}$, where $\langle \cdot, \cdot \rangle$ is the inner product on V. The matrices $T(g)$ of the operators $\mathbf{T}(g)$ with respect to an ON basis $\{\mathbf{v}_i\}$ are unitary matrices

$$\overline{T(g)}_{ji} = T(g^{-1})_{ij} = [T(g)^{-1}]_{ij}.$$

Hence, they form a unitary matrix rep of G. The following theorem shows that for finite groups at least, we can always restrict ourselves to unitary reps.

THEOREM 3.1. *Let* $\mathbf{T}$ *be a rep of* G *on the inner product space* V. *Then* $\mathbf{T}$ *is equivalent to a unitary rep on* V.

Proof. First we define a new inner product $(\cdot, \cdot)$ on V with respect to which $\mathbf{T}$ is unitary. For $\mathbf{u}, \mathbf{v} \in V$ let

(3.14) $$(\mathbf{u}, \mathbf{v}) = \frac{1}{n(G)} \sum_{g \in G} \langle \mathbf{T}(g)\mathbf{u}, \mathbf{T}(g)\mathbf{v} \rangle.$$

(Thus $(\mathbf{u}, \mathbf{v})$ is an average of the numbers $\langle \mathbf{T}(g)\mathbf{u}, \mathbf{T}(g)\mathbf{v} \rangle$ taken over the group.) It is easy to check that $(\cdot, \cdot)$ is an inner product on V. Furthermore,

$$\begin{aligned}
(\mathbf{T}(h)\mathbf{u}, \mathbf{T}(h)\mathbf{v}) &= \frac{1}{n(G)} \sum_{g \in G} \langle \mathbf{T}(gh)\mathbf{u}, \mathbf{T}(gh)\mathbf{v} \rangle \\
&= \frac{1}{n(G)} \sum_{g' \in G} \langle \mathbf{T}(g')\mathbf{u}, \mathbf{T}(g')\mathbf{v} \rangle \\
&= (\mathbf{u}, \mathbf{v}),
\end{aligned}$$

where the next to last equality follows from the fact that if g runs through the elements of G exactly once, then so does gh. Clearly, $\mathbf{T}$ is unitary with respect to the inner product $(\cdot, \cdot)$. Now let $\{\mathbf{u}_i\}$ be an ON basis of V with respect to $(\cdot, \cdot)$ and let $\{\mathbf{v}_i\}$ be an ON basis with respect to $\langle \cdot, \cdot \rangle$. Define the nonsingular linear operator $\mathbf{S} : V \to V$ by $\mathbf{S}\mathbf{u}_i = \mathbf{v}_i, 1 \le i \le n$. Then for $\mathbf{w} = \sum \alpha_i \mathbf{u}_i$ and $\mathbf{x} = \sum \beta_j \mathbf{u}_j$ we find

$$\langle \mathbf{S}\mathbf{w}, \mathbf{S}\mathbf{x} \rangle = \sum_{i,j} \alpha_i \overline{\beta_j} \langle \mathbf{S}\mathbf{u}_i, \mathbf{S}\mathbf{u}_j \rangle$$
$$= \sum_i \alpha_i \overline{\beta_i} = (\mathbf{w}, \mathbf{x})$$

so $\langle \mathbf{w}, \mathbf{x} \rangle = (\mathbf{S}^{-1}\mathbf{w}, \mathbf{S}^{-1}\mathbf{x})$ and

$$\langle \mathbf{S}\mathbf{T}(g)\mathbf{S}^{-1}\mathbf{w}, \mathbf{S}\mathbf{T}(g)\mathbf{S}^{-1}\mathbf{x} \rangle = (\mathbf{T}(g)\mathbf{S}^{-1}\mathbf{w}, \mathbf{T}(g)\mathbf{S}^{-1}\mathbf{x})$$
$$= (\mathbf{S}^{-1}\mathbf{w}, \mathbf{S}^{-1}\mathbf{x}) = \langle \mathbf{w}, \mathbf{x} \rangle.$$

Thus, the rep $\mathbf{T}'(g) = \mathbf{S}\mathbf{T}(g)\mathbf{S}^{-1}$ is unitary on V. ☐

It follows that we can always assume that a rep $\mathbf{T}$ on V is unitary.

Now we study the decomposition of a finite-dimensional rep of a finite group G into irreducible components.

DEFINITION. A subspace W of V is **invariant** under $\mathbf{T}$ if $\mathbf{T}(g)\mathbf{w} \in W$ for every $g \in G$, $\mathbf{w} \in W$.

If W is invariant we can define a rep $\mathbf{T}' = \mathbf{T}|W$ of G on W by

$$\mathbf{T}'(g)\mathbf{w} = \mathbf{T}(g)\mathbf{w}, \quad \mathbf{w} \in W.$$

This rep is called the **restriction** of $\mathbf{T}$ to W. If $\mathbf{T}$ is unitary so is $\mathbf{T}'$.

DEFINITION. The rep $\mathbf{T}$ is **reducible** if there is a proper subspace W of V which is invariant under $\mathbf{T}$. Otherwise, $\mathbf{T}$ is **irreducible (irred)**.

A rep is irred if the only invariant subspaces of V are $\{\boldsymbol{\theta}\}$, the zero vector, and V itself.

Every reducible rep $\mathbf{T}$ can be decomposed into irred reps in an almost unique manner. In proving this, we can assume that $\mathbf{T}$ is unitary.

If W is a proper subspace of the inner product space V and

$$(3.15) \qquad W^\perp = \{\mathbf{v} \in V : \langle \mathbf{v}, \mathbf{w} \rangle = 0, \quad \text{all } \mathbf{w} \in W\}$$

is the subspace of all vectors perpendicular to W, it is easy to show that $V = W \oplus W^\perp, (V$ is the **direct sum** of W and $W^\perp$). That is, every $\mathbf{v} \in V$ can be written uniquely in the form

$$\mathbf{v} = \mathbf{w} + \mathbf{w}', \quad \mathbf{w} \in W, \quad \mathbf{w}' \in W^\perp.$$

THEOREM 3.2. *If $\mathbf{T}$ is a reducible unitary rep of G on V and W is a proper invariant subspace of W, then $W^\perp$ is also a proper invariant subspace of V. In this case we write $\mathbf{T} = \mathbf{T}' \oplus \mathbf{T}''$ and say that $\mathbf{T}$ is the **direct sum** of $\mathbf{T}'$ and $\mathbf{T}''$, where $\mathbf{T}', \mathbf{T}''$ are the (unitary) restrictions of $\mathbf{T}$ to $W, W^\perp$, respectively.*

Proof. We must show $\mathbf{T}(g)\mathbf{u} \in W^\perp$ for every $g \in G, \mathbf{u} \in W^\perp$. Now for every $\mathbf{w} \in W$,

$$\langle \mathbf{T}(g)\mathbf{u}, \mathbf{w} \rangle = \langle \mathbf{u}, \mathbf{T}(g^{-1})\mathbf{w} \rangle = 0$$

since $\mathbf{T}(g^{-1})\mathbf{w} \in W$. The first equality follows from (3.13) and unitarity. Thus $\mathbf{T}(g)\mathbf{u} \in W^\perp$. $\square$

Suppose $\mathbf{T}$ is reducible and V_1 is a proper invariant subspace of V of smallest dimension. Then, necessarily, the restriction $\mathbf{T}_1$ of $\mathbf{T}$ to V_1 is irred and we have the direct sum decomposition $V = V_1 \oplus V_1^\perp$, where $V_1^\perp$ is invariant under $\mathbf{T}$. If $V_1^\perp$ is not irred we can find a proper irred subspace V_2 of smallest dimension such that $V_1^\perp = V_2 \oplus V_2^\perp$, by repeating the above argument. We continue in this fashion until eventually we obtain the direct sum decomposition

$$V = V_1 \oplus V_2 \oplus \cdots \oplus V_\ell \quad \text{or} \quad \mathbf{T} = \mathbf{T}_1 \oplus \mathbf{T}_2 \oplus \cdots \oplus \mathbf{T}_\ell$$

where the V_i are mutually orthogonal proper invariant subspaces of V which transform irreducibly under the restrictions $\mathbf{T}_i$ of $\mathbf{T}$ to V_i. (The decomposition process comes to an end after a finite number of steps because V is finite-dimensional.) Some of the $\mathbf{T}_i$ may be equivalent. If a_1 of the reps $\mathbf{T}_i$ are equivalent to $\mathbf{T}_1, a_2$ to $\mathbf{T}_2, \cdots, a_k$ to $\mathbf{T}_k$ and $\mathbf{T}_1, \cdots, \mathbf{T}_k$ are pairwise nonequivalent, we write

$$(3.16) \qquad \mathbf{T} = \sum_{j=1}^{k} \oplus a_j \mathbf{T}_j.$$

THEOREM 3.3. *Every finite-dimensional unitary rep of a finite group can be decomposed into a direct sum of irred unitary reps.*

The above decomposition is not unique since the irred subspaces $V_1, \cdots, V_\ell$ are not uniquely determined. However, it can be shown that the integers a_j in (3.16) are uniquely determined, [B7], [G1], [M5].

3.3 Shur's lemmas. The following two theorems (Shur's lemmas) are crucial for the analysis of irred reps.

THEOREM 3.4. *Let $\mathbf{T}, \mathbf{T}'$ be irred reps of the group G on the finite-dimensional vector spaces V, V' respectively and let $\mathbf{A}$ be a nonzero linear transformation mapping V into V', such that*

$$(3.17) \qquad \mathbf{T}'(g)\mathbf{A} = \mathbf{A}\mathbf{T}(g)$$

for all $g \in G$. Then $\mathbf{A}$ is a nonsingular linear transformation of V onto V', so $\mathbf{T}$ and $\mathbf{T}'$ are equivalent.

Proof. Let $N_\mathbf{A}$ be the **null space** and $R_\mathbf{A}$ the **range** of $\mathbf{A}$:

$$N_\mathbf{A} = \{\mathbf{v} \in V : \mathbf{A}\mathbf{v} = \boldsymbol{0}\}, \quad R_A = \{\mathbf{v}' \in V' : \mathbf{v}' = \mathbf{A}\mathbf{v} \text{ for some } \mathbf{v} \in V\}.$$

The subspace N_A of V is invariant under $\mathbf{T}$ since $\mathbf{A}\mathbf{T}(g)\mathbf{v} = \mathbf{T}'(g)\mathbf{A}\mathbf{v} = \boldsymbol{\theta}$ for all $g \in G, \mathbf{v} \in N_A$. Since $\mathbf{T}$ is irred, N_A is either V or $\{\boldsymbol{\theta}\}$. The first possibility implies $\mathbf{A}$ is the zero operator, which contradicts the hypothesis. Therefore, $N_{\mathbf{A}} = \{\boldsymbol{\theta}\}$. The subspace $R_{\mathbf{A}}$ of V' is invariant under $\mathbf{T}'$ because $\mathbf{T}'(g)\mathbf{A}\mathbf{v} = \mathbf{A}\mathbf{T}(g)\mathbf{v} \in R_{\mathbf{A}}$ for all $\mathbf{v} \in V$. But $\mathbf{T}'$ is irred so $R_{\mathbf{A}}$ is either V' or $\{\boldsymbol{\theta}\}$. If $R_{\mathbf{A}} = \{\boldsymbol{\theta}\}$ then $\mathbf{A}$ is the zero operator, which is impossible. Therefore $R_{\mathbf{A}} = V'$ which implies that $\mathbf{T}$ and $\mathbf{T}'$ are equivalent. $\square$

COROLLARY 3.1. *Let* $\mathbf{T}, \mathbf{T}'$ *be nonequivalent finite-dimensional irred reps of* G. *If* $\mathbf{A}$ *is a linear transformation from* V *to* V' *which satisfies (3.17) for all* $g \in G$, *then* $\mathbf{A}$ *is the zero operator.*

While Theorems 3.1 – 3.4 and Corollary 3.1 apply also for real vector spaces, the following results are true only for complex reps.

THEOREM 3.5. *Let* $\mathbf{T}$ *be a rep of the group* G *on the finite-dimensional complex vector space* V, $(\dim V \geq 1)$. *Then* $\mathbf{T}$ *is irred if and only if the only transformations* $\mathbf{A} : V \to V$ *such that*

$$(3.18) \qquad \mathbf{T}(g)\mathbf{A} = \mathbf{A}\mathbf{T}(g)$$

for all $g \in G$ *are* $\mathbf{A} = \lambda \mathbf{E}$ *where* $\lambda \in \mathcal{C}$ *and* $\mathbf{E}$ *is the identity operator on* V.

Proof. It is well known that a linear operator on a finite-dimensional **complex** vector space always has at least one eigenvalue. Let λ be an eigenvalue of an operator $\mathbf{A}$ which satisfies (3.18) and define the eigenspace C_λ by

$$C_\lambda = \{\mathbf{v} \in V : \mathbf{A}\mathbf{v} = \lambda\mathbf{v}\}.$$

Clearly C_λ is a subspace of V and $\dim C_\lambda > 0$. Furthermore, C_λ is invariant under $\mathbf{T}$ because

$$\mathbf{A}\mathbf{T}(g)\mathbf{v} = \mathbf{T}(g)\mathbf{A}\mathbf{v} = \lambda\mathbf{T}(g)\mathbf{v}$$

for $\mathbf{v} \in C_\lambda$, $g \in G$, so $\mathbf{T}(g)\mathbf{v} \in C_\lambda$. If $\mathbf{T}$ is irred then $C_\lambda = V$ and $\mathbf{A}\mathbf{v} = \lambda\mathbf{v}$ for all $\mathbf{v} \in V$.

Conversely, suppose $\mathbf{T}$ is reducible. Then there exists a proper invariant subspace V_1 of V and by Theorem 3.2, a proper invariant subspace V_2 such that $V = V_1 \oplus V_2$. Any $\mathbf{v} \in V$ can be written uniquely as $\mathbf{v} = \mathbf{v}_1 + \mathbf{v}_2$ with $\mathbf{v}_j \in V_j$. We define the projection operator $\mathbf{P}$ on V by $\mathbf{P}\mathbf{v} = \mathbf{v}_1 \in V_1$. Then $\mathbf{P}\mathbf{T}(g)\mathbf{v} = \mathbf{T}(g)\mathbf{P}\mathbf{v} = \mathbf{T}(g)\mathbf{v}_1$ (verify this), and $\mathbf{P}$ is clearly not a multiple of $\mathbf{E}$. $\square$

Choosing a basis for V and a basis for V' we can immediately translate Shur's lemmas into statements about irred matrix reps.

COROLLARY 3.2. *Let* T *and* T' *be* $n \times n$ *and* $m \times m$ *complex irred matrix reps of the group* G, *and let* A *be an* $m \times m$ *matrix such that*

$$(3.19) \qquad T'(g)A = AT(g)$$

for all $g \in G$. If T and T' are nonequivalent then A equals the zero matrix. (In particular, this is true for $n \neq m$.) If $T \equiv T'$ then $A \equiv \lambda E_n$ where $\lambda \in \mathcal{C}$ and E_n is the $n \times n$ identity matrix.

Note that the proofs of Shur's lemmas use only the concept of irreducibility and the fact that the rep spaces are finite-dimensional. The homomorphism property of reps and the fact that G is finite are not needed.

3.4 Orthogonality relations for finite group representations. Now let G be a finite group again and select one irred rep $\mathbf{T}^{(\mu)}$ of G in each equivalence class of irred reps Then every irred rep is equivalent to some $\mathbf{T}^{(\mu)}$ and the reps $\mathbf{T}^{(\mu_1)}, \mathbf{T}^{(\mu_2)}$ are nonequivalent if $\mu_1 \neq \mu_2$. The parameter μ indexes the equivalence classes of irred reps. (We will soon show that there are only a finite number of these classes.) Introduction of a basis in each rep space $V^{(\mu)}$ leads to a matrix rep $T^{(\mu)}$. The $T^{(\mu)}$ form a complete set of irred $n_\mu \times n_\mu$ matrix reps of G, one from each equivalence class. Here $n_\mu = \dim V^{(\mu)}$. If we wish, we can choose the $T^{(\mu)}$ to be unitary.

The following procedure leads to an extremely useful set of relations in rep theory, the **orthogonality relations**. Given two irred matrix reps $T^{(\mu)}, T^{(\nu)}$ of G choose an arbitrary $n_\mu \times n_\nu$ matrix B and form the $n_\mu \times n_\nu$ matrix

$$(3.20) \qquad A = N^{-1} \sum_{g \in G} T^{(\mu)}(g) B T^{(\nu)}(g^{-1})$$

where $N = n(G)$. (Here, A is just the average of the matrices $T^{(\mu)}(g) B T^{(\nu)}(g^{-1})$ over the group G.) We will show that A satisfies

$$(3.21) \qquad T^{(\mu)}(h) A = A T^{(\nu)}(h)$$

for all $h \in G$. Indeed,

$$
\begin{aligned}
T^{(\mu)}(h) A &= N^{-1} \sum_{g \in G} T^{(\mu)}(h) T^{(\mu)}(g) B T^{(\nu)}(g^{-1}) \\
&= N^{-1} \sum_{g \in G} T^{(\mu)}(hg) B T^{(\nu)}((hg)^{-1}) T^{(\nu)}(h) \\
&= A T^{(\nu)}(h).
\end{aligned}
$$

We have used the fact that as g runs over each of the elements of G exactly once, so does $g' = hg$. This result and Corollary 3.1 imply that if $\mu \neq \nu$ then A is the zero matrix, whereas if $\mu = \nu$ then $A = \lambda E_{n_\mu}$ for some $\lambda \in \mathcal{C}$. Hence, $A = \lambda(\mu, B) \delta_{\mu\nu} E_{n_\mu}$ where $\delta_{\mu\nu}$ is the Kronecker delta, and the coefficient λ depends on μ and B.

To derive all possible consequences of this identity it is enough to let B run through the $n_\mu \times n_\nu$ matrices $B^{(\ell,m)} = (B_{jk}^{(\ell,m)})$, where

$$B_{jk}^{(\ell,m)} = \delta_{j\ell} \delta_{km}, \quad 1 \leq j, \ell \leq n_\mu, \quad 1 \leq k, m \leq n_\nu.$$

Making these substitutions, we obtain

$$(3.22) \qquad \sum_{g \in G} T_{i\ell}^{(\mu)}(g) T_{ms}^{(\nu)}(g^{-1}) = N \lambda \delta_{\mu\nu} \delta_{is}, \quad 1 \leq i, \ell \leq n_\mu, \quad 1 \leq m, s \leq n_\nu.$$

Here, λ may depend on μ, ℓ and m, but not on i or s. To evaluate λ, set $\nu = \mu, s = i$, and sum on i to obtain

$$n_\nu N \lambda = \sum_{g \in G} \sum_{i=1}^{n_\mu} T_{mi}^{(\mu)}(g^{-1}) T_{i\ell}^{(\mu)}(g) = \sum_{g \in G} T_{mi}^{(\mu)}(e)$$
$$= N \delta_{mi}$$

since $N = n(G)$ and $T_{mi}^{(\mu)}(e) = \delta_{mi}$. Therefore, $\lambda = \delta_{mi}/n_\mu$. We can simplify (3.22) slightly if we assume (as we can) that all of the matrix reps $T^{(\nu)}(g)$ are unitary. Then

$$T_{ms}^{(\nu)}(g^{-1}) = \overline{T}_{sm}^{(\nu)}(g)$$

and (3.22) reduces to

(3.23)
$$\sum_{g \in G} T_{i\ell}^{(\mu)}(g) \overline{T}_{sm}^{(\nu)}(g) = \frac{N}{n_\mu} \delta_{is} \delta_{\ell m} \delta_{\mu\nu}.$$

Expressions (3.22), (3.23) are the **orthogonality relations** for matrix elements of irred reps of G. We can write these relations in a basis-free manner. Suppose the irred unitary reps $\mathbf{T}^{(\mu)}, \mathbf{T}^\nu$ act on the rep spaces $V^{(\mu)}, V^{(\nu)}$ with ON bases $\{\mathbf{v}_i^{(\mu)}\}, \{\mathbf{v}_s^{(\nu)}\}$, respectively. Then we have $T_{i\ell}^{(\mu)}(g) = \langle \mathbf{T}^{(\mu)}(g)\mathbf{v}_\ell^{(\mu)}, \mathbf{v}_i^{(\mu)} \rangle$ with a similar expression for $\mathbf{T}^{(\nu)}$. Expanding arbitrary vectors $\mathbf{f}, \mathbf{h} \in V^{(\mu)}$ in the basis $\{\mathbf{v}_\ell^{(\mu)}\}$ and $\mathbf{f}', \mathbf{h}' \in V^{(\nu)}$ in the basis $\{\mathbf{v}_s^{(\nu)}\}$ and using (3.23) we find

(3.24)
$$\sum_{g \in G} \langle \mathbf{T}^{(\mu)}(g)\mathbf{f}, \mathbf{h} \rangle \langle \mathbf{f}', \mathbf{T}^{(\nu)}(g)\mathbf{h}' \rangle = \langle \mathbf{f}', \mathbf{h} \rangle \langle \mathbf{f}, \mathbf{h}' \rangle \delta_{\mu\nu} \frac{N}{n_\mu}.$$

That is, the left hand side of (3.24) vanishes unless $\mu = \nu$, in which case the inner products on the right hand side correspond to the space $V^{(\mu)} = V^{(\nu)}$.

To better understand the orthogonality relations it is convenient to consider the elements x of the group ring R_G as complex-valued functions $x(g)$ on the group G. The relation between this approach and the definition of R_G as given in example is provided by the correspondence

(3.25)
$$x = \sum_{g \in G} x(g) \cdot g \longleftrightarrow x(g).$$

The elements of the N-tuple $(x(g_1), \cdots, x(g_N))$, where g_i ranges over G, can be regarded as the components of $x \in R_G$ in the natural basis provided by the elements of G. Furthermore the one-to-one mapping (3.25) leads to the relations

(3.26)
$$x + y \leftrightarrow x(g) + y(g), \quad \alpha x \leftrightarrow \alpha x(g)$$
$$xy \leftrightarrow xy(g) = \sum_{g' \in G} x(g')y(g'^{-1}g)$$

where the expression defining $xy(g)$ is called the **convolution product** of $x(g)$ and $y(g)$. The ring of functions just constructed is algebraically isomorphic to R_G with the isomorphism given by (3.26). Under this isomorphism the element $h = 1 \cdot h \in R_G$ is mapped into the function

$$h(g) = \begin{cases} 1 \text{ if } g = h \\ 0 \text{ otherwise}. \end{cases}$$

Now consider the right regular rep on R_G. Writing

$$\mathbf{R}(h)x = \sum_{g \in G} [\mathbf{R}(h)x](g) \cdot g = xh^{-1} = \sum_g x(gh) \cdot g,$$

we obtain

$$(3.27) \qquad\qquad [\mathbf{R}(h)x](g) = x(gh), \quad h \in G,$$

as the action of $\mathbf{R}(h)$ on our new model of R_G. From Theorem 3.1, there is an inner product on the N-dimensional vector space R_G with respect to which the right regular rep $\mathbf{R}$ is unitary. Indeed, the following inner product works:

$$(3.28) \qquad\qquad (x, y) = N^{-1} \sum_{g \in G} x(g) \overline{y(g)}, \quad x, y \in R_G.$$

Now note that for fixed μ, i, j with $1 \leq i, j \leq n_\mu$ the matrix element $T_{ij}^{(\mu)}(g)$ defines a function on G, hence an element of R_G. Furthermore, comparing (3.28) with (3.23), we see that the functions

$$(3.29) \qquad\qquad \varphi_{ij}^{(\mu)}(g) = n_\mu^{1/2} T_{ij}^{(\mu)}(g), \quad 1 \leq i, j \leq n_\mu,$$

where μ ranges over all equivalence classes of irred reps of G, form an ON set in R_G. Since R_G is N-dimensional the ON set can contain at most N elements. Thus there are only a finite number, say ξ, of nonequivalent irred reps of G. Each irred matrix rep μ yields n_μ^2 vectors of the form (3.29). The full ON set $\{\varphi_{ij}^{(\mu)}\}$ spans a subspace of R_G of dimension

$$(3.30) \qquad\qquad n_1^2 + n_2^2 + \cdots + n_\xi^2 \leq N.$$

The inequality (3.30) is a strong restriction on the possible number and dimensions of irred reps of G. This result can be further strengthened by showing that the ON set $\{\varphi_{ij}^{(\mu)}\}$ is actually a **basis** for R_G. Since the dimension N of R_G is equal to the number of basis vectors, we obtain the equality

$$(3.31) \qquad\qquad n_1^2 + n_2^2 + \cdots + n_\xi^2 = N.$$

To prove this result, let V be the subspace of R_G spanned by the ON set $\{\varphi_{ij}^{(\mu)}\}$. From (3.29) and the homomorphism property of the matrices $T^{(\mu)}(g)$ there follows

$$(3.32) \qquad\qquad [\mathbf{R}(h)\varphi_{ij}^{(\mu)}](g) = \varphi_{ij}^{(\mu)}(gh) = \sum_{k=1}^{n_\mu} T_{kj}^{(\mu)}(h)\varphi_{ik}^{(\mu)}(g) \in V.$$

Thus V is invariant under $\mathbf{R}$. According to Theorem 3.2, $V^\perp$ is also invariant under R and $R_G = V \oplus V^\perp$. (Here $V^\perp$ is defined with respect to the inner product (3.28).) If $V^\perp \neq \{\boldsymbol{\theta}\}$ then it contains a subspace W transforming under some irred rep $\mathbf{T}^{(\nu)}$ of G. Thus, there exists an ON basis $x_1, \cdots, x_{n_\nu}$ for W such that

$$(3.33) \qquad [\mathbf{R}(g)x_i](h) = x_i(hg) = \sum_{j=1}^{n_\mu} T_{ji}^{(\nu)}(g)x_j(h), \quad 1 \le i \le n_\mu.$$

Setting $h = e$ in (3.33) we find

$$x_i(g) = \sum_j x_j(e)T_{ji}^{(\nu)}(g) = \sum_j x_j(e)\varphi_{ji}^{(\nu)}(g)/n_\nu^{1/2},$$

so $x_i \in V$. Thus $W \subseteq V \cap V^\perp$. This is possible only if $W = \{\boldsymbol{\theta}\}$. Therefore, $V^\perp = \{\boldsymbol{\theta}\}$ and $V = R_G$.

THEOREM 3.6. *The functions*

$$\{\varphi_{ij}^{(\mu)}(g)\}, \quad \mu = 1, \cdots, \xi, \quad 1 \le i, j \le n_\mu,$$

form an ON basis for R_G. Every function $x \in R_G$ can be written uniquely in the form

$$(3.34) \qquad x(g) = \sum_{i,j,\mu} a_{ij}^\mu \varphi_{ij}^{(\mu)}(g), \quad a_{ij}^\mu = (x, \varphi_{ij}^{(\mu)}).$$

The series (3.34) is the "generalized Fourier expansion" for the functions $x \in R_G$ and the a_{ij}^μ are the "generalized Fourier coefficients". Furthermore we have the "generalized Plancherel formula"

$$(3.35) \qquad (x, y) = \sum_{i,j,\mu} (x, \varphi_{ij}^{(\mu)})(\varphi_{ij}^{(\mu)}, y).$$

We can also write expansion (3.34) in a basis-free form through the use of (3.29) and the homomorphism property of the matrices $T^{(\mu)}(g)$:

$$\begin{aligned}
x(g) &= \sum_{i,j,\mu} n_\mu (x, T_{ij}^{(\mu)}) T_{ij}^{(\mu)}(g) \\
&= \frac{1}{N} \sum_{i,j,\mu} n_\mu \sum_{h \in G} x(h) \overline{T}_{ij}^{(\mu)}(h) T_{ij}^{(\mu)}(g) \\
&= \frac{1}{N} \sum_{i,j,\mu} n_\mu \sum_{h \in G} x(h) T_{ji}^{(\mu)}(h^{-1}) T_{ij}^{(\mu)}(g) \\
&= \frac{1}{N} \sum_{j,\mu} n_\mu \sum_{h \in G} x(h) T_{jj}^{(\mu)}(h^{-1}g) \\
&= \sum_{j,\mu} n_\mu (x, T_{jj}^{(\mu)}(g^{-1}\cdot)) \\
&= \sum_\mu n_\mu (x, \operatorname{tr} T^{(\mu)}(g^{-1}\cdot))
\end{aligned}$$

Here tr $T^{(\mu)}(h)$ *is the* **trace** *of the matrix* $T^{(\mu)}(h)$.

Relation (3.35) can be written in the basis-free form

$$(x,y) = \sum_{\mu} n_{\mu}(s_x^{(\mu)}, y)$$

where $s_x^{(\mu)}(h) = (x, \text{tr } T^{(\mu)}(h^{-1}\cdot))$.

3.5 Representations of Abelian groups. The representation theory of Abelian (not necessarily finite) groups is especially simple.

LEMMA 3.1. *Let G be an Abelian group and let* **T** *be a finite-dimensional irred rep of G on a complex vector space V. Then* **T**, *hence V, is one-dimensional.*

Proof. Suppose **T** is irred on V and dim $V > 1$. There must exist a $g \in G$ such that $\mathbf{T}(g)$ is not a multiple of the identity operator, for otherwise V would be reducible. Let λ be an eigenvalue of $\mathbf{T}(g)$ and let C_λ be the eigenspace

$$C_\lambda = \{\mathbf{v} \in V : \mathbf{T}(g)\mathbf{v} = \lambda\mathbf{v}\}.$$

Clearly, C_λ is a proper subspace of V. If $h \in G$ and $\mathbf{w} \in C_\lambda$ then

$$\mathbf{T}(g)(\mathbf{T}(h)\mathbf{w}) = \mathbf{T}(h)(\mathbf{T}(g)\mathbf{w}) = \lambda(\mathbf{T}(h)\mathbf{w}),$$

since G is Abelian, so C_λ is invariant under the operator $\mathbf{T}(h)$. Therefore, **T** is reducible. Contradiction! □

Example. $Z_N, N > 1$

This is the Abelian group of order N, example (6) in §3.1, with addition of integers mod N as group multiplication. Since $n_j = 1$ for each irred rep of Z_N, it follows from (3.31) that Z_N has exactly $\xi = N$ distinct irred reps. Since the N elements of Z_N can be represented as $g_0^k, k = 0, 1, 2, \cdots, N-1$ where $g_0^N = e$, we have $[\mathbf{T}^{(\mu)}(g_0)]^N = \mathbf{T}^{(\mu)}(g_0^N) = \mathbf{T}^{(\mu)}(e) = 1$ for each irred rep $T^{(\mu)}$. Thus $T^{(\mu)}(g_0)$ is an Nth root of unity and it uniquely determines $\mathbf{T}^{(\mu)}(g)$ for all $g \in Z_N$. Since there are exactly N such roots, which we label as

$$\mathbf{T}^{(\mu)}(g_0) = \exp(2\pi i\mu/N), \quad \mu = 0, 1, \cdots, N-1,$$

the possible irred reps are

$$(3.36) \qquad \mathbf{T}^{(\mu)}(g_0^k) = \exp(2\pi i k\mu/N) \quad k, \mu = 0, 1, \cdots, N-1.$$

It follows from Theorem 3.6 that every function $x \in R_{Z_N}$ has the finite Fourier expansion

$$(3.37) \qquad x(k) = \sum_{\mu=0}^{N-1} \hat{x}(\mu) \exp(2\pi i k\mu/N)$$

where

$$\hat{x}(\mu) = (x, T^{(\mu)}) = \frac{1}{N} \sum_{\ell=0}^{N-1} x(\ell) \exp(-2\pi i \ell \mu / N).$$

Furthermore, the orthogonality relations are

(3.38) $$(T^{(\nu)}, T^{(\mu)}) = \delta_{\nu\mu} = \frac{1}{N} \sum_{\ell=0}^{N-1} \exp[2\pi i \ell (\nu - \mu)/N)]$$

and the Plancherel formula reads

$$\frac{1}{N} \sum_{k=0}^{N-1} x(k)\overline{y}(k) = \sum_{\mu=0}^{N-1} \hat{x}(\mu)\overline{\hat{y}}(\mu), \quad x, y \in R_{Z_N}.$$

Since finite non-Abelian groups do not occur in the sequel, we refer the reader to standard textbooks for examples of unitary irred reps of these groups, [B7], [M5].

3.6 Exercises.

3.1 The **center** of a group G is the subgroup

$$C = \{h \in G : hg = gh, \ \forall g \in G\}.$$

Compute the center of the Heisenberg group H_R and the center of the affine group G_A.

3.2 Let V be an n-dimensional complex vector space with basis $\{\mathbf{v}_1, \cdots, \mathbf{v}_n\}$. The **matrix** of the operator $\mathbf{T} \in GL(V)$ with respect to this basis is $T = (T_{ij})$ where $\mathbf{T}\mathbf{v}_j = \sum_{i=1}^{n} T_{ij}\mathbf{v}_i$, $1 \leq j \leq n$. Verify that the correspondence $\mathbf{T} \leftrightarrow T$ is an isomorphism of the groups $GL(V)$ and $GL(n)$.

3.3 Show that the map $g \to \mathbf{R}(g)$ is a rep of the group G on the group ring R_G where $\mathbf{R}(g)x = xg^{-1}$ for $g \in G$, $x \in R_G$.

3.4 Verify explicitly that the Hermitian form $(\mathbf{u}, \mathbf{v})$, (3.14), defines an inner product on the vector space V. (Among other things, you must show that $(\mathbf{u}, \mathbf{u}) = 0$ if and only if $\mathbf{u} = \boldsymbol{0}$.)

3.5 Let $T_1(g)$ and $T_2(g)$ be $n \times n$ matrix reps of the group G with **real** matrix elements. These reps are **real equivalent** if there is a real nonsingular matrix S such that $T_1(g)S = ST_2(g)$ for all $g \in G$. Show that T_1 and T_2 are complex equivalent if and only if they are real equivalent. (Hint: Write $S = A + iB$, where A and B are real, and show that $A + tB$ is invertible for some real number t.)

3.6 Show that the matrix elements of two real irred reps of a group G which are not real equivalent satisfy an orthogonality relation. Show that every real irred rep is real equivalent to a rep by real orthogonal matrices.

§4. REPRESENTATION THEORY FOR INFINITE GROUPS

4.1 Linear Lie groups. We will now indicate how some of the basic results in the rep theory of finite groups can be extended to infinite groups. A fundamental tool in the rep theory of finite groups is the averaging of a function or operator over the group by taking a sum. We will now introduce a new class of groups, the linear Lie groups, in which one can (explicitly) integrate over the group manifold and, at least for compact linear Lie groups, can prove close analogs of Theorems 3.1–3.5.

Let W be an open connected set containing $\mathbf{e} = (0, \cdots, 0)$ in the space F_n of all (real or complex) n-tuples $\mathbf{g} = (g_1, \cdots, g_n)$. (The reader can assume W is an open sphere with center $\mathbf{e}$.)

DEFINITION. An n-dimensional **local linear Lie group** G is a set of $m \times m$ nonsingular matrices $A(\mathbf{g}) = A(g_1, \cdots, g_n)$, defined for each $\mathbf{g} \in W$, such that

(1) $A(\mathbf{e}) = E_m$ (the identity matrix).
(2) The matrix elements of $A(\mathbf{g})$ are analytic functions of the parameters $g_1, \cdots, g_n$ and the map $\mathbf{g} \to A(\mathbf{g})$ is one-to-one.
(3) The n matrices $\frac{\partial A(\mathbf{g})}{\partial g_j}$, $j = 1, \cdots, n$, are linearly independent for each $\mathbf{g} \in W$. That is, these matrices span an n-dimensional subspace of the m^2-dimensional space of all $m \times m$ matrices.
(4) There exists a neighborhood W' of $\mathbf{e}$ in F_n, $W' \subseteq W$, with the property that for every pair of n-tuples $\mathbf{g}, \mathbf{h}$ in W' there is an n-tuple $\mathbf{k}$ in W satisfying

$$(4.1) \qquad\qquad A(\mathbf{g})A(\mathbf{h}) = A(\mathbf{k})$$

where the operation on the left is matrix multiplication.

From the implicit function theorem one can show that (4) implies that there exists a nonzero neighborhood V of $\mathbf{e}$ such that $\mathbf{k} = \boldsymbol{\varphi}(\mathbf{g}, \mathbf{h})$ for all $\mathbf{g}, \mathbf{h} \in V$ where $\boldsymbol{\varphi}$ is an analytic vector-valued function of its an arguments and $\mathbf{g}, \mathbf{h}, \mathbf{k}$ are related by (4.1), [HS], [R4].

Let G be a local linear group of $m \times m$ matrices. We will now construct a (connected, global) **linear Lie group** $\tilde{G}$ containing G. Algebraically, $\tilde{G}$ is the abstract subgroup of $GL(m, \mathcal{C})$ generated by the matrices of G. That is, $\tilde{G}$ consists of all possible products of finite sequences of elements in G. In addition, the elements of $\tilde{G}$ can be parametrized analytically. If $B \in \tilde{G}$ we can introduce coordinates in a neighborhood of B by means of the map $\mathbf{g} \to BA(\mathbf{g})$ where $\mathbf{g}$ ranges over a suitably small neighborhood Z of $\mathbf{e}$ in F_n. In particular, the coordinates of B will be $\mathbf{e} = (0, \cdots, 0)$. Proceeding in this way for each $B \in \tilde{G}$ we can cover $\tilde{G}$ with local coordinate systems as "coordinate patches". The same group element C will have many different sets of coordinates, depending on which coordinate patch containing C we happen to consider. Suppose C lies in the intersection of coordinate patches around B_1 and B_2, respectively. Then C will have coordinates $\mathbf{g}_1, \mathbf{g}_2$ respectively, where $C = B_1 A(\mathbf{g}_1) = B_2 A(\mathbf{g}_2)$. Since

$$A(\mathbf{g}_1) = B_1^{-1} B_2 A(\mathbf{g}_2), \quad A(\mathbf{g}_2) = B_2^{-1} B_1 A(\mathbf{g}_1)$$

it follows that in a suitably small neighborhood of $\mathbf{e}$: (a) the coordinates $\mathbf{g}_2$ are analytic functions $\mathbf{g}_2 = \boldsymbol{\rho}(\mathbf{g}_1)$ of the coordinates $\mathbf{g}_1$, (b) $\boldsymbol{\rho}$ is one-to-one, and (c) the Jacobian of the coordinate transformation is nonzero. This makes $\tilde{G}$ into an analytic manifold. (In addition to the coordinate neighborhoods described above, we can always add more coordinate neighborhoods to $\tilde{G}$ provided they satisfy conditions (a) – (c) on the overlap with any of the original coordinate systems.) We leave it to the reader to show that $\tilde{G}$ is **connected**. That is, any two elements A, B in $\tilde{G}$ can be connected by an analytic curve $C(t)$ lying entirely in $\tilde{G}$.

In general an n-dimensional (global) **linear Lie group** K is an abstract matrix group which is also an n-dimensional local linear group G. Clearly, $K \sqsupseteq \tilde{G}$. Indeed, $\tilde{G}$ is the connected component of K containing the identity matrix. The group K need not be connected.

Examples (3), (5) ($GL(n, R)$, $SL(n, R)$, $O(n)$, $GL(n, \mathcal{C})$, $SL(n, \mathcal{C})$, $U(n)$), (7), (8) (H_R) from §3.1 are all linear Lie groups. To illustrate, we can write $A \in GL(n, R)$ for A "close" to the identity matrix as $A_{jk} = \delta_{jk} + g_{jk}, 1 \le j, k \le n$, and verify that conditions (1) – (4) of the definition of a local linear Lie group are satisfied for the coordinates $\mathbf{g} = \{g_{jk}\} \in R_{n^2}$. It follows that $GL(n, R)$ is a real global n^2-dimensional linear Lie group.

4.2 Invariant measures on Lie groups. Let G be a real n-dimensional global Lie group of $m \times m$ matrices. A function $f(B)$ on G is **continuous** at $B \in G$ if it is a continuous function of the parameters $(g_1, \cdots, g_n)$ in a local coordinate system for G at B. (Clearly, if f is continuous with respect to one local coordinate system at B it is continuous with respect to all coordinate systems.) If f is continuous at every $B \in G$ then it is a **continuous function** on G. We shall show how to define an infinitesimal volume element dA in G with respect to which the associated integral over the group is left-invariant, i.e.,

$$(4.2) \qquad \int_G f(BA)dA = \int_G f(A)dA, \quad B \in G,$$

where f is a continuous function on G such that either of the integrals converges. In terms of local coordinates $\mathbf{g} = (g_1, \cdots, g_n)$ at A,

$$(4.3) \qquad dA = w(\mathbf{g})dg_1 \cdots dg_n = w(\mathbf{g})d\mathbf{g},$$

where the positive continuous function w is called a **weight function**. If $\mathbf{k} = (\mathbf{k}_1, \cdots, \mathbf{k}_n)$ is another set of local coordinates at A then,

$$dA = \tilde{w}(\mathbf{k})dk_1 \cdots dk_n, \quad \tilde{w}(k) = w(\mathbf{g}(\mathbf{k}))|\det(\partial g_i / \partial k_j)|,$$

where the determinant is the Jacobian of the coordinate transformation. (For a precise definition of integrals on manifolds see [S4].)

Two examples of left-invariant integrals are well known. The group R' (example (3) in §3.1) with elements

$$(4.4) \qquad A(x) = \begin{pmatrix} 1 & x \\ 0 & 1 \end{pmatrix}, \quad x \in R$$

is isomorphic to the real line. The continuous functions on R' are just the continuous functions $f(x)$ on the real line. Here, dx is a left-invariant measure. Indeed by a simple change of variable we have

$$\int_{-\infty}^{\infty} f(y + x)dx = \int_{-\infty}^{\infty} f(x)dx, \quad y \in R$$

where f is any continuous function on R such that the integral converges. Since R is Abelian, dx is also right-invariant.

A second example is the circle group $U(1) = \{e^{i\phi}\}$, the Abelian group of 1×1 unitary matrices. The continuous functions on $U(1)$ can be written $f(\phi)$, where f is continuous for $0 \leq \phi \leq 2\pi$ and periodic with period 2π. The measure $d\phi$ is left-invariant (right invariant) since

$$\int_0^{2\pi} f(\alpha + \phi)d\phi = \int_0^{2\pi} f(\phi)d\phi.$$

We now show how to construct a left-invariant measure for the n-dimensional real linear Lie group G. Let $A(\mathbf{g})$ be a parametrization of G in a neighborhood of the identity element and chosen such that $A(\boldsymbol{\theta}) = E_m$. Set $\mathcal{C}_j = \partial_{g_j} A(\mathbf{g})|_{\mathbf{g}=\theta}$, $1 \leq j \leq n$. Then by property (3) of a local linear Lie group, the $m \times m$ matrices $\{\mathcal{C}_j\}$ are linearly independent. Given any other parametrization $A'(\mathbf{h})$ near the identity and such that $A'(\mathbf{h}^0) = E_m$ we have $A'(\mathbf{h}) = A(\mathbf{g}(\mathbf{h}))$ for some analytic function $\mathbf{g}(\mathbf{h})$ and $\mathcal{C}'_j = \partial_{h_j} A'(\mathbf{h})|_{\mathbf{h}=\mathbf{h}^0} = \sum_{\ell=1}^n \partial_{g_\ell} A(\mathbf{g})|_{\mathbf{g}=\theta} \left(\frac{\partial g_\ell}{\partial h_j}(\mathbf{h}^0)\right) = \sum_{\ell=1}^n \alpha_\ell^{(j)} \mathcal{C}_\ell$. Here the $\{\mathcal{C}'_j\}$ must be linearly independent so $\det(\alpha_\ell^{(j)}) \neq 0$. In other words, the **tangent space** at the identity element of G is n-dimensional and once we chose a fixed basis $\{\mathcal{C}_j\}$ for the tangent space, any coordinate system in a neighborhood of E_m will determine n linearly independent n-tuples $\alpha^{(1)}, \cdots, \alpha^{(n)}$ which generate a parallelepiped with volume

$$V = |\det(\alpha_k^{(i)})| > 0$$

in the tangent space. Now suppose $A'(\mathbf{h})$ is a parametrization of G in a neighborhood of the group element A_0 and such that $A'(\mathbf{h}^0) = A_0$. Then the matrices $\mathcal{C}'_j = \partial_{h_j} A'(\mathbf{h})|_{\mathbf{h}=\mathbf{h}^0}$, $j = 1, \cdots, n$, are linearly independent. Furthermore, the matrices in a neighborhood of $E_m \in G$ can be represented as $A_0^{-1} A'(\mathbf{h}) = A(\mathbf{g}(\mathbf{h}))$. Differentiating this expression with respect to h_j and setting $\mathbf{h} = \mathbf{h}^0$ we find

$$A_0^{-1} \mathcal{C}'_j = \sum_{\ell=1}^n \alpha_\ell^{(j)} \mathcal{C}_\ell$$

for n linearly independent n-tuples $\alpha^{(1)}, \cdots \alpha^{(n)}$. This defines a parallelepiped in the tangent space to G at A_0 as the volume of its image in the tangent space at E_m:

$$V_{A_0}^{(\mathbf{h})} = |\det(\alpha_\ell^{(j)})| > 0.$$

By construction, our volume element is left-invariant. Indeed if $B \in G$ then

$$(BA_0)^{-1} \frac{\partial}{\partial h_j} [BA_0(\mathbf{h})]|_{\mathbf{h}=\boldsymbol{\theta}} = A_0^{-1} B^{-1} B \mathcal{C}'_j = A_0^{-1} \mathcal{C}'_j = \sum_{\ell=1}^n \alpha_\ell^{(j)} \mathcal{C}_\ell,$$

so $V_{BA_0} = V_{A_0}$. We define the measure $d_\ell A$ on G by

(4.5) $$d_\ell A = V_A(g) dg_1 \cdots dg_n.$$

Expression (4.5) actually makes sense independent of local coordinates. If $\mathbf{k} = (k_1, \cdots, k_n)$ is another local coordinate system at A then

$$A^{-1} \frac{\partial A}{\partial k_\ell} = \sum_j A^{-1} \mathcal{C}'_j \frac{\partial g_j}{\partial k_\ell} = \sum_{j,s} \frac{\partial g_j}{\partial k_\ell} \alpha_s^{(j)} \mathcal{C}_s$$

so

$$V_a(k) = \left| \det \left(\sum_j \frac{\partial g_j}{\partial k_\ell} \alpha_s^{(j)} \right) \right| = \left| \det \left(\frac{\partial g_j}{\partial k_\ell} \right) \right| \cdot |\det(\alpha_s^{(j)})|.$$

Thus

(4.6) $$\begin{aligned} V_A(\mathbf{k}) dk_1 \cdots dk_n &= V_A(\mathbf{g}) |\det(\partial g_j / \partial k_\ell)| dk_1 \cdots dk_n \\ &= V_A(\mathbf{g}) dg_1 \cdots dg_n. \end{aligned}$$

This shows that the integral

$$\int_G f(A) d_\ell A = \int_G f(g_1, \cdots, g_n) V_A(\mathbf{g}) dg_1 \cdots dg_n$$

is well-defined, provided it converges. Furthermore,

(4.7) $$\begin{aligned} \int_G f(BA) d_\ell A &= \int_G f(BA(\mathbf{g})) V_A(\mathbf{g}) d\mathbf{g} = \int_G f(BA(\mathbf{g})) V_{BA}(\mathbf{g}) d\mathbf{g} \\ &= \int_G f(A) V_A(\mathbf{g}) d\mathbf{g} = \int_G f(A) d_\ell A, \end{aligned}$$

where the third equality follows from the fact that BA runs over G if A does.

By analogous procedures one can also define a right-invariant measure $d_r A$ in G. Writing

$$(\partial A / \partial g_j) A^{-1} = B_j = \sum_k \beta_k^{(j)} \mathcal{C}_k,$$

we define

(4.8) $$W_A(\mathbf{g}) = |\det(\beta_k^{(j)})|, \quad d_r A = W_A(\mathbf{g}) dg_1 \cdots dg_n.$$

The reader can verify that $d_r A$ is right-invariant on G.

Since $A(A^{-1} \partial A / \partial g_j) A^{-1} = (\partial A / \partial g_j) A^{-1}$, we have

(4.9) $$W_A(\mathbf{g}) = |\det \tilde{A}| \cdot V_A(\mathbf{g}),$$

where $\tilde{A}$ is the automorphism $\mathcal{C} \to A\mathcal{C}A^{-1}$ of the tangent space at the identity. (That is, we consider $\tilde{A}$ as an $n \times n$ matrix acting on the n-dimensional tangent space.) Thus, if $\det \tilde{A} = 1$ then $d_\ell A = d_r A$ and there exists a two-sided invariant measure on G.

It can be shown that a much larger class of groups (the locally compact topological (groups) possesses left-invariant (right-invariant) measures. Furthermore, the left-invariant (right-invariant) measure of a group is unique up to a constant factor. That is, if δA and $\delta' A$ are left-invariant measures on G then there exists a constant $c > 0$ such that $\delta A = c\delta' A$, [G1], [N2].

To illustrate our construction consider the Heisenberg matrix group H_R, example (8) in §3.1. Here H_R is a 3-dimensional Lie group which can be covered by a single coordinate patch $\mathbf{x} = (x_1, x_2, x_3)$.

$$A(\mathbf{x}) = \begin{pmatrix} 1 & x_1 & x_3 \\ 0 & 1 & x_2 \\ 0 & 0 & 1 \end{pmatrix}$$

Clearly, the matrices

$$\mathcal{C}_1 = \begin{pmatrix} 0 & 1 & 0 \\ 0 & 0 & 0 \\ 0 & 0 & 0 \end{pmatrix}, \mathcal{C}_2 = \begin{pmatrix} 0 & 0 & 0 \\ 0 & 0 & 1 \\ 0 & 0 & 0 \end{pmatrix}, \mathcal{C}_3 = \begin{pmatrix} 0 & 0 & 1 \\ 0 & 0 & 0 \\ 0 & 0 & 0 \end{pmatrix}$$

form a basis for the tangent space at the identity element of H_R. Now

$$A^{-1}dA = \begin{pmatrix} 0 & dx_1 & dx_3 - x_1 dx_2 \\ 0 & 0 & dx_2 \\ 0 & 0 & 0 \end{pmatrix}$$
$$= \mathcal{C}_1 dx_1 + (\mathcal{C}_2 - x_1\mathcal{C}_3)dx_2 + \mathcal{C}_3 dx_3$$

where dA is the differential of $A(\mathbf{x})$. Thus,

$$V_A(\mathbf{x}) = |\det \begin{pmatrix} 1 & 0 & 0 \\ 0 & 1 & -x_1 \\ 0 & 0 & 1 \end{pmatrix}| = 1$$

and $d_\ell A = dx_1 dx_2 dx_3$. Similarly,

$$(dA)A^{-1} = \begin{pmatrix} 0 & dx_1 & -x_2 dx_1 + dx_3 \\ 0 & 0 & dx_2 \\ 0 & 0 & 0 \end{pmatrix}$$
$$= (\mathcal{C}_1 - x_2\mathcal{C}_3)dx_1 + \mathcal{C}_2 dx_2 + \mathcal{C}_3 dx_3,$$

so

$$W_A(\mathbf{x}) = |\det \begin{pmatrix} 1 & 0 & -x_2 \\ 0 & 1 & 0 \\ 0 & 0 & 1 \end{pmatrix}| = 1$$

and $d_r A = dx_1 dx_2 dx_3$.

The affine group G_A, example (7) of §3.1 is 2-dimensional and can again be covered by a single coordinate patch:

$$A(a, b) = \begin{pmatrix} a & b \\ 0 & 1 \end{pmatrix}, \quad a > 0.$$

A basis for the tangent space at the identity is

$$\mathcal{C}_1 = \begin{pmatrix} 1 & 0 \\ 0 & 0 \end{pmatrix}, \quad \mathcal{C}_2 = \begin{pmatrix} 0 & 1 \\ 0 & 0 \end{pmatrix},$$

and

$$A^{-1}dA = \begin{pmatrix} da/a & db/a \\ 0 & 0 \end{pmatrix} = \frac{1}{a}(\mathcal{C}_1 da + \mathcal{C}_2 db).$$

Therefore

$$V_A(a, b) = |\det \begin{pmatrix} \frac{1}{a} & 0 \\ 0 & \frac{1}{a} \end{pmatrix}| = \frac{1}{a^2}$$

and $d_\ell A = da\, db/a^2$. On the other hand,

$$(dA)A^{-1} = \begin{pmatrix} da/a & -bda/a & +db \\ 0 & 0 \end{pmatrix}$$
$$= \frac{1}{a}(\mathcal{C}_1 - b\mathcal{C}_2)da + \mathcal{C}_2 db.$$

Therefore

$$W_A(a, b) = |\det \begin{pmatrix} \frac{1}{a} & -\frac{b}{a} \\ 0 & 1 \end{pmatrix}| = \frac{1}{a}$$

and $d_r A = da\, db/a$. In this case the left- and right-invariant measures are distinct.

4.3 Orthogonality relations for compact Lie groups. Now we are ready to extend the orthogonality relations to representations of a larger class of groups, the compact linear Lie groups. We say that a sequence of $m \times m$ complex matrices $\{A^{(j)}\}$ is a **Cauchy sequence** if each of the sequences of matrix elements $\{A_{ik}^{(j)}\}, 1 \leq i, k \leq m$ is Cauchy. Clearly, every Cauchy sequence of matrices converges to a unique matrix $A = (A_{ik}), A_{ik} = \lim_{j \to \infty} A_{ik}^{(j)}$. A set U of $m \times m$ matrices is **bounded** if there exists a constant $K > 0$ such that $|A_{ik}| \leq K$ for $1 \leq i, k \leq m$ and all $A \in U$. The set U is **closed** provided every Cauchy sequence in U converges to a matrix in U.

DEFINITION. A (global) group of $m \times m$ matrices is **compact** if it is a bounded, closed subset of the set of all $m \times m$ matrices.

As an example we show that the orthogonal group $O(3, R)$ is compact. If $A \in O(3, R)$ then $A^t A = E_3$, i.e.,

$$\sum_{j=1}^{3} A_{j\ell} A_{jk} = \delta_{\ell k}.$$

Setting $\ell = k$ we obtain $\sum_i (A_{ik})^2 = 1$, so $|A_{ik}| \leq 1$ for all i, k. Thus the matrix elements of A are bounded. Let $\{A^{(j)}\}$ be a Cauchy sequence in $O(3, R)$ with limit A. Then $E_3 = \lim_{j \to \infty} (A^{(j)})^t A^{(j)} = A^t A$ so $A \in O(3, R)$ and $O(3, R)$ is compact.

Now suppose G is a real, compact, linear Lie group of dimension n. It follows from the Heine-Borel Theorem [R4] that the group manifold of G can be covered by a finite number of bounded coordinate patches. Thus, for any continuous function $f(A)$ on G, the integral

$$\int_G f(A) d_\ell A = \int_G f(A) V_A(\mathbf{g}) d\mathbf{g}$$

will converge (since the domain of integration is bounded.) In particular the integral

(4.10)
$$V_G = \int_G 1 d_\ell A,$$

called the **volume** of G, converges. The preceding remarks also hold for the right-invariant measure $d_r A$. Moreover, we can show $d_\ell A = d_r A$ for compact groups.

THEOREM 4.1. *If G is a compact linear Lie group then $d_\ell A = d_r A$.*

Proof. By (4.9), $d_r A = |\det \tilde{A}| d_\ell A$, where $\tilde{A}$ is the mapping $\mathcal{C} \to A\mathcal{C}A^{-1}$ of the tangent space at the identity of G. (We can think of $\tilde{A}$ as an $m^2 \times m^2$ matrix rep of G.) Since G is compact the matrices $A, A^{-1} \in G$ are uniformly bounded. Thus the matrices $\tilde{A}$ are bounded and there exists a constant $M > 0$ such that $|\det \tilde{A}| \leq M$ for all $A \in G$. Now fix A and suppose $|\det \tilde{A}| = s \neq 1$. Then

$$|\det \tilde{A}^j| = |\det \tilde{A}|^j = s^j, \quad j = 0 \pm 1, \pm 2, \cdots.$$

Choosing j appropriately we get $s^j > M$, which is impossible. Therefore $s = 1$ for all $A \in G$ and $d_\ell A = d_r A$. $\square$

For G compact we write $dA = d_\ell A = d_r A$ where the measure dA is both left- and right-invariant.

Using the invariant measure for compact groups we can mimic the proofs of most of the results for finite groups obtained in Chapter 3. In particular we can show that any finite-dimensional rep of a compact group can be decomposed into a direct sum of irred reps and can obtain orthogonality relations for the matrix elements.

For finite groups K these results were proved using the average of a function over K. If f is a function on K then the **average** of f over K is

$$AV(f(k)) = \frac{1}{n(K)} \sum_{k \in K} f(k).$$

If $h \in K$ then

(4.11)
$$AV(f(hk)) = AV(f(kh)) = AV(f(k)).$$

Furthermore

$$(4.12) \qquad \mathcal{AV}(a_1 f_1(k) + a_2 f_2(k)) = a_1 \mathcal{AV}(f_1(k)) + a_2 \mathcal{AV}(f_2(k)), \quad \mathcal{AV}(1) = 1.$$

Properties (4.11) and (4.12) are sufficient to prove most of the fundamental results on the reps of finite groups. Now let G be a compact linear Lie group and let f be a continuous function on G. We define

$$(4.13) \qquad \mathcal{AV}(f(A)) = \frac{1}{V_G} \int_G f(A)dA = \int_G f(A)\delta A$$

where dA is the invariant measure on G, $V_G = \int_G 1 dA$ is the volume of G, and $\delta A = V_G^{-1} dA$ is the **normalized** invariant measure. Then

$$(4.14) \qquad \begin{aligned} \mathcal{AV}(f(BA)) &= \int_G f(BA)\delta A = \int_G f(A)\delta A = \mathcal{AV}(f(A)), \\ \mathcal{AV}(f(AB)) &= \mathcal{AV}(f(A)), \quad \mathcal{AV}(1) = \int_G \delta A = 1, \quad B \in G. \end{aligned}$$

since δA is both left- and right-invariant. Thus, $\mathcal{AV}(f(A))$ also satisfies properties (4.11), (4.12).

In order to mimic the finite group constructions we need to limit ourselves to **continuous** reps of G, i.e., reps $\mathbf{T}$ such that the operators $\mathbf{T}(A)$ are continuous functions of the group parameters of $A \in G$.

THEOREM 4.2. *Let* $\mathbf{T}$ *be a continuous rep of the compact linear Lie group* G *on the finite-dimensional inner product space* V. *Then* $\mathbf{T}$ *is equivalent to a unitary rep on* V.

Proof. Let $\langle \cdot, \cdot \rangle$ be the inner product on V. We define another inner product $(\cdot, \cdot)$ on V, with respect to which $\mathbf{T}$ is unitary. For $\mathbf{u}, \mathbf{v} \in V$ define

$$(4.15) \qquad (\mathbf{u}, \mathbf{v}) = \int_G \langle \mathbf{T}(A)\mathbf{u}, \mathbf{T}(A)\mathbf{v} \rangle \delta A = \mathcal{AV}[\langle \mathbf{T}(A)\mathbf{u}, \mathbf{T}(A)\mathbf{v} \rangle].$$

(The integral converges since the integrand is continuous and the domain of integration is finite.) It is straightforward to check that $(\cdot, \cdot)$ is an inner product. In particular, the positive definite property follows from the fact that the weight function is strictly positive (except possibly on a set of Lebesgue measure 0.) Now

$$(\mathbf{T}(B)\mathbf{u}, \mathbf{T}(B)\mathbf{v}) = \mathcal{AV}[\langle \mathbf{T}(AB)\mathbf{u}, \mathbf{T}(AB)\mathbf{v} \rangle] = \mathcal{AV}[\langle \mathbf{T}(A)\mathbf{u}, \mathbf{T}(A)\mathbf{v} \rangle] = (\mathbf{u}, \mathbf{v}),$$

so $\mathbf{T}$ is unitary with respect to $(\cdot, \cdot)$. The remainder of the proof is identical with that of Theorem 3.1. $\square$

Thus with no loss of generality, we can restrict ourselves to the study of unitary reps.

THEOREM 4.3. *If* **T** *is a unitary rep of* G *on* V *and* W *is an invariant subspace of* V *then* $W^{\perp}$ *is also an invariant subspace under* **T**.

THEOREM 4.4. *Every finite-dimensional, continuous, unitary rep of a compact linear Lie group can be decomposed into a direct sum of irred unitary reps.*

The proofs of these theorems are identical with corresponding proofs for finite groups.

Let $\{\mathbf{T}^{(\mu)}\}$ be a complete set of nonequivalent finite-dimensional unitary irred reps of G, labeled by the parameter μ. (Here we consider only reps of G on **complex** vector spaces.) Initially we have no way of telling how many distinct values μ can take. (It will turn out that μ takes on a countably infinite number of values, so that we can choose $\mu = 1, 2, \cdots$.) We introduce an ON basis in each rep space $V^{(\mu)}$ to obtain a unitary $n_\mu \times n_\mu$ matrix rep $T^{(\mu)}$ of G.

Now we mimic the construction of the orthogonality relations for finite groups. Given the matrix reps $T^{(\mu)}, T^{(\nu)}$, choose an arbitrary $n_\mu \times n_\nu$ matrix C and form the $n_\mu \times n_\nu$ matrix

$$D = \mathcal{A}V[T^{(\mu)}(A)CT^{(\nu)}(A^{-1})] = \int_G T^{(\mu)}(A)CT^{(\nu)}(A^{-1})\delta A.$$

Just as in the corresponding construction for finite groups, one can easily verify that

$$T^{(\mu)}(B)D = DT^{(\nu)}(B)$$

for all $B \in G$. Recall that the Shur lemmas are valid for finite-dimensional reps of all groups, not just finite groups. Thus if $\mu \neq \nu$, i.e., $T^{(\mu)}$ not equivalent to $T^{(\nu)}$, then D is the zero matrix. If $\mu = \nu$ then $D = \lambda E_{n_\mu}$ for some $\lambda \in \mathcal{C}$:

$$D(C, \mu, \nu) = \lambda(\mu, C)\delta_{\mu\nu}E_{n_\mu}.$$

Letting C run over all $n_\mu \times n_\nu$ matrices, we obtain the independent identities

(4.16) $$\int_G T_{i\ell}^{(\mu)}(A)T_{ks}^{(\nu)}(A^{-1})\delta A = \lambda(\mu, \ell, k)\delta_{\mu\nu}\delta_{is},$$

for the matrix elements $T_{i\ell}^{(\mu)}(A)$. To evaluate λ we set $\nu = \mu, s = i$ and sum on i:

$$\sum_{i=1}^{n_\mu} \lambda = n_\mu \lambda = \int_G \sum_{i=1}^{n_\mu} T_{ki}^{(\mu)}(A^{-1})T_{i\ell}^{(\mu)}(A)\delta A = \delta_{k\ell}.$$

Therefore $\lambda = \delta_{k\ell}/n_\mu$. Since the matrices $T^{(\mu)}(A)$ are unitary, (4.16) becomes

$$\int_G T_{i\ell}^{(\mu)}(A)\overline{T_{sk}^{(\nu)}}(A)\delta A = (\delta_{is}/n_\mu)\delta_{\ell k}\delta_{\mu\nu}, \quad 1 \leq i, \ell \leq n_\mu, \ 1 \leq s, k \leq n_\nu.$$

These are the **orthogonality relations** for matrix elements of irred reps of G.

4.4 The Peter-Weyl theorem. In the case of finite groups K we were able to relate the orthogonality relations to an inner product on the group ring R_K. We can consider R_K as the space of all functions $f(k)$ on K. Then

$$\langle f_1, f_2 \rangle = \frac{1}{n(K)} \sum_{k \in K} f_1(k)\overline{f_2}(k)$$

defines an inner product on R_K with respect to which the functions $\{n_\mu^{1/2} T_{i\ell}^{(\mu)}(g)\}$ form an ON basis. We extend this idea to a compact linear Lie group G as follows: Let $L_2(G)$ be the space of all functions on G which are (Lebesgue) square-integrable:

(4.18) $$L_2(G) = \{f(A) : \int_G |f(A)|^2 \delta A < \infty\}.$$

With respect to the inner product

(4.19) $$\langle f_1, f_2 \rangle = \int_G f_1(A)\overline{f_2}(A)\delta A,$$

$L_2(G)$ is a Hilbert space, [G1], [N2]. Note that every continuous function on G belongs to $L_2(G)$. Let

(4.20) $$\varphi_{ij}^{(\mu)}(A) = n_\mu^{1/2} T_{ij}^{(\mu)}(A).$$

It follows from (4.17) and (4.19) that $\{\varphi_{ij}^{(\mu)}\}$, where $1 \leq i, j \leq n_\mu$ and μ ranges over all equivalence classes of irred reps, forms an ON set in $L_2(G)$.

For finite groups we know that the set $\{\varphi_{ij}^{(\mu)}\}$ is an ON **basis** for the group ring, and every function f on the group can be written as a unique linear combination of these basis functions. Similarly one can show that for G compact the set $\{\varphi_{ij}^{(\mu)}\}$ is an ON basis for $L_2(G)$. Thus, every $f \in L_2(G)$ can be expanded uniquely in the (generalized) **Fourier series**

(4.21) $$f(A) \sim \sum_\mu \sum_{i,k=1}^{n\mu} c_{ik}^\mu \varphi_{ik}^{(\mu)}(A)$$

where

(4.22) $$c_{ik}^\mu = \langle f, \varphi_{ik}^{(\mu)} \rangle.$$

Furthermore we have the **Plancherel equality**

$$\langle f_1, f_2 \rangle = \sum_\mu \sum_{i,k=1}^{n\mu} c_{ik}^{\mu(1)} \overline{c}_{ik}^{\mu(2)}.$$

(For $f_1 \equiv f_2$ this is called **Parseval's equality**.) Convergence of the right-hand side of (4.21) to the left-hand side is meant in the sense of the Hilbert space norm. We will not here take up the question of pointwise convergence. See, however, [DS2], [DM2], [K2], [R1].

THEOREM 4.5. *(Peter-Weyl). If G is a compact linear Lie group, the set $\{\varphi_{ij}^{(\mu)}\}$ is an ON basis for $L_2(G)$.*

The proof of this theorem depends heavily on facts about symmetric completely continuous operators in Hilbert space and will not be given here. For the details see [G1] or [N2].

COROLLARY 4.1. *A compact linear Lie group G has a countably infinite (not finite) number of equivalence classes of irred reps $\{\mathbf{T}^{(\mu)}\}$. Thus, we can label the reps so that $\mu = 1, 2, \cdots$.*

Proof. The functions $\{\varphi_{jk}^{(\mu)}\}$ form an ON basis for $L_2(G)$. Since $L_2(G)$ is a separable, infinite-dimensional Hilbert space there are a countably infinite number of basis vectors [G1], [N2]. $\square$

We illustrate the **Peter-Weyl theorem** for an important example, the circle group $U(1)$, example 5 in §3.1. Since $U(1) = \{e^{2\pi i\phi}\}$, $(i = \sqrt{-1})$, is compact and Abelian its irred matrix reps are continuous functions $x(\phi)$ such that

$$(4.23) \qquad x(\phi_1 + \phi_2) = x(\phi_1)x(\phi_2), \quad \phi_1, \phi_2 \in R,$$

and $x(\phi + 1) = x(\phi)$. The functional equation (4.23) has only the solutions $x(\phi) = e^{a\phi}$ and the periodicity of x implies $a = 2\pi i m$, where m is an integer. Therefore, there are an infinite number of irreducible unitary representations of $U(1)$:

$$x_m(\phi) = e^{2\pi i m\phi}, \quad m = 0, \pm 1, \pm 2, \cdots.$$

The normalized invariant measure on $U(1)$ is $d\phi$. The space $L_2(U(1))$ is just the space $L_2([0,1])$ consisting of all measurable functions $f(\phi)$ with period 1 such that $\int_0^1 |f(\phi)|^2 d\phi < \infty$. By the Peter-Weyl theorem the functions $\{e^{2\pi i m\phi}\}$ form an ON basis for $L_2([0,1])$. Every $f \in L_2([0,1])$ can be expressed uniquely in the form

$$(4.24) \qquad f(\phi) \sim \sum_{m=-\infty}^{\infty} c_m e^{2\pi i m\phi}, \quad c_m = \int_0^1 f(\phi)e^{-2\pi i m\phi} d\phi.$$

Furthermore,

$$(4.25) \qquad \int_0^1 |f(\phi)|^2 d\phi = \sum_{m=-\infty}^{\infty} |c_m|^2.$$

Here (4.24) is the well-known Fourier series expansion of a periodic function and (4.25) is Parseval's equality.

4.5 The rotation group and spherical harmonics. A second very important example of the orthogonality relations for compact linear Lie groups and the Peter-Weyl theorem in the rotation group $SO(3) = SO(3, R)$. Here we will give some basic facts about the irreducible representations of $SO(3)$ and refer to the literature for most of the proofs, [B7], [GMS], [M5], [V].

Recall that $SO(3)$ has a convenient realization as the group of all 3×3 real matrices A such that $A^t A = E_3$ and $\det A = 1$, example 5, §3.1. This is the natural realization of $SO(3)$ as the group of all rotations in R_3 which leave the origin fixed. One convenient parametrization of $SO(3)$ is in terms of the **Euler angles**. Recall that a rotation through angle φ about the z axis is given by

$$R_z(\varphi) = \begin{pmatrix} \cos\varphi & -\sin\varphi & 0 \\ \sin\varphi & \cos\varphi & 0 \\ 0 & 0 & 1 \end{pmatrix} \in SO(3)$$

and rotations through angle φ about the x and y axis are given by

$$R_x(\varphi) = \begin{pmatrix} 1 & 0 & 0 \\ 0 & \cos\varphi & -\sin\varphi \\ 0 & \sin\varphi & \cos\varphi \end{pmatrix} \in SO(3),$$

$$R_y(\varphi) = \begin{pmatrix} \cos\varphi & 0 & \sin\varphi \\ 0 & 1 & 0 \\ -\sin\varphi & 0 & \cos\varphi \end{pmatrix} \in SO(3),$$

respectively. Differentiating each of these curves in $SO(3)$ with respect to φ and setting $\varphi = 0$ we find the following linearly independent matrices in the tangent space at the identity:

$$(4.26) \quad \mathcal{L}_z = \begin{pmatrix} 0 & -1 & 0 \\ 1 & 0 & 0 \\ 0 & 0 & 0 \end{pmatrix}, \quad \mathcal{L}_x = \begin{pmatrix} 0 & 0 & 0 \\ 0 & 0 & -1 \\ 0 & 1 & 0 \end{pmatrix}, \quad \mathcal{L}_y = \begin{pmatrix} 0 & 0 & 1 \\ 0 & 0 & 0 \\ -1 & 0 & 0 \end{pmatrix}.$$

One can check from the definition $A^t A = E_3$ that the tangent space at the identity is at most three-dimensional, so the matrices (4.26) form a basis for this space.

The Euler angles φ, θ, ψ for $A \in SO(3)$ are given by

$$A(\varphi, \theta, \psi) = R_z(\varphi) R_x(\theta) R_z(\psi)$$

$$(4.27) \quad = \begin{pmatrix} \cos\varphi\cos\psi - \sin\varphi\sin\psi\cos\theta, & -\cos\varphi\sin\psi - \sin\varphi\cos\psi\cos\theta, & \sin\varphi\sin\theta \\ \sin\varphi\cos\psi + \cos\varphi\sin\psi\cos\theta, & -\sin\varphi\sin\psi + \cos\varphi\cos\psi\cos\theta, & -\cos\varphi\sin\theta \\ \sin\psi\sin\theta, & \cos\psi\sin\theta, & \cos\theta \end{pmatrix}.$$

It can be shown that every $A \in SO(3)$ can be represented in the form (4.27) where the Euler angles run over the domain

$$(4.28) \qquad 0 \le \varphi < 2\pi, \quad 0 \le \theta \le \pi, \quad 0 \le \psi < 2\pi.$$

The representation of A by Euler angles is unique except for the cases $\theta = 0, \pi$ where only the sum $\varphi + \psi$ is determined by A, but this exceptional set is only one-dimensional and doesn't contribute to an integral over the group manifold. The invariant measure on $SO(3)$ can be computed directly from the formulas of §4.2. Let $A(\varphi, \theta, \psi) \in SO(3)$. Then

$$A^{-1}\frac{\partial A}{\partial \varphi} = \sin\psi\sin\theta\mathcal{L}_x + \cos\psi\sin\theta\mathcal{L}_y + \cos\theta\mathcal{L}_z,$$

$$A^{-1}\frac{\partial A}{\partial \theta} = \cos\psi\mathcal{L}_x - \sin\psi\mathcal{L}_y,$$

$$A^{-1}\frac{\partial A}{\partial \psi} = \mathcal{L}_z.$$

Thus,

$$V_A(\varphi, \theta, \psi) = \left| \det \begin{pmatrix} \sin\psi\sin\theta & \cos\psi\sin\theta & \cos\theta \\ \cos\psi & -\sin\psi & 0 \\ 0 & 0 & 1 \end{pmatrix} \right| = \sin\theta$$

and

(4.29)
$$dA = \sin\theta\, d\varphi\, d\theta\, d\psi.$$

Since $SO(3)$ is compact, dA is both left- and right-invariant. The volume of $SO(3)$ is

(4.30)
$$V_{SO(3)} = \int_{SO(3)} dA = \int_0^{2\pi} d\psi \int_0^{2\pi} d\varphi \int_0^\pi \sin\theta\, d\theta = 8\pi^2.$$

The irred unitary reps of $SO(3)$ are denoted $\mathbf{T}^{(\ell)}, \ell = 0, 1, 2, \cdots$, where $\dim \mathbf{T}^{(\ell)} = 2\ell + 1$. (In particular $T^{(0)}(A) = 1$ and $T^{(1)}(A) = A$.) Expressed in terms of an ON basis for the rep space $V^{(\ell)}$ consisting of simultaneous eigenfunctions for the operators $\mathbf{T}^{(\ell)}(R_z(\varphi))$, the matrix elements are

(4.31)
$$T^\ell_{km}(\varphi, \theta, \psi) = i^{k-m} \left[\frac{(\ell+m)!(\ell-k)!}{(\ell+k)!(\ell-m)!}\right]^{1/2} \times$$

$$e^{i(k\varphi+m\psi)} \frac{[\sin\theta]^{m-k}(1+\cos\theta)^{\ell+k-m}}{2^\ell \Gamma(m-k+1)} {}_2F_1\left(\begin{matrix} -\ell-k, m-1; \frac{\cos\theta-1}{\cos\theta+1} \\ m-k+1 \end{matrix}\right)$$

$$= i^{k-m} \left[\frac{(\ell+m)!(\ell-k)!}{(\ell+k)!(\ell-m)!}\right]^{1/2} e^{i(k\varphi+m\psi)} P_\ell^{-k,m}(\cos\theta),$$

$$-\ell \le k, m \le \ell.$$

Here ${}_2F_1\left(\begin{smallmatrix} a,b \\ c,z \end{smallmatrix}\right)$ is the Gaussian hypergeometric function and $\Gamma(z)$ is the gamma function [EMOT1], [V], [WW]. A generating function for the matrix elements is

$$g(A, z) = \frac{(\beta z + \bar{\alpha})^{\ell-m}(\alpha z - \bar{\beta})^{\ell+m}}{[(\ell-m)!(\ell+m)!]^{1/2}} = \sum_{k=-\ell}^{\ell} T^\ell_{km}(A) \frac{(-1)^{k-m} z^{\ell+k}}{[(\ell-k)!(\ell+k)!]^{1/2}}$$

where

$$\alpha = e^{i(\varphi+\psi)/2} \cos\frac{\theta}{2}, \quad \beta = i e^{i(\phi-\psi)/2} \sin\frac{\theta}{2}.$$

The group property

$$T_{km}^\ell(A_1 A_2) = \sum_{j=-\ell}^{\ell} T_{kj}^\ell(A_1) T_{jm}^\ell(A_2)$$

defines an **addition theorem** obeyed by the matrix elements. The unitary property of the operator $\mathbf{T}^{(\ell)}(A)$ implies

$$T_{km}^\ell(A^{-1}) = \overline{T_{mk}^\ell}(A),$$

or in Euler angles,

$$(-1)^{m-k} P_\ell^{-k,m}(\cos\theta) = \frac{(\ell+k)!(\ell-m)!}{(\ell-k)!(\ell+m)!} P_\ell^{-m,k}(\cos\theta).$$

Also, $|T_{km}^\ell(A)| \le 1$ or

$$|P_\ell^{-k,m}(\cos\theta)| \le \left[\frac{(\ell+k)!(\ell-m)!}{(\ell+m)!(\ell-k)!}\right]^{1/2}, 0 \le \theta \le \pi.$$

The matrix elements $T_{om}^\ell(\varphi,\theta,\psi)$, are proportional to the spherical harmonics $Y_\ell^m(\theta,\psi)$. Indeed

$$T_{om}^\ell(\varphi,\theta,\psi) = i^m \left(\frac{4\pi}{2\ell+1}\right)^{1/2} Y_\ell^m(\theta,\psi) = i^m \left[\frac{(\ell-m)!}{(\ell+m)!}\right]^{1/2} P_\ell^m(\cos\theta) e^{im\psi}$$

where the $P_\ell^m(\cos\theta)$ are the associated Legendre functions [EMOT1], [GMS], [M4], [M5]. Moreover,

$$T_{oo}^\ell(\varphi,\theta,\psi) = P_\ell(\cos\theta)$$

where $P_\ell(\cos\theta)$ is the Legendre polynomial.

According to the general theory of §4.3, the matrix elements $T_{km}^\ell(A)$ satisfy the orthogonality relations

(4.32) $$\int_{SO(3)} T_{km}^\ell(A)\overline{T_{k'm'}^{\ell'}}(A) dA = \frac{8\pi^2}{2\ell+1}\delta_{kk'}\delta_{mm'}\delta_{\ell\ell'}.$$

Thus

$$\int_0^{2\pi} d\psi \int_0^{2\pi} d\varphi \int_0^\pi d\theta\, T_{km}^\ell(\varphi,\theta,\psi)\overline{T_{k'm'}^{\ell'}}(\varphi,\theta,\psi) \sin\theta = \frac{8\pi^2}{2\ell+1}\delta_{kk'}\delta_{mm'}\delta_{\ell\ell'}.$$

The ψ and φ integrations are trivial, while the θ integration gives

$$\int_0^\pi P_\ell^{k,m}(\cos\theta) P_{\ell'}^{k,m}(\cos\theta)\sin\theta d\theta = \frac{2}{2\ell+1}\frac{(\ell-k)!(\ell-m)!}{(\ell+k)!(\ell+m)!}\delta_{\ell\ell'}.$$

For $k = m = 0$ these are the orthogonality relations for the Legendre polynomials. (Note: By definition, $P_\ell^{0,-m}(\cos\theta) = P_\ell^m(\cos\theta), P_\ell^{0,0}(\cos\theta) = P_\ell(\cos\theta)$, where P_ℓ^m, P_ℓ are Legendre functions.)

By the Peter-Weyl theorem, the functions

$$\varphi_{km}^\ell(\varphi,\theta,\psi) = (2\ell+1)^{1/2}T_{km}^\ell(\varphi,\theta,\psi),$$
$$-\ell \leq k, m \leq \ell, \quad \ell = 0, 1, 2, \cdots$$

constitute an ON basis for $L_2(SO(3))$. If $f \in L_2(SO(3))$ then

$$(4.33) \qquad f(\varphi,\theta,\psi) = \sum_{\ell=0}^{\infty} \sum_{k,m=-\ell}^{\ell} a_{km}^\ell \varphi_{km}^\ell(\varphi,\theta,\psi)$$

where

$$(4.34) \qquad a_{km}^\ell = (f, \varphi_{km}^\ell) = \frac{1}{8\pi^2} \int_0^{2\pi} d\psi \int_0^{2\pi} d\varphi \int_0^{\pi} d\theta \times$$
$$\times f(\varphi,\theta,\psi)\overline{\varphi_{km}^\ell}(\varphi,\theta,\psi)\sin\theta.$$

Some particular cases of (4.33) are of special interest. Suppose $f(\theta,\psi) \in L_2(SO(3))$ is independent of the variable φ. If we think of (θ,ψ) as latitude and longitude, we can consider f as a function on the unit sphere $S_2 \cong SO(3)/U(1)$, square-integrable with respect to the area measure on S_2. Since the φ-dependence of $\varphi_{km}^\ell(\varphi,\theta,\psi)$ is $e^{ik\varphi}$, it follows from (4.34) that $a_{km}^\ell = 0$ unless $k = 0$; the only possible nonzero coefficients are a_{om}^ℓ. Now

$$\varphi_{om}^\ell(\varphi,\theta,\psi) = (4\pi)^{1/2}Y_\ell^m(\theta,\psi)$$

where Y_ℓ^m is a spherical harmonic. Thus,

$$(4.35) \qquad f(\theta,\psi) = \sum_{\ell=0}^{\infty} \sum_{m=-\infty}^{\infty} c_m^\ell Y_\ell^m(\theta,\psi)$$

where

$$c_m^\ell = \int_0^{2\pi} d\psi \int_0^{\pi} d\theta f(\theta,\psi)\overline{Y_\ell^m}(\theta,\psi)\sin\theta, \quad (Y_\ell^m, Y_{\ell'}^{m'}) = \delta_{\ell\ell'}\delta_{mm'}.$$

This is the expansion of a function on the sphere as a linear combination of spherical harmonics. Again, (4.35) converges in the norm of $L_2(SO(3))$, not necessarily pointwise.

If $f(\theta) \in L_2(SO(3))$ is a function of θ alone then the coefficients a_{km}^ℓ are zero unless $k = m = 0$. Here,

$$\varphi_{oo}^\ell(\varphi,\theta,\psi) = (2\ell+1)^{1/2}P_\ell(\cos\theta), \quad \ell = 0, 1, 2, \cdots$$

where

$$P_\ell(x) = {}_2F_1\left(\begin{matrix} \ell+1, & -\ell, & \frac{1-x}{2} \\ & 1 & \end{matrix}\right) = \left(\frac{1+x}{2}\right)^\ell {}_2F_1\left(\begin{matrix} -\ell, & -\ell, & \frac{1-x}{x+1} \\ & 1 & \end{matrix}\right)$$

is a Legendre polynomial of order ℓ. The coefficient of x^ℓ in the expansion of $P_\ell(x)$ is nonzero and $P_\ell(1) = 1$. The expansion of $f(\theta)$ becomes

(4.37)

$$f(\theta) = \sum_{\ell=\theta}^{\infty} c_\ell P_\ell(\cos\theta),$$

$$c_\ell = \frac{1}{2}(2\ell+1)\int_0^\pi f(\theta)P_\ell(\cos\theta)\sin\theta d\theta,$$

$$\int_0^\pi P_\ell(\cos\theta)P_k(\cos\theta)\sin\theta d\theta = \frac{2\delta_{k\ell}}{2\ell+1}.$$

4.6 Fourier transforms and their relation to Fourier series. Abelian groups G, (not necessarily compact) are another class of groups concerning which one can make general statements about the decomposition of $L_2(G)$ in terms of unitary representations [B7], [DM2], [G1], [K2], [N2], [R1], [V]. We will not go into this theory but consider only a single, very important, example where the results are familiar to everyone: The group R of real numbers t with addition of numbers as group multiplication. Here R is isomorphic to the matrix group R', examples (1) and (3), Section §3.1, and dt is the invariant measure. The unitary irred reps of R are one-dimensional, hence continuous functions $\chi(t)$ such that

(4.38) $$\chi(t_1 + t_2) = \chi(t_1)\chi(t_2), \quad t_1, t_2 \in R.$$

This functional equation has only the solutions $\chi(t) = e^{at}$ and the unitarity requirement implies $a = 2\pi i\omega$ where ω is real. Given a function $s \in L_2(R)$ we have

$$s(t) = \int_{-\infty}^{\infty} S(\omega)e^{2\pi i\omega t} d\omega$$

where the **Fourier coefficients** $S(\omega)$ are defined by

$$S(\omega) = \int_{-\infty}^{\infty} s(t)e^{-2\pi i\omega t} dt$$
$$= (s, \chi_\omega),$$

and $\chi_\omega(t) = e^{2\pi i\omega t}$ is the irred rep. Parseval's equality is

$$\int_{-\infty}^{\infty} |s(t)|^2 dt = \int_{-\infty}^{\infty} |S(\omega)|^2 d\omega.$$

Formally, the orthogonality relations are

$$(\chi_\omega, \chi_{\omega'}) = \int_{-\infty}^{\infty} \exp[2\pi i t(\omega - \omega')] dt = \delta(\omega - \omega')$$

where $\delta(\omega)$ is the Dirac delta function. Note that the sum over irred reps, familiar for compact groups, is here replaced by an integral over irred reps.

It is illuminating to compare the Fourier transform on R with the corresponding results for $U(1)$ and Z_n. Assume that $f \in L_2(R)$ belongs to the Schwartz class, i.e., f is in $C^\infty(R)$ and there exist constants $C_{n,q}$ (depending on f) such that $|t^n \frac{d^q}{dt^q} f| \le C_{n,q}$ on R for each $n, q = 0, 1, 2, \cdots$. Then the projection operator P maps f to a continuous function in $L_2([0,1])$ with period one:

$$(4.39) \qquad P[f](x) = \sum_{m=-\infty}^{\infty} f(x+m).$$

Expanding $P[f](x)$ into a Fourier series we find

$$P[f](x) = \sum_{m=-\infty}^{\infty} c_n e^{2\pi i n x}$$

where

$$c_n = \int_0^1 P[f](x) e^{-2\pi i n x} dx = \int_{-\infty}^{\infty} f(x) e^{-2\pi i n x} dx = \hat{f}(n)$$

and $\hat{f}(\omega)$ is the Fourier transform of $f(x)$. Thus,

$$(4.40) \qquad \sum_{n=-\infty}^{\infty} f(x+n) = \sum_{n=-\infty}^{\infty} \hat{f}(n) e^{2\pi i n x},$$

and we see that $P[f](x)$ tells us the value of $\hat{f}$ at the integer points $\omega = n$, but not in general at the non-integer points. (For $x = 0$, equation (4.40) is known as the Poisson summation formula, [DM2]. If we think of f as a signal, we see that **periodization** (4.39) of f results in a loss of information. However, if f vanishes outside of $[0,1)$ then $P[f](x) \equiv f(x)$ for $0 \le x < 1$ and

$$(4.41) \qquad f(x) = \sum_n \hat{f}(n) e^{2\pi i n x}, \quad 0 \le x < 1$$

without error. Now suppose (4.41) holds, so that there is no periodization error. For an integer $N > 1$ we sample the signal at the points $a/N, a = 0, 1, \cdots, N-1$:

$$(4.42) \qquad f\left(\frac{a}{N}\right) = \sum_n \hat{f}(n) e^{2\pi i n a/N}, \quad 0 \le a < N.$$

From the Euclidean algorithm we have $n = b + cN$ where $0 \le b < N$ and b, c are integers. Thus

$$(4.43) \qquad f\left(\frac{a}{N}\right) = \sum_{b=0}^{N-1} \left[\sum_c \hat{f}(b+cN) \right] e^{2\pi i a b/N}, \quad 0 \le a < N.$$

Note that the quantity in brackets is the projection of $\hat{f}$ at integer points to a periodic function of period N. Furthermore, the expansion (4.43) is essentially the finite Fourier expansion (3.37). However, simply sampling the signal at the points a/N tells us only $\sum_c \hat{f}(b+cN)$ not (in general) $\hat{f}(b)$. This is known as **aliasing error**.

Although we will not work out the details in these notes, a similar approach to the foregoing (with periodizing and aliasing error) is appropriate and useful for Fourier analysis on the Heisenberg and affine groups.

4.7 Exercises.

4.1 Construct a real irred two-dimensional rep of the circle group $U(1)$.

4.2 Show how to decompose any real finite-dimensional rep of $U(1)$ as a direct sum of real irred reps.

4.3 Verify that the measure $d_r A$, (4.8), is right-invariant on the group G.

4.4 Prove: If G is a compact linear Lie group then $d(A^{-1}) = dA$, i.e., $\int_G f(A^{-1})dA = \int_G f(B)dA$. Hint: Show that $V_{A^{-1}}(\mathbf{g}) = |\det(-\tilde{A})|V_A(\mathbf{g})$ where $\tilde{A}$ is the automorphism $\mathcal{A} \to A\mathcal{A}A^{-1}$ of $n \times n$ matrices $\mathcal{A}$.

4.5 Assuming that $\chi(t)$ is a continuously differentiable function of t, use a differential equations argument to show that the only nonzero solutions of the functional equation

$$\chi(t_1 + t_2) = \chi(t_1)\chi(t_2), \quad t_1, t_2 \in R$$

are $\chi(t) = e^{at}$, where a is a constant.

4.6 Assuming only that that $\chi(t)$ is a real continuous function of t, show that the only nonzero solutions of the functional equation

$$\chi(t_1 + t_2) = \chi(t_1)\chi(t_2), \quad t_1, t_2 \in R$$

are $\chi(t) = e^{at}$, where a is a constant. Hint: Set $\chi(t) = e^{\phi(t)}$ so that $\phi(t_1 + t_2) = \phi(t_1) + \phi(t_2)$. Then determine $\phi(t)$ for t rational from $\phi(1)$.

4.7 Use the Poisson summation formula for the Gauss kernel $f(x) = (2\pi t)^{-\frac{1}{2}} e^{-\frac{x^2}{2t}}$ to derive the identity

$$\frac{1}{\sqrt{2\pi t}} \sum_{n=-\infty}^{\infty} e^{-\frac{n^2}{2t}} = \sum_{n=-\infty}^{\infty} e^{-2\pi^2 n^2 t}.$$

Hint: $\hat{f}(\omega) = e^{-2\pi^2 \omega^2 t}$.

§5. REPRESENTATIONS OF THE HEISENBERG GROUP

5.1 Induced representations of H_R. Recall that the Heisenberg group H_R can be realized as the linear Lie group of matrices

$$(5.1) \qquad A(\mathbf{x}) = \begin{pmatrix} 1 & x_1 & x_3 \\ 0 & 1 & x_2 \\ 0 & 0 & 1 \end{pmatrix}$$

with group product

$$(x_1, x_2, x_3) \cdot (y_1, y_2, y_3) = (x_1 + y_1, x_2 + y_2, x_3 + y_3 + x_1 y_2),$$

and invariant measure

$$d_\ell A = d_r A = dx_1 dx_2 dx_3.$$

To motivate the construction of the unitary irred reps of H_R we review the essentials of the Frobenius construction of induced representations. Let G be a linear Lie group and H a linear Lie subgroup of G. If $\mathbf{T}$ is a rep of G on the vector space W we can obtain a rep $\mathbf{T}_H$ of H by restricting $\mathbf{T}$ to H,

$$\mathbf{T}_H(B) = \mathbf{T}(B), \quad B \in H.$$

On the other hand the method of Frobenius allows one to construct a rep of G from a rep of H. Let $\mathbf{T}$ be a finite-dimensional unitary rep of H on the inner product space V. Denote by U^G the vector space of all functions $\mathbf{f}(A)$ with domain G and range contained in V where addition and scalar multiplication of functions are the vector operations. Here, for a fixed $A \in G$, $\mathbf{f}(A)$ is a vector in V. Let V^G be the subspace of U^G defined by

$$(5.2) \qquad V^G = \{\mathbf{f} \in U^G : \mathbf{f}(BA) = \mathbf{T}(B)\mathbf{f}(A) \text{ for all } B \in H, A \in G\}.$$

We define a rep $\mathbf{T}^G$ of G on V^G by

$$(5.3) \qquad [\mathbf{T}^G(A)]\mathbf{f}(A') = \mathbf{f}(A'A), \quad A, A' \in G, \quad \mathbf{f} \in V^G.$$

It is clear that V^G is invariant under G and the operators $\mathbf{T}^G(A)$ satisfy the homomorphism property. Here, $\mathbf{T}^G$ is called an **induced representation**. If $(\cdot, \cdot)$ is the inner product on V and H is compact we can initially define the inner product $\langle \cdot, \cdot \rangle$ on V^G by

$$\langle \mathbf{f}_1, \mathbf{f}_2 \rangle = \int_G (\mathbf{f}_1(A), \mathbf{f}_2(A)) d_r A$$

where $d_r A$ is the right-invariant measure on G. Then we restrict the operators $\mathbf{T}^G$ to the subspace $V'^G \subseteq V^G$ of functions $\mathbf{f}$ such that $\langle \mathbf{f}, \mathbf{f} \rangle < \infty$. If $A' \in G$ and $\mathbf{f} \in V'^G$ we have

$$\langle \mathbf{T}^G(A')\mathbf{f}_1, \mathbf{T}^G(A')\mathbf{f}_2 \rangle = \int_G (\mathbf{f}_1(AA'), \mathbf{f}_2(AA')) d_r A$$

$$= \int_G (\mathbf{f}_1(A), \mathbf{f}_2(A)) d_r A = \langle \mathbf{f}_1, \mathbf{f}_2 \rangle$$

so $\mathbf{T}^G$ is unitary. However, note from (5.2) that $(\mathbf{f}_1(BA), \mathbf{f}_2(BA)) = (\mathbf{f}_1(A), \mathbf{f}_2(A))$ for all $B \in H$, i.e., the inner product $(\cdot, \cdot)$ is constant on the right cosets HA. Thus if H is noncompact the above integral will be undefined (or V'^G will contain only the zero function). If the coset space $X = H \backslash G$ admits a G-invariant measure, i.e., a measure $d\mu(x)$ such that $d\mu(xA) = d\mu(x)$ for all $x \in X, A \in G$, then we can define the inner product on V'^G as

$$(5.4) \qquad \langle \mathbf{f}_1, \mathbf{f}_2 \rangle = \int_X (\mathbf{f}_1(x), \mathbf{f}_2(x)) d\mu(x)$$

and the operators $\mathbf{T}^G : \mathbf{f}(x) \rightarrow \mathbf{T}(B)\mathbf{f}(xA)$ will still be unitary. Here, we choose an $A'_x \in G$ in each right coset $HA' \leftrightarrow x$, so that $A'_x A = BA'_{xA}$ and $\mathbf{f}(x) \equiv \mathbf{f}(A'_x)$. (Note: It is always possible to find local coordinates $\mathbf{f} = (g_1, \cdots, g_n)$ on G such that $\mathbf{h} = (g_1, \cdots, g_m)$, $m \leq n$, are local coordinates on H and $\mathbf{x} = (g_{m+1} \cdots, g_n)$ are local coordinates on X.) In general, no such invariant measure $d\mu$ exists, but it does exist in many cases. In particular, if both G and H are **unimodular**, i.e., if the left-invariant and right-invariant measures for each of these groups are the same, and H is a closed subgroup of G then it can be shown that an (unique up to a constant multiplier) invariant measure $d\mu$ exists and that in local coordinates

$$dA = V_G(\mathbf{g})dg_1 \cdots dg_n, \quad A \in G$$
$$dB = V_H(\mathbf{h})dg_1 \cdots dg_m, \quad B \in H$$

and

$$d\mu(\mathbf{x}) = V_x(\mathbf{x})dg_{m+1} \cdots dg_n$$

where $dA = d\mu dB$. Thus $V_x(\mathbf{x}) = V(\mathbf{g})/V_h(\mathbf{h})$.

5.2 The Schrödinger representation. Let us consider the case where $G = H_R$ and H is the subgroup of matrices $\{A(0, x_2, x_3)\}$. Since $A(0, x_2, x_3)A(0, x'_2, x'_3) = A(0, x_2 + x'_2, x_3 + x'_3)$ it is clear that H is Abelian and that the operators $\mathbf{T}_\lambda[A(0, x_2, x_3)] = e^{2\pi i \lambda x_3}$ define a one-dimensional unitary irred rep of H. Furthermore the left-invariant and right-invariant measure on H is $dx_2 dx_3$. Since

$$A(x_1, x_2, x_3) = A(0, x_2, x_3)A(x_1, 0, 0)$$

it is clear that the coset space $H \backslash G$ can be parametrized by the coordinate $x_1 = t$, and the (scalar) functions in V'^G can be taken as $f(t)$. If $A(x_1, x_2, x_3)$ acts on this function, it transforms to

$$f(t + x_1, x_2, x_3 + tx_2) \equiv \mathbf{T}_\lambda(B)f(t + x_1)$$

where $B = A(0, x_2, x_3 + tx_2)$, (5.2). Thus the action of H_R restricted to the coset space is

$$(5.5) \qquad \mathbf{T}^\lambda(A(\mathbf{x}))f(t) \equiv \mathbf{T}^\lambda[x_1, x_2, x_3]f(t) = e^{2\pi i \lambda(x_3 + tx_2)} f(t + x_1).$$

One can verify directly that $\mathbf{T}^\lambda$ is a rep, i.e.,

$$\mathbf{T}^\lambda(A_1)\mathbf{T}^\lambda(A_2) = \mathbf{T}^\lambda(A_1 A_2)$$

and that it is unitary with respect to the inner product

(5.6) $$\langle f_1, f_2 \rangle = \int_{-\infty}^{\infty} f_1(t)\overline{f_2}(t)dt.$$

For $\lambda = 0$ this rep is reducible but (as we will show later) for $\lambda \neq 0$ it is irred in the sense that there is no nontrivial closed subspace of $L_2(R)$ which is invariant under the operators $\mathbf{T}^\lambda(A)$, $A \in H_R$. This rep is called the **Schrödinger representation** of the Heisenberg group.

Note that the mapping

$$A(x_1, x_2, x_3) \xrightarrow{\rho} (x_1, x_2)$$

is a homomorphism of H_R onto the Abelian group $R_2 : (x_1, x_2) \cdot (y_1, y_2) = (x_1 + y_1, x_2 + y_2)$. The unitary irred reps of R_2 are clearly of the form $\chi_{\alpha_1,\alpha_2}(x_1, x_2) = e^{2\pi i(\alpha_1 x_1 + \alpha_2 x_2)}$ for real constants α_1, α_2. It follows that the matrices $T^{\alpha_1,\alpha_2}(A(\mathbf{x})) = e^{2\pi i(\alpha_1 x_1 + \alpha_2 x_2)}$ define one-dimensional unitary irred reps of H_R. It can be shown that $\mathbf{T}^\lambda$ and T^{α_1,α_2} are the only irred unitary reps of H_R, [S2].

5.3 Square integrable representations. Assuming for the time being that $\mathbf{T}^\lambda$ is irred for real $\lambda \neq 0$ let us see if there is an analog for H_R of the orthogonality and completeness relations for matrix elements which hold for linear compact Lie groups, and for compact topological groups in general. Noncompact groups G can have infinite dimensional irred unitary reps. That is, the rep space is a separable infinite dimensional Hilbert space $\mathcal{H}$ with inner product $\langle \cdot, \cdot \rangle$. The rep operators $\mathbf{T}(g), g \in G$ are **unitary** i.e., each $\mathbf{T}(g)$ is a linear mapping of $\mathcal{H}$ onto itself which preserves inner product: $\langle \mathbf{T}(g)f_1, \mathbf{T}(g)f_2 \rangle = \langle f_1, f_2 \rangle$, for all $f_1, f_2 \in \mathcal{H}$. The rep is **irreducible** if there is no proper closed subspace of H which is invariant under the operator $\mathbf{T}(g), g \in G$. The reps $\mathbf{T}, \mathbf{T}'$ of G on the Hilbert spaces $\mathcal{H}, \mathcal{H}'$, respectively, are **equivalent** if there is a bounded invertible linear operator $\mathbf{S} : \mathcal{H} \to \mathcal{H}'$ such that $\mathbf{ST}(g) = \mathbf{T}'(g)\mathbf{S}$ for all $g \in G$, i.e., $\mathbf{ST}(g)\mathbf{S}^{-1} = \mathbf{T}'(g)$.

With these definitions the analog of Theorems 3.2, 3.4 are true and can be proven with ease. Moreover, the following analog of Theorem 3.5 holds:

THEOREM 5.1. *Let $\mathbf{T}$ be a unitary rep of the group G on the separable Hilbert space $\mathcal{H}$. Then $\mathbf{T}$ is irred if and only if the only bounded operator $\mathbf{S}$ on $\mathcal{H}$ satisfying*

(5.7) $$\mathbf{T}(g)\mathbf{S} = \mathbf{ST}(g)$$

is $\mathbf{S} = \lambda\mathbf{E}$ where $\mathbf{E}$ is the identity operator on $\mathcal{H}$.

Sketch of the proof. Part of this result is easy to prove. Suppose $\mathbf{S} = \lambda\mathbf{E}$ is the only solution of (5.7) and let $\mathcal{M}$ be a closed subspace of $\mathcal{H}$ which is invariant under $\mathbf{T}$: $\mathbf{T}(g)\mathbf{f} \in \mathcal{M}$ for all $g \in G, \mathbf{f} \in \mathcal{M}$. Then $\mathcal{M}^\perp$ is also invariant under $\mathbf{T}$ and $\mathcal{H} = \mathcal{M} \oplus \mathcal{M}^\perp$. Thus for every $\mathbf{f} \in \mathcal{H}$ we have the unique decomposition $\mathbf{f} = \mathbf{f}_1 + \mathbf{f}_2$, $\mathbf{f}_1 \in \mathcal{M}$, $\mathbf{f}_2 \in \mathcal{M}^\perp$ where $\mathbf{T}(g)\mathbf{f}_1 \in \mathcal{M}$ and $\mathbf{T}(g)\mathbf{f}_2 \in \mathcal{M}^\perp$. Now define the **projection operator** $\mathbf{P}$ by $\mathbf{Pf} = \mathbf{f}_1$. Clearly $\mathbf{P}$ is a bounded operator on $\mathcal{H}$ and $\mathbf{T}(g)\mathbf{P} = \mathbf{PT}(g)$ for all $g \in G$. By hypothesis, $\mathbf{P} = \lambda\mathbf{E}$. If $\lambda = 0$ then $\mathcal{M}$ is the zero subspace; if $\lambda \neq 0$ then $\lambda = 1$ and $\mathcal{M} = \mathcal{H}$. Thus $\mathcal{M}$ is not a proper invariant subspace and $\mathbf{T}$ is irred.

The converse is somewhat more difficult. Suppose $\mathbf{T}$ is irred and $\mathbf{S}$ is a bounded linear operator satisfying (5.7). Then the adjoint $\mathbf{S}^*$ of $\mathbf{S}$ also satisfies (5.7) so, without loss of generality, one can assume that $\mathbf{S}$ is self-adjoint. Then, by the spectral theorem for self-adjoint operators [AG1], [AG2], [G1], [N2], [RN], the projection operators $\mathbf{P}_\lambda$ in the spectral family associated with $\mathbf{S}$ must all commute with each $\mathbf{T}(g)$. By hypothesis then, each $\mathbf{P}_\lambda$ is either the zero operator or the identity operator. Hence $\mathbf{S} = \lambda\mathbf{E}$ for some real number λ. □

Note: At this point it is appropriate to mention that if the unitary irred reps $\mathbf{T}, \mathbf{T}'$ of a group G are equivalent, then they are **unitary equivalent**, i.e., there is a unitary operator $\mathbf{U}$ such that $\mathbf{UT}(g) = \mathbf{T}'(g)\mathbf{U}$ for all $g \in G$. (Here $\mathbf{U}$ maps the Hilbert space $\mathcal{H}$ onto the Hilbert space $\mathcal{H}'$ and preserves inner product: $\langle \mathbf{Uf}_1, \mathbf{Uf}_2 \rangle' = \langle \mathbf{f}_1, \mathbf{f}_2 \rangle$ for all $\mathbf{f}_1, \mathbf{f}_2 \in \mathcal{H}$. In particular, this means that $\mathbf{U}^* = \mathbf{U}^{-1}$ where the **adjoint** $\mathbf{S}^*$: $\mathcal{H}' \to \mathcal{H}$ of a bounded operator $\mathbf{S}$: $\mathcal{H} \to \mathcal{H}'$ is defined by $\langle \mathbf{Sf}, \mathbf{f}' \rangle' = \langle \mathbf{f}, \mathbf{S}^*\mathbf{f}' \rangle$ for all $f \in \mathcal{H}, f' \in \mathcal{H}'$.) Indeed, if $\mathbf{T} \sim \mathbf{T}'$ then there is a nonzero bounded invertible operator $\mathbf{S}$: $\mathcal{H} \to \mathcal{H}'$ such that $\mathbf{ST}(g) = \mathbf{T}(g)\mathbf{S}$. Taking the adjoint of both sides of this equation and using the fact that $\mathbf{T}^*(g) = \mathbf{T}^{-1}(g), \mathbf{T}'^*(g) = \mathbf{T}'^{-1}(g)$ we have $\mathbf{S}^*\mathbf{T}'^{-1}(g) = \mathbf{T}^{-1}(g)\mathbf{S}^*$. Eliminating $\mathbf{T}'(g)$ from the two equations we find $\mathbf{ST}(g)\mathbf{S}^{-1} = \mathbf{S}^{*-1}\mathbf{T}(g)\mathbf{S}^*$ or $(\mathbf{S}^*\mathbf{S})\mathbf{T}(g) = \mathbf{T}(g)(\mathbf{S}^*\mathbf{S})$. Since $\mathbf{T}$ is irred it follows from Theorem 5.1 that $\mathbf{S}^*\mathbf{S} = \alpha\mathbf{E}$ where $\mathbf{E}$ is the identity operator and α is a constant. Since $\alpha\|\mathbf{f}\|^2 = \langle \mathbf{S}^*\mathbf{Sf}, \mathbf{f} \rangle = \langle \mathbf{Sf}, \mathbf{Sf} \rangle = \|\mathbf{Sf}\|^2 > 0$ for $\mathbf{f} \neq \boldsymbol{0}$, then $\alpha = \beta^2$ is a positive constant. Setting $\mathbf{U} = \beta^{-1}\mathbf{S}$ we have $\mathbf{U}^*\mathbf{U} = \alpha^{-1}\mathbf{S}^*\mathbf{S} = \mathbf{E}$, so $\mathbf{U}$ is unitary and $\mathbf{UT}(g) = \mathbf{T}'(g)\mathbf{U}$.

Based on these results we can mimic the proof of the orthogonality relations (4.17) for any linear Lie group which is unimodular, i.e., such that $d_r A = d_\ell A$. There are two differences, however: (1) In order that the integrals (4.17) exist we must limit ourselves to those irred reps $\mathbf{T}^{(\mu)}$ of G whose matrix elements $T_{jk}^{(\mu)}(A)$ are square integrable. (2) We cannot normalize the measure in general, since the volume of G may be infinite. Thus we obtain the result

$$(5.8) \qquad \int_G T_{i\ell}^{(\mu)}(A)\overline{T}_{sk}^{(\nu)}(A)\,dA = \frac{\delta_{is}}{d(\mu)}\delta_{\ell k}\delta_{\mu\nu}$$

where $T_{i\ell}^{(\mu)}(A)$ are the matrix elements of $\mathbf{T}^{(\mu)}$ with respect to some ON basis and $\mathbf{T}^{(\mu)}, \mathbf{T}^{(\nu)}$ are unitary irred reps of G whose matrix elements are square integrable. The constant $d(\mu) > 0$ is called the **degree** of $\mathbf{T}^{(\mu)}$. (It can be shown that if one matrix element $T_{ij}^{(\mu)}(A)$ of $\mathbf{T}^{(\mu)}$ is square integrable, then all possible matrix elements $\langle \mathbf{T}^{(\mu)}(A)\mathbf{f}, \mathbf{g} \rangle$ are square integrable.) Furthermore, (5.8) can be written in a basis-free form analogous to (3.24).

THEOREM 5.2. *For* $\lambda \neq 0$ *the representation* $\mathbf{T}^\lambda$ *of* H_R *on* $L_2(R)$ *is irreducible.*

Proof. We will present the basic ideas of the proof, omitting some of the technical details. Our aim will be to show that if $\mathbf{L}$ is a bounded operator on $L_2(R)$ which commutes with the operators $\mathbf{T}^\lambda(A)$ for some $\lambda \neq 0$ and all $A \in H_R$ then $\mathbf{L} = \kappa\mathbf{E}$ for some constant κ, where $\mathbf{E}$ is the identity operator. (It follows from this that $\mathbf{R}_\lambda$ is irred, for if $\mathcal{M}$ were a proper closed subspace of $L_2(R)$, invariant under $\mathbf{T}^\lambda$, then the self-adjoint projection operator $\mathbf{P}$ on $\mathcal{M}$ would commute with the operators $\mathbf{T}^\lambda(A)$. This is impossible since $\mathbf{P}$ could not be a scalar multiple of $\mathbf{E}$.)

Suppose the bounded operator $\mathbf{L}$ satisfies

$$\mathbf{L}\mathbf{T}^\lambda[x_1, x_2, x_3] = \mathbf{T}^\lambda[x_1, x_2, x_3]\mathbf{L}$$

for all real x_i. First consider the case $x_1 = x_3 = 0$: $\mathbf{L}$ commutes with the operation of multiplication by functions of the form $e^{i\lambda bt}$ for real b. Clearly $\mathbf{L}$ must also commute with multiplication by finite sums of the form $\sum_{b_j} c_j e^{2\pi i\lambda b_j t}$ and, by using the well-known fact that trigonometric polynomials are dense in the space of measurable functions, $\mathbf{L}$ must commute with multiplication by any bounded function $f(t)$ on $(-\infty, \infty)$. Now let Q be a bounded closed internal in $(-\infty, \infty)$ and let $\chi_Q \in L_2(R)$ be the **characteristic function** of Q:

$$\chi_Q(t) = \begin{cases} 1 & \text{if } t \in Q \\ 0 & \text{if } t \notin Q \end{cases}.$$

Let $f_Q \in L_2(R)$ be the function $f_Q = \mathbf{L}\chi_Q$. Since $\chi_Q^2 = \chi_Q$ we have $f_Q(t) = \mathbf{L}\chi_Q(t) = \mathbf{L}\chi_Q^2(t) = \chi_Q(t)\mathbf{L}\chi_Q(t) = \chi_Q(t)f_Q(t)$ so f_Q is nonzero only for $t \in Q$. Furthermore, if Q' is a closed interval with $Q' \subseteq Q$ and $f_{Q'} = \mathbf{L}\chi_{Q'}$ then $f_{Q'}(t) = \mathbf{L}\chi_{Q'}\chi_Q(t) = \chi_{Q'}(t)\mathbf{L}\chi_Q(t) = \chi_{Q'}(t)f_Q(t)$ so $f_{Q'}(t) = f_Q(t)$ for $t \in Q'$ and $f_{Q'}(t) = 0$ for $t \notin Q'$. It follows that there is a unique function $f(t)$ such that $\chi_{\tilde{Q}}f \in L_2(R)$ and $\chi_{\tilde{Q}}(t)f(t) = \mathbf{L}\chi_{\tilde{Q}}(t)$ for any closed bounded interval $\tilde{Q}$ in $(-\infty, \infty)$. Now let φ be a C^∞ function which is zero in the exterior of $\tilde{Q}$. Then $\mathbf{L}\varphi(t) = \mathbf{L}(\varphi\chi_{\tilde{Q}}(t)) = \varphi(t)\mathbf{L}\chi_{\tilde{Q}}(t) = \varphi(t)f(t)\chi_{\tilde{Q}}(t) = f(t)\varphi(t)$, so $\mathbf{L}$ acts on φ by multiplication by the function $f(t)$. Since as $\tilde{Q}$ runs over all finite subintervals of $(-\infty, \infty)$ the functions φ are dense in $L_2(R)$, it follows that $\mathbf{L} = f(t)\mathbf{E}$.

Now we use the hypothesis that $\mathbf{L}\mathbf{T}^\lambda[x_1, 0, 0]\varphi(t) = \mathbf{T}^\lambda[x_1, 0, 0]\mathbf{L}\varphi(t)$ for all x_1 and $\varphi \in L_2(R)$: $f(t)\varphi(t + x_1) = f(t + x_1)\varphi(t + x_1)$. Thus $f(t) = f(t + x_1)$ almost everywhere, which implies that $f(t)$ is a constant. $\square$

5.4 Orthogonality of radar cross-ambiguity functions.

The results of §5.3 do not directly apply to the unitary rep $\mathbf{T}^\lambda$ of H_R because this rep fails to be square integrable. Indeed it is evident from (5.5) that the x_3-dependence of the matrix element $\langle \mathbf{T}^\lambda(\mathbf{x})f_1, f_2 \rangle$ is of the form $e^{2\pi i\lambda x_3}$, so the integral

$$\iiint \langle \mathbf{T}^\lambda(\mathbf{x})f_1, f_2 \rangle \langle \mathbf{T}^\lambda(\mathbf{x})f_3, f_4 \rangle \, dx_1 \, dx_2 \, dx_3$$

will diverge. All is not lost because we can factor out the **center** of H_R and consider only the factor space. The center $\mathcal{C}$ consists of all elements of H_R which commute with every element of H_R. Clearly

$$\mathcal{C} = \{A(0,0,x_3) = A(x_3), \quad x_3 \in R\}.$$

Now $\mathbf{T}^\lambda$ is irred and $\mathbf{T}^\lambda(x_3)\mathbf{T}^\lambda(\mathbf{x}) = \mathbf{T}^\lambda(\mathbf{x})\mathbf{T}^\lambda(x_3)$ for all $A(\mathbf{x})$. From Theorem 5.1, $\mathbf{T}^\lambda(x_3)$ must be a multiple of the identity operator on $L_2(R)$. Indeed we know that $\mathbf{T}^\lambda(x_3) = e^{2\pi i \lambda x_3}\mathbf{E}$. Now

$$A(\mathbf{x}) = A(x_1, x_2, 0)A(x_3),$$

and since $\mathbf{T}^\lambda$ is irred, the unitary operators $\mathbf{T}^\lambda(x_1, x_2, 0)$, $x_1, x_2 \in R$, must act irreducibly on $L_2(R)$. Furthermore the measure $dx_1 dx_2$ is (two-sided) invariant under the action of H_R. Thus we can repeat the arguments leading to (5.8) for reps $\mathbf{T}^{(\mu)} = \mathbf{T}^{(\nu)} \equiv \mathbf{T}^\lambda$ and measure $dx_1 dx_2$ to obtain (with the assumption, correct as we shall see, that the matrix elements $T_{jk}^\lambda(A)$ are square integrable):

$$(5.9) \qquad \iint_{-\infty}^{\infty} T_{j\ell}^\lambda(x_1, x_2, 0)\overline{T_{sk}^\lambda}(x_1, x_2, 0)dx_1 dx_2 = \frac{\delta_{js}\delta_{\ell k}}{d(\lambda)}.$$

In fact, the structure of H_R is so simple that one can use ordinary Fourier analysis to evaluate the left-hand side of (5.9) for arbitrary matrix elements. The result is

$$(5.10) \qquad \iint_{-\infty}^{\infty} dx_1 dx_2 \langle \mathbf{T}^\lambda(x_1, x_2)f_1, f_2\rangle\langle f_4, \mathbf{T}^\lambda(x_1, x_2)f_3\rangle = \langle f_1, f_3\rangle\langle f_4, f_2\rangle$$

where

$$\langle \mathbf{T}^\lambda(x_1, x_2)f_1, f_2\rangle = \int_{-\infty}^{\infty} e^{2\pi i \lambda t x_2} f_1(t+x_1)\overline{f_2}(t)dt.$$

(This shows explicitly that the matrix elements $\langle \mathbf{T}^\lambda(x_1, x_2)f_1, f_2\rangle$ are square integrable.)

At this point we recall that the narrow-band cross-ambiguity function can be written in the form

$$(5.11) \qquad \psi_{nm}(-x_1, x_2/2) = e^{\pi i x_1 x_2} \int_{-\infty}^{\infty} f_n(t+x_1)\overline{f_m}(t)e^{2\pi i t x_2}dt$$

so that the cross-ambiguity function differs from the matrix element $\langle \mathbf{T}^1(x_1, x_2, 0)f_n, f_m\rangle$ of H_R by the simple multiplicative factor $e^{\pi i x_1 x_2}$. Thus the results of group representation theory can be brought to bear on the radar ambiguity function. (**For the purposes of computation of the ambiguity function, the phase factor $e^{\pi i x_1 x_2}$ is of no concern and we henceforth will identify the matrix element itself with the ambiguity function.**)

Note the special case of (5.10) where $f_j \equiv f$ and $\|f\| = 1$:

$$(5.10') \qquad \iint_{-\infty}^{\infty} dx_1 dx_2 |\langle \mathbf{T}^\lambda(x_1, x_2)f, f\rangle|^2 = 1.$$

For $\lambda = 1$ this is the **radar uncertainty relation**. (The maximum of $| < \mathbf{T}^\lambda(x_1, x_2)f, f > |$ is 1 and occurs for $x_1 = x_2 = 0$. However, in view of (5.10') the graph of this function cannot be too "peaked" around the maximum.)

From (5.10) we see that if $\{f_n, \ n = 0, 1, 2 \cdots \}$ is an ON basis for $L_2(R)$ then the matrix elements $\{\langle \mathbf{T}^\lambda(x_1, x_2)f_n, f_m \rangle\}$ form an ON set in $L_2(R^2)$. This set is actually an ON **basis** for $L_2(R^2)$, see Exercise 5.2. **This allows us to expand a moving target distribution function in the narrow-band case as a series in this ON basis, [W5].**

5.5 The Heisenberg commutation relations. To motivate our next example we make a brief digression to study the **Heisenberg commutation relations** of quantum mechanics. Recall that the matrices

$$\mathcal{C}_1 = \begin{pmatrix} 0 & 1 & 0 \\ 0 & 0 & 0 \\ 0 & 0 & 0 \end{pmatrix}, \quad \mathcal{C}_2 = \begin{pmatrix} 0 & 0 & 0 \\ 0 & 0 & 1 \\ 0 & 0 & 0 \end{pmatrix}, \quad \mathcal{C}_3 = \begin{pmatrix} 0 & 0 & 1 \\ 0 & 0 & 0 \\ 0 & 0 & 0 \end{pmatrix}$$

form a basis for the tangent space at the identity element of H_R. Defining the **commutator** $[\cdot, \cdot]$ of two 3×3 matrices A and B by $[A, B] = AB - BA$, we see that

$$(5.12) \qquad [\mathcal{C}_1, \mathcal{C}_2] = \mathcal{C}_3, \quad [\mathcal{C}_1, \mathcal{C}_3] = \Theta, \quad [\mathcal{C}_2, \mathcal{C}_3] = \Theta$$

where Θ is the zero matrix. (Indeed, one can show that the tangent space at the identity for any local linear Lie group G is closed under the commutator operation: if A and B belong to the tangent space, then so does $[A, B]$. The tangent space equipped with the commutator operation is called the **Lie algebra** of G. The Lie algebra contains essential information about G. Indeed one can reconstruct the connected component of the identity element in G just from a knowledge of the Lie algebra. The lack of commutivity in the group operations corresponds to the nonvanishing of the Lie algebra commutators. For more details on the relationship between Lie groups and Lie algebras see [G1], [HS], [M4], [M5].)

Recall that $\mathcal{C}_j = \partial_{x_j} A(\mathbf{x})|_{\mathbf{x}=\theta}$. Corresponding to the representation $\mathbf{T}^\lambda$ of H_R we can define the analogous operators $\mathbf{C}_j$ where

$$\mathbf{C}_j f(t) = \partial_{x_j} \mathbf{T}^\lambda(\mathbf{x})f(t)|_{\mathbf{x}=\theta}, \ j = 1, 2, 3$$

and $f \in L_2(R)$ is a C^∞ function with compact support. From (5.5) we see that

$$(5.13) \qquad \mathbf{C}_1 = \frac{d}{dt}, \quad \mathbf{C}_2 = 2\pi i \lambda t, \quad \mathbf{C}_3 = 2\pi i \lambda.$$

Defining the **commutator** of operators $\mathbf{A}, \mathbf{B}$ by $[\mathbf{A}, \mathbf{B}] = \mathbf{AB} - \mathbf{BA}$ we verify that

$$(5.14)$$

$$[\mathbf{C}_1, \mathbf{C}_2] = \left[\frac{d}{dt}, 2\pi i \lambda t\right] = 2\pi i \lambda = \mathbf{C}_3,$$

$$[\mathbf{C}_1, \mathbf{C}_3] = \left[\frac{d}{dt}, 2\pi i \lambda\right] = \mathbf{O},$$

$$[\mathbf{C}_2, \mathbf{C}_3] = [2\pi i \lambda t, 2\pi i \lambda] = \mathbf{O},$$

where $\mathbf{O}$ is the zero operator, in analogy with (5.12). Applying these operators on the domain $\mathcal{D}$ of C^∞ functions with compact support, a dense subdomain of $L_2(R)$, we see that they are skew-adjoint; i.e., $\mathbf{C}_j^* = -\mathbf{C}_j$, $j = 1, 2, 3$, where the adjoint $\mathbf{C}^*$ of $\mathbf{C}$ is defined by

$$(5.15) \qquad \langle \mathbf{C}f_1, f_2 \rangle = \int_{-\infty}^{\infty} (\mathbf{C}f_1(t)) \overline{f_2}(t) dt = \langle f_1, \mathbf{C}^* f_2 \rangle$$

for all $f_1, f_2 \in \mathcal{D}$. (In the case of $\mathbf{C}_1$ we have to integrate by parts.)

The skew-adjoint operators

$$(5.16) \qquad \mathbf{C}_1 = \frac{d}{dt}, \quad \mathbf{C}_2 = 2\pi i \lambda t, \quad \mathbf{C}_3 = 2\pi i \lambda$$

satisfy the **Heisenberg commutation relations** (5.14) and are reminiscent of the **annihilation and creation operators for bosons**, familiar from quantum theory. In this theory there is a separable Hilbert space $\mathcal{H}$, an **annihilation operator** $\mathbf{a}$ and its adjoint the **creation operator $\mathbf{a}^*$** such that

$$(5.17) \qquad [\mathbf{a}^*, \mathbf{a}] = -\mathbf{E}$$

where $\mathbf{E}$ is the identity operator on $\mathcal{H}$. Here $\mathbf{a}$ and $\mathbf{a}^*$, and the relation (5.17) are well-defined on some dense subspace $\mathcal{D}$ of $\mathcal{H}$ and map $\mathcal{D}$ into itself. It is further assumed that the equation $\mathbf{a}\psi = \boldsymbol{\theta}$, $\psi \in \mathcal{H}$, has a unique solution in $\mathcal{D}$, up to a multiplicative factor. The normalized solution ψ_0, $\|\psi_0\| = 1$ is called the **vacuum state**. Finally it is assumed that the closure of the subspace generated by applying $\mathbf{a}$ and $\mathbf{a}^*$ to ψ_0 recursively, is $\mathcal{H}$ itself. With these assumptions one can construct explicitly an ON basis for $\mathcal{H}$, a basis of eigenvectors of the **number of particles operator** $\mathbf{N} = \mathbf{a}^* \mathbf{a}$.

To make the commutation relations of the C-operators agree with (5.17) and to assure that $\mathbf{a}^*$ is the adjoint of $\mathbf{a}$ we have, in essence, only one choice:

$$(5.18) \qquad \begin{aligned} \mathbf{a}^* &= \frac{1}{\sqrt{2}} \left(\frac{d}{dt} - t \right) = \frac{1}{\sqrt{2}} \left(\mathbf{C}_1 + \frac{i}{2\pi\lambda} \mathbf{C}_2 \right) \\ \mathbf{a} &= \frac{1}{\sqrt{2}} \left(-\frac{d}{dt} - t \right) = \frac{1}{\sqrt{2}} \left(-\mathbf{C}_1 + \frac{i}{2\pi\lambda} \mathbf{C}_2 \right) \\ \mathbf{E} &= -\frac{i}{2\pi\lambda} \mathbf{C}_3. \end{aligned}$$

To find the vacuum state $\psi_0(t)$ we solve the equation $\mathbf{a}\psi_0 = \boldsymbol{\theta}$. The solution of this first order differential equation is easily seen to be

$$\psi_0(t) = \pi^{-1/4} e^{-t^2/2}$$

where the constant factor is chosen so $\|\psi_0\| = 1$. Since

$$(5.19) \qquad \mathbf{N} = \mathbf{a}^* \mathbf{a} = -\frac{1}{2} \frac{d^2}{dt^2} + \frac{t^2}{2} - \frac{1}{2}$$

we have $\mathbf{N}\boldsymbol{\psi}_0 = \boldsymbol{0}$. (We can take $\mathcal{D}$ to be the space of all functions $p(t)e^{-t^2/2}$ where p is a polynomial.) Now let $\boldsymbol{\psi}$ be a normalized eigenvector of $\mathbf{N}$ with eigenvalue μ. The commutation relations (5.17) imply

$$(5.20) \qquad \begin{aligned} \mathbf{N}(\mathbf{a}^*\boldsymbol{\psi}) &= (\mu+1)\mathbf{a}^*\boldsymbol{\psi}, \\ \mathbf{N}(\mathbf{a}\boldsymbol{\psi}) &= (\mu-1)\mathbf{a}\boldsymbol{\psi}. \end{aligned}$$

Thus $\mathbf{a}^*$ and $\mathbf{a}$ are raising and lowering operators: Given an eigenvector with eigenvalue μ we can obtain a ladder of eigenvectors with eigenvalues $\mu+n$, n an integer. Now

$$(5.21) \qquad \langle \mathbf{a}^*\boldsymbol{\psi}, \mathbf{a}^*\boldsymbol{\psi} \rangle = \langle \mathbf{a}\mathbf{a}^*\boldsymbol{\psi}, \boldsymbol{\psi} \rangle = \langle (\mathbf{a}^*\mathbf{a}+\mathbf{E})\boldsymbol{\psi}, \boldsymbol{\psi} \rangle = \mu+1.$$

Thus for $\mu \geq 0$, $\|\mathbf{a}^*\boldsymbol{\psi}\| = \sqrt{\mu+1} > 0$, so the process of constructing eigenvectors of $\mathbf{N}$ by applying recursively the creation operator to the vacuum state can be continued indefinitely. Indeed from (5.20) and (5.21) we can define normalized eigenvectors $\boldsymbol{\psi}_n$ with eigenvalues n recursively by

$$(5.22) \qquad \mathbf{a}^*\boldsymbol{\psi}_n = (n+1)^{1/2}\boldsymbol{\psi}_{n+1}, \quad n = 0, 1, 2, \cdots.$$

The commutation relations imply the formulas

$$(5.23) \qquad \mathbf{N}\boldsymbol{\psi}_n = n\boldsymbol{\psi}_n, \quad \mathbf{a}\boldsymbol{\psi}_n = n^{1/2}\boldsymbol{\psi}_{n-1}.$$

Substituting expressions (5.18) into (5.22) and (5.23), we obtain a second-order differential equation and two recurrence formulas for the special functions $\boldsymbol{\psi}_n(t)$. We can obtain a generating function for the $\boldsymbol{\psi}_n$ from the first-order operator $\mathbf{a}^* = \frac{1}{\sqrt{2}}\left(\frac{d}{dt}-t\right)$. Note that $\mathbf{a}^* = \frac{1}{\sqrt{2}}e^{t^2/2}\left(\frac{d}{dt}\right)e^{-t^2/2}$. Hence by Taylor's theorem,

$$\begin{aligned} e^{\alpha \mathbf{a}^*}\boldsymbol{\psi}(t) &= \sum_{k=0}^{\infty}\frac{\alpha^k}{k!}(\mathbf{a}^*)^k = e^{t^2/2}\sum_{k=0}^{\infty}\frac{\left(\frac{\alpha}{\sqrt{2}}\right)^k}{k!}\left(\frac{d}{dt}\right)^k[e^{-t^2/2}\boldsymbol{\psi}(t)] \\ &= e^{t^2/2}e^{-(t+\frac{\alpha}{\sqrt{2}})^2/2}\boldsymbol{\psi}(t+2^{-1/2}\alpha) \\ &= \exp(-\frac{\alpha^2}{4}-2^{-1/2}\alpha t)\boldsymbol{\psi}(t+2^{-1/2}\alpha) \end{aligned}$$

for any analytic function $\boldsymbol{\psi}$. On the other hand, from (5.22) we have

$$e^{\alpha \mathbf{a}^*}\boldsymbol{\psi}_n(t) = \sum_{k=0}^{\infty}\left[\frac{(n+k)!}{n!}\right]^{1/2}\frac{\alpha^k}{k!}\boldsymbol{\psi}_{n+k}(t).$$

Comparing these equations, we find the identity

$$(5.24) \qquad \exp\left(-\frac{\beta^2}{2}-\beta t\right)\boldsymbol{\psi}_n(t+\beta) = \sum_{k=0}^{\infty}\left[\frac{2^k(n+k)!}{n!}\right]^{1/2}\frac{\beta^k}{k!}\boldsymbol{\psi}_{n+k}(t).$$

In the special case $n = 0$, the generating function yields

(5.25) $$\pi^{-1/4} \exp\left(-\beta^2 - 2\beta t - \frac{1}{2}t^2\right) = \sum_{k=0}^{\infty} \frac{2^{k/2}\beta^k}{(k!)^{1/2}} \boldsymbol{\psi}_k(t).$$

Comparing this with the well-known generating function

$$\exp(-\beta^2 + 2\beta t) = \sum_{k=0}^{\infty} \frac{\beta^k}{k} H_k(t)$$

for the Hermite polynomials $H_k(t)$, [EMOT1], [M4], [M5], [M6], [V], [WW], we obtain

(5.26) $$\boldsymbol{\psi}_k(t) = \pi^{-1/4}(k!)^{-1/2}(-1)^k 2^{-k/2} e^{-t^2/2} H_k(t).$$

The above series converge for all t and β. Since the $\{\boldsymbol{\psi}_n(t)\}$ form an ON set in $L_2(R)$ we easily obtain the formula

$$\int_{-\infty}^{\infty} H_n(t)H_k(t)e^{-t^2} dt = \pi^{1/2} 2^n n! \delta_{nk}.$$

We sketch a proof of the fact that the $\{\boldsymbol{\psi}_n(t)\}$ form an ON basis for $L_2(R)$. It is enough to show that this set is *dense* in $L_2(R)$, i.e., if

(5.27) $$< g, \boldsymbol{\psi}_n >= 0, \quad \text{for } g \in L_2(R), \quad n = 0, 1, 2, \cdots,$$

then $g(t) = 0$ almost everywhere. Since $H_n(t)$ is a polynomial of order n in t with the coefficient of t^n nonzero, conditions (5.27) are equivalent to

$$\int_{-\infty}^{\infty} e^{-\frac{t^2}{2}} t^n g(t) \, dt = 0, \quad n = 0, 1, 2, \cdots.$$

Now consider the function

$$G(z) = \int_{-\infty}^{\infty} e^{izt} e^{-\frac{t^2}{2}} g(t) \, dt.$$

Since $g \in L_2(R)$, $G(z)$ is an (entire) analytic function of z and its derivatives can be obtained by differentiating under the integral sign:

$$\frac{d^n G(z)}{dz^n} = G^{(n)}(z) = i^n \int_{-\infty}^{\infty} e^{izt} t^n e^{-\frac{t^2}{2}} g(t) \, dt.$$

Now

$$G(z) = \sum_{n=0}^{\infty} \frac{G^{(n)}(0)}{n!} z^n \equiv 0.$$

Since $G(z)$ is the Fourier transform of $e^{-\frac{t^2}{2}}$, it follows that $g(t) = 0$ almost everywhere.

5.6 The Bargmann-Segal Hilbert space. Our next task is to compute the matrix elements $T_{jk}^\lambda(A) = (\mathbf{T}^\lambda(A)\boldsymbol{\psi}_k, \boldsymbol{\psi}_j)$ with respect to this basis. However, a computation of these matrix elements using the direct evaluation of the integral is not very enlightening. A better approach is to use a simpler model of the representation $\mathbf{T} \equiv \mathbf{T}^\lambda$, motivated by another solution of the Heisenberg commutation relations $[\mathbf{a}^*, \mathbf{a}] = -\mathbf{E}$. Rather than the solution (5.18) we can try

$$(5.28) \qquad\qquad \mathbf{a}^* = z, \quad \mathbf{a} = \frac{d}{dz}.$$

This will work provided we can define a Hilbert space $\mathcal{F}$ on which $\mathbf{a}$ and $\mathbf{a}^*$ act and such that $\mathbf{a}^*$ is the adjoint of $\mathbf{a}$. (We will construct an ON basis $\{\mathbf{j}_n\}$ for $\mathcal{F}$ corresponding to the basis $\{\boldsymbol{\psi}_n\}$.) Since $\mathbf{a}\mathbf{j}_0(z) = 0$ implies that $\mathbf{j}_0(z)$ is constant, in order to mimic successfully the construction (5.22), (5.23) we see that the $\mathbf{j}_n$ must be proportional to z^n, so the elements of $\mathcal{F}$ must be functions $\mathbf{j}(z)$. If z were a real variable it would not be possible to find an inner product $(\mathbf{f}_1, \mathbf{f}_2) = \int \mathbf{f}_1(z)\overline{\mathbf{f}_2}(z)\rho(z)dz$ with respect to which $\mathbf{a}^*$ is the adjoint of $\mathbf{a}$. However, if we take z to be a complex variable and search for an inner product of the form $(\mathbf{f}_1, \mathbf{f}_2) = \iint_{-\infty}^{\infty} \mathbf{f}_1(z)\overline{\mathbf{f}_2}(z)\rho(z, \bar{z})dxdy$ we will be successful. (Here $z = x + iy$ and the region of integration is the plane R_2. We assume that the weight function ρ is nonnegative.) Integrating by parts in the formula $(\mathbf{f}_1, \mathbf{a}\mathbf{f}_2) = (\mathbf{a}^*\mathbf{f}_1, \mathbf{f}_2)$, assuming that the boundary terms vanish and that the formula holds identically in $\mathbf{f}_1, \mathbf{f}_2$ we obtain the condition $-\partial_{\bar{z}}\rho(z, \bar{z}) = z\rho(z, \bar{z})$. Thus $\rho(z, \bar{z}) = \pi^{-1}e^{-z\bar{z}}$ where we have chosen the constant π^{-1} so that $(1, 1) = 1$. It follows from the analogs of (5.22), (5.23) that

$$(5.29) \qquad \begin{aligned} \mathbf{a}\mathbf{j}_n &= \sqrt{n}\,\mathbf{j}_{n-1}, \quad \mathbf{a}^*\mathbf{j}_n = \sqrt{n+1}\,\mathbf{j}_{n+1}, \\ \mathbf{N}\mathbf{j}_n &= n\mathbf{j}_n, \quad (\mathbf{j}_n, \mathbf{j}_m) = \delta_{nm} \\ n, m &= 0, 1, 2, \cdots \end{aligned}$$

where $\mathbf{j}_{-1} \equiv \boldsymbol{0}$ and

$$(5.30) \qquad\qquad \mathbf{j}_n(z) = \frac{z^n}{\sqrt{n!}}, \quad n = 0, 1, 2, \cdots.$$

For any two functions $\mathbf{f}(z) = \sum_{k=0}^{\infty} a_k z^k$ and $\mathbf{h}(z) = \sum_{k=0}^{\infty} b_k z^k$ in $\mathcal{F}$ we find

$$(5.31) \qquad\qquad (\mathbf{f}, \mathbf{h}) = \sum_{k=0}^{\infty} k!a_k\bar{b}_k$$

and

$$(5.32) \qquad\qquad ||\mathbf{f}||^2 = (\mathbf{f}, \mathbf{f}) = \sum_{k=0}^{\infty} k!|a_k|^2.$$

From (5.32), $\mathbf{f}$ belongs to the Hilbert space $\mathcal{F}$ if and only if $\sum_{k=0}^{\infty} k!|a_k|^2 < \infty$. Clearly if $\mathbf{f} \in \mathcal{F}$ then there is a constant $C \geq 0$ such that $|a_k| \leq C/\sqrt{k!}$ for all k. By the ratio test, the series $\sum_k z^k/\sqrt{k!}$ converges for all complex z, so $\mathbf{f}$ is an entire function, i.e., the power series expansion for $\mathbf{f}$ has an infinite radius of convergence.

The space $\mathcal{F}$ was introduced by Segal and studied in detail by Bargmann [B2]. We mention here some of the special properties of $\mathcal{F}$.

Define the function $\mathbf{e}_b \in \mathcal{F}$ for some complex constant b by $\mathbf{e}_b(z) = \exp(\bar{b}z) = \sum_{k=0}^{\infty} (\bar{b}z)^k/k!$. It follows from (5.31) that for any $\mathbf{f} \in \mathcal{F}$,

$$(\mathbf{f}, \mathbf{e}_b) = \sum_{k=0}^{\infty} a_k b^k = \mathbf{f}(b),$$

$$(\mathbf{e}_b, \mathbf{e}_b) = \mathbf{e}_b(b) = e^{\bar{b}b}.$$

Thus $\mathbf{e}_b \in \mathcal{F}$ acts like a delta function!

From the Schwarz inequality, $|\mathbf{f}(b)| = |(\mathbf{f}, \mathbf{e}_b)| \leq ||\mathbf{e}_b|| \cdot ||\mathbf{f}|| = e^{\bar{b}b/2}||\mathbf{f}||$. Thus, if $\mathbf{f}, \mathbf{h} \in \mathcal{F}$, then $|\mathbf{f}(b) - \mathbf{h}(b)| \leq e^{\bar{b}b/2}||f - h||$ which shows that convergence in the norm of $\mathcal{F}$ implies pointwise convergence, uniform on any compact set in $\mathcal{C}$.

Now we construct a representation of H_R on $\mathcal{F}$, using the annihilation and creation operators (5.28). Comparing with (5.18) we see that the standard basis for the Lie algebra of H_R is

(5.33)
$$\mathbf{C}_1 = \frac{1}{\sqrt{2}}(\mathbf{a}^* - \mathbf{a}) = \frac{1}{\sqrt{2}}\left(z - \frac{d}{dz}\right)$$
$$\mathbf{C}_2 = -\frac{2\pi\lambda i}{\sqrt{2}}(\mathbf{a}^* + \mathbf{a}) = -\frac{2\pi\lambda i}{\sqrt{2}}\left(z + \frac{d}{dz}\right)$$
$$\mathbf{C}_3 = 2\pi\lambda i \mathbf{E}.$$

Mimicking the derivation of (5.24) by exponentiating these operators, we obtain the following candidates for operators defining a unitary rep of H_R on $\mathcal{F}$:

(5.34)
$$\mathbf{T}'(x_1,0,0)\mathbf{f}(z) = \exp\left(-\frac{x_1^2}{4} + 2^{-1/2}x_1 z\right)\mathbf{f}(z - 2^{-1/2}x_1),$$
$$\mathbf{T}'(0,x_2,0)\mathbf{f}(z) = \exp\left(-\pi^2\lambda^2 x_2^2 - 2^{1/2}i\pi\lambda x_2 z\right)\mathbf{f}(z - 2^{1/2}i\pi\lambda x_2),$$
$$\mathbf{T}'(0,0,x_3)\mathbf{f}(z) = e^{2\pi\lambda i x_3}\mathbf{f}(z).$$

Since $A(\mathbf{x}) = A(0,x_2,0)A(x_1,0,0)A(0,0,x_3)$ we construct the operators $\mathbf{T}'(\mathbf{x}) \equiv \mathbf{T}'(A(\mathbf{x}))$ by
(5.35)
$$\mathbf{T}'(\mathbf{x})\mathbf{f}(z) = [\mathbf{T}'(0,x_2,0)\mathbf{T}'(x_1,0,0)\mathbf{T}'(0,0,x_3)\mathbf{f}](z)$$
$$= \exp\left[-\frac{(x_1^2 + 4\pi^2\lambda^2 x_2^2)}{4} + \frac{(x_1 - 2i\pi\lambda x_2)}{\sqrt{2}}z - i\pi\lambda x_1 x_2 + 2i\pi\lambda x_3\right] \times$$
$$\mathbf{f}\left(z - \frac{[x_1 + 2i\pi\lambda x_2]}{\sqrt{2}}\right).$$

It is straightforward to check that the operators $\mathbf{T}'(\mathbf{x})$ define a unitary rep of H_R on $\mathcal{F}$. As one would expect, this rep is equivalent to the rep $\mathbf{T} \equiv \mathbf{T}^\lambda$ of H_R on $L_2(R)$.

To establish the equivalence we construct the unitary operator $\mathbf{A} : L_2(R) \to \mathcal{F}$ which maps the ON basis vector $\boldsymbol{\psi}_k(t)$ of $L_2(R)$, (5.26), to the ON basis vector $\mathbf{j}_k(z) = z^k/\sqrt{k!}$ of $\mathcal{F}$:

$$[\mathbf{A}\boldsymbol{\psi}](z) = \int_{-\infty}^{\infty} A(z,t)\boldsymbol{\psi}(t)dt, \quad \boldsymbol{\psi} \in L_2(R),$$

(5.36)

$$A(z,t) = \sum_{k=0}^{\infty} \mathbf{j}_k(z)\boldsymbol{\psi}_k(t) = \pi^{-1/4} \exp[-(z^2 + t^2)/2 - \sqrt{2}zt].$$

The last identity follows from (5.25) with $z = \sqrt{2}\beta$. Since $A(z,\cdot) \in L_2(R)$ for each $z \in \mathcal{C}$ the integral in (5.36) is always defined. Now if $\boldsymbol{\psi} \in L_2(R)$ then it can be expanded uniquely in the form $\boldsymbol{\psi} = \sum_{n=0}^{\infty} c_n \boldsymbol{\psi}_n$. From (5.36) and the orthogonality of the basis $\{\boldsymbol{\psi}_k\}$ for $L_2(R)$ we have

$$\mathbf{f} = \mathbf{A}\boldsymbol{\psi} = \sum_{n=0}^{\infty} c_n \mathbf{j}_n \in \mathcal{F}$$

where the $\{\mathbf{j}_k\}$ form an ON basis for $\mathcal{F}$. Clearly the operator $\mathbf{A}$ is unitary. A correct expression for $\mathbf{A}^{-1} : \mathcal{F} \to L_2(R)$ is a bit more involved:

$$\mathbf{A}^{-1}\mathbf{f}(t) = \lim_{\substack{\mu \to 1 \\ \mu < 1}} \int \overline{A(\mu z, t)}\mathbf{f}(z)\rho(z, \bar{z})dx\,dy,$$

see [B2], [M4]. (It is necessary to insert the parameter $\mu < 1$ because $A(z,t)$ for fixed t does not belong to $\mathcal{F}$.)

Now we can verify explicitly the relations

$$\mathbf{T}'(x_1, 0, 0)\mathbf{A} = \mathbf{A}\mathbf{T}(x_1, 0, 0),$$
$$\mathbf{T}'(0, x_2, 0)\mathbf{A} = \mathbf{A}\mathbf{T}(0, x_2, 0),$$
$$\mathbf{T}'(0, 0, x_3)\mathbf{A} = \mathbf{A}\mathbf{T}(0, 0, x_3),$$

where $\mathbf{T} \equiv \mathbf{T}^{\lambda}$ is given by (5.5). Since $\mathbf{A}$ is invertible, it follows that $\mathbf{T}'(\mathbf{x}) = \mathbf{A}\mathbf{T}(\mathbf{x})\mathbf{A}^{-1}$, so $\mathbf{T}'$ is equivalent to $\mathbf{T}$.

Since our chosen ON basis for $\mathcal{F}$ consists simply of powers of z, it is relatively easy to compute the matrix elements $T_{k\ell}(\mathbf{x}) = (\mathbf{T}'(\mathbf{x})\mathbf{j}_\ell, \mathbf{j}_k)$. (It is immediate that $(\mathbf{T}'(\mathbf{x})\mathbf{j}_\ell, \mathbf{j}_k) = \langle \mathbf{T}(x)\boldsymbol{\psi}_\ell, \boldsymbol{\psi}_k \rangle$ so these matrix elements are exactly the same as those which could be computed using the ON basis $\{\boldsymbol{\psi}_k\}$ for $L_2(R)$.) We define the generating function

(5.37) $$G(\mathbf{x}; u, v) = (\mathbf{T}'(\mathbf{x})e_{\bar{u}}, e_v) = \sum_{n,m=0}^{\infty} (\mathbf{T}'(\mathbf{x})\mathbf{j}_m, \mathbf{j}_n)\frac{u^m v^n}{\sqrt{m!n!}}.$$

Due to the delta function property of $e_{\bar{v}}$ we obtain

(5.38)
$$(\mathbf{T}'(\mathbf{x})e_{\bar{u}}, e_v) = [\mathbf{T}(\mathbf{x})e_{\bar{u}}](v)$$
$$= \exp\left[-\frac{(x_1^2 + 4\pi^2 \lambda^2 x_2^2)}{4} + \frac{(x_1 - 2\pi \lambda i x_2)}{\sqrt{2}}v - \pi \lambda i x_1 x_2 + 2\pi \lambda i x_3 \right]$$
$$\times \exp\left[u\left(v - \frac{[x_1 + 2\pi \lambda i x_2]}{\sqrt{2}} \right) \right].$$

Introducing polar coordinates

$$x_1 = r\cos\theta, \quad 2\pi\lambda x_2 = r\sin\theta$$

and equating coefficients of $u^m v^n$ in (5.37) and (5.38) we obtain the explicit expression

(5.39)
$$T_{k\ell}(\mathbf{x}) = \exp\left[2\pi\lambda i x_3 + i(k-\ell)\theta - \pi\lambda i x_1 x_2\right] e^{-r^2/4} \left(\frac{k!}{\ell!}\right)^{1/2} \times$$
$$\left(\frac{r}{2^{1/2}}\right)^{\ell-k} L_k^{(\ell-k)}(r^2/2)$$

where $L_k^{(\alpha)}(x)$ is the associated Laguerre polynomial [EMOT1], [M3], [M4]. An alternate expression is

(5.40)
$$T_{k\ell}(\mathbf{x}) = \exp\left[2\pi\lambda i x_3 + i(k-\ell)\theta - \pi\lambda i x_1 x_2\right] e^{-r^2/4} \left(\frac{\ell!}{k!}\right)^{1/2} \times$$
$$\left(-\frac{r}{2^{1/2}}\right)^{k-\ell} L_\ell^{(k-\ell)}(r^2/2).$$

Note that the skew adjoint operator $-i\mathbf{N} = -iz\frac{d}{dz}$ is well defined on $\mathcal{F}$ and can be exponentiated to yield the unitary operator $\mathbf{U}'(\alpha)$:

(5.41)
$$\mathbf{U}'(\alpha)\mathbf{f}(z) = \exp\left(-i\alpha z\frac{d}{dz}\right)\mathbf{f}(z) = \mathbf{f}(e^{-i\alpha}z), \quad \mathbf{f} \in \mathcal{F}.$$

Here the eigenvectors of $\mathbf{U}$ acting on $\mathcal{F}$ are just the ON basis vectors $\mathbf{j}_n$:

(5.42)
$$\mathbf{U}'(\alpha)\mathbf{j}_n = e^{-in\alpha}\mathbf{j}_n.$$

(One can extend the rep $\mathbf{T}'$ of the three parameter group H_R to the four-parameter **oscillator group** generated by $\mathbf{T}'(\mathbf{x})$ and $\mathbf{U}(\alpha)$. See [M3], [M4] for the details.) Using the unitary transformation $\mathbf{A}$ one can transform the $\mathbf{U}'(\alpha)$ to unitary operators on $L_2(R)$:

(5.43)
$$\mathbf{U}(\alpha)\boldsymbol{\psi}(t) = \lim_{n\to\infty} \int_{-n}^{n} \frac{e^{i\epsilon(\pi/4 - \beta/2)}}{(2\pi|\sin\alpha|)^{1/2}} \times$$
$$\exp\left[i\cot\alpha(t^2 + \tau^2)/2 - it\tau/\sin\alpha\right]\boldsymbol{\psi}(\tau)d\tau.$$

Here, $\mathbf{U}(\alpha) = \mathbf{A}^{-1}\mathbf{U}'(\alpha)\mathbf{A}$ and $\alpha = 2k\pi + \epsilon\beta$, k an integer, $\epsilon = \pm 1$, $0 < \beta < \pi$. (See [B2] for details.) Note that $\mathbf{U}(\pi/2)$ is just the (ordinary) Fourier transform on $L_2(R)$. Thus, the Fourier transform is embedded in a one parameter group of transformations. the infinitesimal generator of this one-parameter group is the second-order differential operator

$$-i\mathbf{N} = i\left(\frac{1}{2}\frac{d^2}{dt^2} - \frac{t^2}{2} + \frac{1}{2}\right).$$

Furthermore, $\mathbf{U}(\alpha)\psi_n = e^{-in\alpha}\psi_n$.

It follows from the orthogonality relations (5.10) that the matrix elements (5.39) form an ON set in $L_2(R^2)$. (See [M3] or [M4] for a direct proof of this fact.) These orthogonality relations reduce to the following orthogonality relations for associated Laguerre polynomials:

$$(5.44) \qquad \int_0^\infty L_m^{(k)}(r^2)L_n^{(k)}(r^2)e^{-r^2}r^{2k+1}dr = \frac{(n+k)!}{2n!}\delta_{mn},$$

valid for all integers $m, n \geq 0$ and all integers k such that $n+k \geq 0, m+k \geq 0$. (By switching between (5.39) and (5.40) depending on whether $k - \ell \leq 0$ or $k - \ell > 0$ we can always take $k \geq 0$ in (5.44).) Since for each $k \geq 0$ the associated Laguerre polynomials are known to be complete in $L_2(0, \infty)$ with weight function $e^{-r^2}r^{2k+1}$, it follows that the matrix elements $\{T_{mn}\}$ form an ON basis for $L_2(R^2)$, with the usual Lebesgue weight function 1. (Indeed, this follows from Exercise 5.2, without any knowledge of the completeness properties of the Laguerre polynomials.) Thus, the matrix elements of $\mathbf{T}$ with respect to any ON basis for $L_2(R)$ will form an ON basis for $L_2(R^2)$. **(In other words, the cross-ambiguity functions with respect to an ON basis of signals $\{s_k\}$ form an ON basis for $L_2(R^2)$.)**

5.7 The lattice representation of H_R. There is another realization of the irred unitary rep $\mathbf{T}^\lambda$ that we shall find useful: an induced rep of H_R from the subgroup H' where

$$H' = \left\{ A(a_1, a_2, y_3) = \begin{pmatrix} 1 & a_1 & y_3 \\ 0 & 1 & a_2 \\ 0 & 0 & 1 \end{pmatrix} \right\},$$

a_1, a_2 are integers and $y_3 \in R$. Note that the operators $\mathbf{T}_0(a_1, a_2, y_3) = e^{2\pi i y_3}$ define a one-dimensional unitary rep of H'. (Here, the fact that a_1, a_2 are integers is crucial in verifying that $\mathbf{T}_0$ is a rep of H'.) We will study the rep $\tilde{\mathbf{T}}$ of H_R induced from the rep $\mathbf{T}_0$ of H', (5.2 - 5.4). Here $\tilde{\mathbf{T}}$ is defined on the space V of functions $\mathbf{f}$ on H_R such that $\mathbf{f}(BA) = \mathbf{T}_0(B)\mathbf{f}(A)$ for all $B \in H', A \in H_R$, i.e.,

$$(5.45) \qquad \mathbf{f}(a_1 + x_1, \ a_2 + x_2, \ y_3 + x_3 + a_1 x_2) = e^{2\pi i y_3}\mathbf{f}(x_1, x_2, x_3).$$

The operators $\tilde{\mathbf{T}}(A), A \in H_R$ act on V according to

$$(5.46) \qquad [\tilde{\mathbf{T}}(A)\mathbf{f}](A') = \mathbf{f}(A'A).$$

We see from (5.45) that for any $A(x_1, x_2, x_3)$ we can always choose $B(a_1, a_2, y_3)$ such that $BA = A'(x_1', x_2', 0)$ where $0 \leq x_1' < 1, 0 \leq x_2' < 1$. Thus $\mathbf{f}$ can be restricted to $X = H'\backslash H_R$ with coordinates $(x_1', x_2', 0)$. Moreover, setting $x_3 = 0, y_3 = -a_1 x_2$ in (5.45) we have the periodicity condition

$$(5.46) \qquad \varphi(a_1 + x_1, a_2 + x_2) = e^{-2\pi i a_1 x_2}\varphi(x_1, x_2)$$

where $\varphi(x_1, x_2) = \mathbf{f}(x_1, x_2, 0)$. Conversely, given φ satisfying (5.46) we can define a unique $\mathbf{f}$ satisfying (5.45) by

$$\mathbf{f}(x_1, x_2, x_3) = \varphi(x_1, x_2)e^{2\pi i x_3}.$$

The H_R-invariant inner product on X is $dx_1 dx_2$:

$$(5.47) \qquad \langle \varphi_1, \varphi_2 \rangle = \int_0^1 \int_0^1 \varphi_1(x_1, x_2)\overline{\varphi_2}(x_1, x_2)dx_1 dx_2,$$

and the operator $\tilde{\mathbf{T}}[\mathbf{y}] \equiv \tilde{\mathbf{T}}(A(y_1, y_2, y_3))$ acts on these functions by

$$(5.48) \qquad (\tilde{\mathbf{T}}[\mathbf{y}]\varphi)(x_1, x_2) = \exp[2\pi i(y_3 + x_1 y_2)]\varphi(x_1 + y_1, x_2 + y_2).$$

To recapitulate, we have defined a unitary rep $\tilde{\mathbf{T}}$ of H_R on the Hilbert space V' of all functions φ satisfying (5.46) and of finite norm with respect to the inner product (5.47). This is known as the **lattice representation** of H_R.

The lattice rep is equivalent to the irred Schrödinger rep $\mathbf{T}^1$, (5.5). To see this consider the periodizing operator (Weil-Brezin-Zak isomorphism)

$$(5.49) \qquad \mathbf{P}\psi(x_1, x_2, x_3) = \sum_{n=-\infty}^{\infty} (\mathbf{T}^1[x_1, x_2, x_3]\psi)(n)$$
$$= e^{2\pi i x_3} \sum_{n=-\infty}^{\infty} e^{2\pi i n x_2}\psi(n + x_1)$$

which is well defined for any $\psi \in L_2(R)$ which belongs to the Schwartz space. It is straightforward to verify that $\mathbf{f} = \mathbf{P}\psi$ satisfies the periodicity condition (5.45), hence $\mathbf{f}$ belongs to V. Now

$$\langle \mathbf{P}\psi(\cdot, \cdot, 0), \mathbf{P}\psi'(\cdot, \cdot, 0) \rangle$$
$$= \int_0^1 dx_1 \int_0^1 dx_2 \sum_{m,n=-\infty}^{\infty} e^{2\pi i(n-m)x_2}\psi(n + x_1)\overline{\psi'(m + x_2)}$$
$$= \int_0^1 dx_1 \sum_{n=-\infty}^{\infty} \psi(n + x_1)\overline{\psi'(n + x_1)} = \int_{-\infty}^{\infty} \psi(t_1)\overline{\psi'(t)}\, dt$$
$$= (\psi, \psi')$$

so $\mathbf{P}$ can be extended to an inner product preserving mapping of $L_2(R)$ into V.

It is clear from (5.49) that if $\varphi(x_1, x_2) = \mathbf{P}\psi(x_1, x_2, 0)$ then we can recover $\psi(x_1)$ by integrating with respect to $x_2 : \psi(x_1) = \int_0^1 \varphi(x_1, y)dy$. Thus we define the mapping $\mathbf{P}^*$ of V' into $L_2(R)$ by

$$(5.50) \qquad \mathbf{P}^*\varphi(t) = \int_0^1 \varphi(t, y)dy, \quad \varphi \in V'.$$

Since $\boldsymbol{\varphi} \in V'$ we have

$$\mathbf{P}^*\boldsymbol{\varphi}(t+a) = \int_0^1 \boldsymbol{\varphi}(t,x)e^{-2\pi i a y}\,dy = \hat{\varphi}_{-a}(t)$$

for a an integer. (Here $\hat{\varphi}_n(t)$ is the nth Fourier coefficient of $\boldsymbol{\varphi}(t,y)$.) The Parseval formula then yields

$$\int_0^1 |\boldsymbol{\varphi}(t,y)|^2\,dy = \sum_{a=-\infty}^{\infty} |\mathbf{P}^*\boldsymbol{\varphi}(t+a)|^2$$

so

$$\langle\boldsymbol{\varphi},\boldsymbol{\varphi}\rangle = \int_0^1\int_0^1 |\boldsymbol{\varphi}(t,y)|^2\,dt\,dy = \int_0^1 \sum_{a=-\infty}^{\infty} |\mathbf{P}^*\boldsymbol{\varphi}(t+a)|^2\,dt$$

$$= \int_{-\infty}^{\infty} |\mathbf{P}^*\boldsymbol{\varphi}(t)|^2\,dt = (\mathbf{P}^*\boldsymbol{\varphi}, \mathbf{P}^*\boldsymbol{\varphi}).$$

and $\mathbf{P}^*$ is an inner product preserving mapping of V' into $L_2(R)$. Moreover, it is easy to verify that

$$\langle\mathbf{P}\boldsymbol{\psi},\boldsymbol{\varphi}\rangle = (\boldsymbol{\psi}, \mathbf{P}^*\boldsymbol{\varphi})$$

for $\boldsymbol{\psi} \in L_2(R)$, $\boldsymbol{\varphi} \in V'$, i.e., $\mathbf{P}^*$ is the adjoint of $\mathbf{P}$. Since $\mathbf{P}^*\mathbf{P} = \mathbf{E}$ on $L_2(R)$ it follows that $\mathbf{P}$ is a unitary operator mapping $L_2(R)$ **onto** V' and $\mathbf{P}^* = \mathbf{P}^{-1}$ is a unitary operator mapping V' **onto** $L_2(R)$.

Finally,

$$(\mathbf{PT}^1[\mathbf{y}]\boldsymbol{\psi})(\mathbf{x}) = e^{2\pi i(x_3+y_3+x_1 y_2)} \sum_{n=-\infty}^{\infty} e^{2\pi i n(x_2+y_2)}\boldsymbol{\psi}(n+x_1+y_1)$$

$$= (\tilde{\mathbf{T}}[\mathbf{y}]\mathbf{P}\boldsymbol{\psi})(\mathbf{x})$$

so $\mathbf{PT}^1[\mathbf{y}] = \tilde{\mathbf{T}}[\mathbf{y}]\mathbf{P}$ and the unitary reps $\mathbf{T}^1$ and $\tilde{\mathbf{T}}$ are equivalent.

5.8 Functions of positive type. Let G be a group. A complex function ρ on G is said to be of **positive type** provided for every finite set $g_1, \cdots, g_{k_0}$ of elements of G and every set of complex numbers $\lambda_1, \cdots, \lambda_{k_0}$ the inequality

(5.51)
$$\sum_{j,\ell} \rho(g_j^{-1}g_\ell)\bar{\lambda}_j\lambda_\ell \geqq 0$$

holds. If $\mathbf{U}$ is a unitary rep of G on the Hilbert space $\mathcal{H}$ then the inner product $\rho(g) = \langle\mathbf{U}(g)\mathbf{f},\mathbf{f}\rangle$ is of positive type on G for every $\mathbf{f} \in \mathcal{H}$. Indeed,

(5.52)
$$\sum_{j,\ell} \rho(g_j^{-1}g_\ell)\bar{\lambda}_j\lambda_\ell = \sum_{j,\ell}\langle\mathbf{U}(g_j^{-1}g_\ell)\mathbf{f}, \mathbf{f}\rangle\bar{\lambda}_j\lambda_\ell$$

$$= \sum_{j,\ell}\langle\mathbf{U}(g_\ell)\mathbf{f}, \mathbf{U}(g_j)\mathbf{f}\rangle\bar{\lambda}_j\lambda_\ell$$

$$= \langle\sum_\ell \lambda_\ell\mathbf{U}(g_\ell)\mathbf{f}, \sum_\ell \lambda_\ell\mathbf{U}(g_\ell)\mathbf{f}\rangle \geqq 0.$$

It follows from this construction, in the case where $G = H_R$, that narrow band ambiguity functions are of positive type on H_R.

It is an important result of abstract harmonic analysis that all functions of positive type on a group arise as diagonal matrix elements of unitary reps, exactly as in the construction (5.52). This result, whose proof we now sketch, sheds light on the structure of the set of ambiguity functions.

Note first that the inequality (5.51) implies that the $k_0 \times k_0$ matrix with elements $H_{j\ell} = \rho(g_j^{-1} g_\ell)$ is Hermitian ($H_{j\ell} = \bar{H}_{\ell j}$) and nonnegative. Thus the k_0 eigenvalues of this matrix are nonnegative; hence the determinant is also nonnegative.

LEMMA 5.1. *Let ρ be a function of positive type on the group G. Then*

1) $\overline{\rho(g^{-1})} = \rho(g)$

2) $\rho(e) \geq |\rho(g)|$

for each $g \in G$.

Proof. For $k_0 = 1$, the inequality (5.51) yields $\rho(e) \geq 0$. Now take $k_0 = 2$, $g_1 = e$, $g_2 = g$. Then

$$(H_{j\ell}) = \begin{pmatrix} \rho(e) & \rho(g) \\ \rho(g^{-1}) & \rho(e) \end{pmatrix}.$$

Since $H_{12} = \bar{H}_{21}$ we have $\rho(g) = \overline{\rho(g^{-1})}$. Further, $\det H = \rho(e)^2 - \rho(g)\rho(g^{-1}) \geq 0$. □

THEOREM 5.3. *Let ρ be a function of positive type on G. Then there is a unitary representation U of G on a Hilbert space $\mathcal{H}$ such that $\rho(g) = \langle U(g)\mathbf{f}, \mathbf{f} \rangle$ for some $\mathbf{f} \in \mathcal{H}$. Furthermore the span of the set $\{U(g)\mathbf{f} : g \in G\}$ is dense in $\mathcal{H}$.*

Proof. We will use the computation (5.52) as a motivation for the construction of $\mathcal{H}$ and U. Let L be the subspace of the group ring R_G consisting of those elements x which take on only a finite number of nonzero values:

$$x = \sum_i x_i \cdot g_i.$$

(Alternatively we can consider x as a function $x : G \to \mathcal{C}$ which is zero except at a finite number of points g_i.) The inner product of two vectors in L is defined as

$$\langle y, x \rangle = \sum_{i,j} \rho(g_i^{-1} g_j) \bar{x}_i y_j$$

where we sum over all points g_ℓ such that either $x(g_\ell)$ or $y(g_\ell)$ is nonzero. It is evident that $\langle \cdot, \cdot \rangle$ is linear in its first argument and from Lemma 5.1 that $\langle y, x \rangle = \overline{\langle x, y \rangle}$ and $\langle x, x \rangle \geq 0$.

Thus, $\langle \cdot, \cdot \rangle$ satisfies all requirements for an inner product, except that we might have $\langle x, x \rangle = 0$ with $x \neq \theta$. It follows from this result that the Cauchy-Schwarz inequality is valid: $|\langle x, y \rangle|^2 \leq \langle x, x \rangle \langle y, y \rangle$.

We use a standard construction to convert $\langle \cdot, \cdot \rangle$ to a true inner product.

Let $N = \{x \in L : \langle x, x \rangle = 0\}$. Then N is a subspace of L. From the Cauchy-Schwarz inequality we have $\langle x, y \rangle = 0$ for $x \in N$, $y \in L$. Thus

$$(5.53) \qquad \langle y_1 + x_1,\, y_2 + x_2 \rangle = \langle y_1, y_2 \rangle$$

for $y_j \in L, x_j \in N$. We can now define $\langle \cdot, \cdot \rangle$ on the factor space L/N whose elements are the sets $\mathbf{y} = y + N = \{y + x : x \in N\}$. The set $\boldsymbol{\theta} = \theta + N$ corresponds to the zero vector. From (5.53) we see that the definition

$$\langle \mathbf{y_1}, \mathbf{y_2} \rangle = \langle y_1 + N, y_2 + N \rangle = \langle y_1, y_2 \rangle$$

is unambiguous. Furthermore $\langle \mathbf{y}, \mathbf{y} \rangle = 0$ only if $\mathbf{y} = \theta + N$, i.e., $\mathbf{y} = \boldsymbol{\theta}$. Thus $\langle \cdot, \cdot \rangle$ is a true inner product on L/N. Now by taking all Cauchy sequences in L/N we can complete L/N to a Hilbert space $\mathcal{H}$.

Given $h \in G, x = \sum_i x_i \cdot g_i \in L$ we define the linear operator $\mathbf{U}(h) : L \to L$ by $\mathbf{U}(h)x = \sum_i x_i \cdot hg_i$. Then

$$\langle \mathbf{U}(h)y, \mathbf{U}(h)x \rangle = \sum_{i,j} \rho(g_i^{-1}h^{-1}hg_j)\bar{x}_i x_j$$

$$= \sum_{i,j} \rho(g_i^{-1}g_j)\bar{x}_i y_j = \langle y, x \rangle$$

so $\mathbf{U}(h)$ is an isometry on L. Furthermore, $\mathbf{U}(h^{-1})[\mathbf{U}(h)x] = x$ for all $x \in L$, so $\mathbf{U}(h)$ is invertible, hence unitary on L. Also, $\mathbf{U}(h_1 h_2)x = \mathbf{U}(h_1)[\mathbf{U}(h_2)x]$ so $\mathbf{U}$ is a unitary rep of G on L. Clearly, the operators $\mathbf{U}(h)$ extend uniquely to a unitary rep of G on $\mathcal{H}$.

Let $f = 1 \cdot e \in L$. Then the vector $\{\mathbf{U}(g)f\}$ span L, for if $x = \sum_i x_i \cdot g_i$ we have $x = \sum_i x_i \mathbf{U}(g_i)f$. Moreover, $\langle \mathbf{U}(g)f, f \rangle = \rho(g)$. In the extension of L/N to $\mathcal{H}, f$ maps to $\mathbf{f} \in \mathcal{H}$ such that the span of $\{\mathbf{U}(g)\mathbf{f}\}$ is dense in $\mathcal{H}$ and $\langle \mathbf{U}(g)\mathbf{f}, \mathbf{f} \rangle = \rho(g)$. $\square$

If G is a linear Lie group, one can show that the function ρ is continuous on G if and only if $\mathbf{u}$ is a continuous rep of G, [G1], [N2].

The unitary rep $\mathbf{U}$ constructed in Theorem 5.3 is unique up to equivalence. Indeed, suppose there are unitary reps $\mathbf{U}_1, \mathbf{U}_2$ on Hilbert spaces $\mathcal{H}_1, \mathcal{H}_2$ such that $\langle \mathbf{U}_1(g)\mathbf{f}_1, \mathbf{f}_1 \rangle_1 = \langle \mathbf{U}_2(g)\mathbf{f}_2, \mathbf{f}_2 \rangle_2$ where the spans of $\{\mathbf{U}_j(g)\mathbf{f}_j\}$ are dense in $\mathcal{H}_j$. Define the map $\mathbf{S} : \mathcal{H}_1 \to \mathcal{H}_2$ by

$$(5.54) \qquad \mathbf{S}\left(\sum_i \lambda_i \mathbf{U}_1(g_i)\mathbf{f}_1\right) = \sum_i \lambda_i \mathbf{U}_2(g_i)\mathbf{f}_2$$

for all finite sums $\mathbf{f}_1' = \sum_i \lambda_i \mathbf{U}_1(g_i)\mathbf{f}_1$. In particular $\mathbf{S}\mathbf{f}_1 = \mathbf{f}_2$. This mapping is well defined because if $\mathbf{f}_1' = \boldsymbol{\theta}$ and $\mathbf{S}\mathbf{f}_1' = \mathbf{f}_2'$ then

$$\langle \mathbf{f}_2', \mathbf{f}_2' \rangle_2 = \sum_{i,j} \langle \lambda_i \mathbf{U}_2(g_i)\mathbf{f}_2, \lambda_j \mathbf{U}_2(g_j)\mathbf{f}_2 \rangle_2$$

$$= \sum_{i,j} \langle \mathbf{U}_2(g_j^{-1}g_i)\mathbf{f}_2, \mathbf{f}_2 \rangle_2 \lambda_i \bar{\lambda}_j$$

$$= \sum_{i,j} \langle \mathbf{U}_1(g_j^{-1}g_i)\mathbf{f}_1, \mathbf{f}_1 \rangle_1 \lambda_i \bar{\lambda}_j = \langle \mathbf{f}_1', \mathbf{f}_1' \rangle_1$$

$$= 0,$$

so $\mathbf{f}_2' = \boldsymbol{\theta}$. (Furthermore this same calculation shows that $\mathbf{S}$ is an isometry.) Thus, $\mathbf{S}$ extends to a unitary transformation from $\mathcal{H}_1$ onto $\mathcal{H}_2$. Finally $\mathbf{SU}_1(g)\mathbf{f}_1' = \mathbf{U}_2(g)\mathbf{f}_2' = \mathbf{U}_2(g)\mathbf{Sf}_1'$ for all finite sums $\mathbf{f}_1'$, so $\mathbf{SU}_1(g) = \mathbf{U}_2(g)\mathbf{S}$, and the reps $\mathbf{U}_1$ and $\mathbf{U}_2$ are equivalent.

COROLLARY 5.1. *Let* $\mathbf{U}$ *be a unitary irred rep of the group* G *on the Hilbert space* $\mathcal{H}$ *such that*

$$\langle \mathbf{U}(g)\mathbf{f}_1, \mathbf{f}_1 \rangle = \langle \mathbf{U}(g)\mathbf{f}_2, \mathbf{f}_2 \rangle$$

for all $g \in G$, *where* $\mathbf{f}_1, \mathbf{f}_2$ *are nonzero elements of* $\mathcal{H}$. *Then* $\mathbf{f}_2 = \lambda \mathbf{f}_1$ *for some* $\lambda \in \mathcal{C}$ *with* $|\lambda| = 1$.

Proof. Since $\mathbf{U}$ is irred, the spans of $\{\mathbf{U}(g)\mathbf{f}_j\}$ are dense in $\mathcal{H}$. From the construction (5.54) with $\mathcal{H}_1 \equiv \mathcal{H}_2$ and $\mathbf{U}_1(g) \equiv \mathbf{U}_2(g)$ we see that there is a unitary operator $\mathbf{S} : \mathcal{H} \to \mathcal{H}$ such that $\mathbf{SU}(g) = \mathbf{U}(g)\mathbf{S}$ for all $g \in G$ and $\mathbf{Sf}_1 = \mathbf{f}_2$. Since $\mathbf{U}$ is irred it follows from Theorem 5.1 that $\mathbf{S} = \lambda \mathbf{E}$, where $|\lambda| = 1$ since $\mathbf{S}$ is unitary. Thus, $\lambda \mathbf{f}_1 = \mathbf{f}_2$. □

Note that the corollary implies that two signals correspond to the same ambiguity function if and only if they differ by a constant factor of absolute value one.

Next we will characterize those functions of positive type on a topological group G that correspond to **irred** unitary reps of G. Consider functions ρ_1, ρ_2 of positive type on G. We say that ρ_1 **dominates** ρ_2 if $\rho_1 - \rho_2$ is of positive type on G (so that $\rho_1 = \rho_2 + (\rho_1 - \rho_2)$ is a sum of two functions of positive type). A function ρ of positive type on G is **indecomposable** if the only functions of positive type dominated by ρ are scalar multiples of ρ. (Clearly, the multiples must be of the form $a\rho$ where $0 < a < 1$.)

THEOREM 5.4. *The function* ρ *of positive type on the topological group* G *is indecomposable if and only if*

$$(5.55) \qquad \rho(g) = \langle \mathbf{U}(g)\mathbf{f}, \mathbf{f} \rangle$$

for all $g \in G$ *where* $\mathbf{U}$ *is a unitary* **irred** *rep of* G *on some Hilbert space* $\mathcal{H}$ *and* $\mathbf{f} \in \mathcal{H}$.

Proof. Suppose ρ is indecomposable. By Theorem 5.3 there is a unitary rep $\mathbf{U}$ of G on the Hilbert space $\mathcal{H}$ and a vector $\mathbf{f} \in \mathcal{H}$ such that the span of $\{\mathbf{U}(g)\mathbf{f}\}$ is dense in $\mathcal{H}$, and $\rho(g) = \langle \mathbf{U}(g)\mathbf{f}, \mathbf{f} \rangle$. Let $\mathcal{K}$ be a nonzero subspace of $\mathcal{H}$ which is invariant under $\mathbf{U} : \mathbf{U}(g)\mathcal{K} \subseteq \mathcal{K}$ for all $g \in G$. Then $\mathcal{K}^\perp$ is also invariant under $\mathbf{U}$ (see Theorem 4.3), and we have the unique decomposition $\mathbf{f} = \mathbf{f}_1 + \mathbf{f}_2$ with $\mathbf{f}_1 \in \mathcal{K}, \mathbf{f}_2 \in \mathcal{K}^\perp$. It follows easily that $\rho(g) = \langle \mathbf{U}(g)\mathbf{f}, \mathbf{f} \rangle = \langle \mathbf{U}(g)\mathbf{f}_1, \mathbf{f}_1 \rangle + \langle \mathbf{U}(g)\mathbf{f}_2, \mathbf{f}_2 \rangle = \rho_1(g) + \rho_2(g)$. Since $\{\mathbf{U}(g)\mathbf{f}_1\}$ is dense in $\mathcal{K}$, $\rho_1(g) = \langle \mathbf{U}(g)\mathbf{f}_1, \mathbf{f}_1 \rangle$ is a function of positive type and ρ dominates ρ_1. Hence $\rho_1 = a\rho$ for some constant $a \neq 0$, so $\langle \mathbf{U}(g)\mathbf{f}_1, \mathbf{f}_1 \rangle = \langle \mathbf{U}(g)\mathbf{f}, \mathbf{f}_1 \rangle = \langle \mathbf{U}(g)\mathbf{f}, a\mathbf{f} \rangle$ or,

$$(5.56) \qquad \langle \mathbf{U}(g)\mathbf{f}, \mathbf{f}_1 - a\mathbf{f} \rangle = 0$$

for all $g \in G$. It follows that $\mathbf{f}_1 = a\mathbf{f}$ so $\mathbf{f} \in \mathcal{K}$, hence $\mathcal{K} = \mathcal{H}$ and $\mathbf{U}$ is irred.

Conversely, suppose there is a unitary irred rep $\mathbf{U}$ of G on $\mathcal{H}$ such that (5.55) holds for some $\mathbf{f} \in \mathcal{H}$. Let ρ_1 be a function of positive type on G that is dominated by ρ. Without loss of generality we can assume that $\mathbf{U}$ is obtained from ρ and the space L/N according to the construction given in the proof of Theorem 5.3. On this same space we can construct the inner product $\langle \cdot, \cdot \rangle_1$ associated with the function ρ_1. Since ρ_1 is dominated by ρ, the Cauchy-Schwarz inequality implies $|\langle x, y \rangle_1|^2 \leq \langle x, x \rangle_1 \langle y, y \rangle_1 \leq ||x||^2 ||y||^2$ for all $x, y \in L/N$. Thus $\langle \cdot, \cdot \rangle_1$ extends to a positive definite Hermitian form on $\mathcal{H}$ such that

(5.57) $$|\langle \mathbf{k}_1, \mathbf{k}_2 \rangle_1|^2 \leq ||\mathbf{k}_1||_1^2 ||\mathbf{k}_2||^2 \leq ||\mathbf{k}_1||^2 ||\mathbf{k}_2||^2.$$

This means that there exists a bounded self-adjoint operator $\mathbf{A}$ on $\mathcal{H}$ such that

(5.58) $$\langle \mathbf{k}_1, \mathbf{k}_2 \rangle_1 = \langle \mathbf{k}_1, \mathbf{A}\mathbf{k}_2 \rangle$$

for all $\mathbf{k}_1, \mathbf{k}_2 \in \mathcal{H}$. (Indeed, it follows from (5.57) that for fixed $\mathbf{k}_2, \langle \mathbf{k}_1, \mathbf{k}_2 \rangle_1$ is a bounded linear functional $\mathcal{H}$. By the Riesz representation theorem [RN] there exists a vector $\mathbf{s}_\mathbf{k} \in \mathcal{H}$ for all $\mathbf{k} \in \mathcal{H}$ such that $\langle \mathbf{k}_1, \mathbf{k} \rangle_1 = \langle \mathbf{k}_1, \mathbf{s}_\mathbf{k} \rangle$. Clearly, the map $\mathbf{k} \to \mathbf{s}_\mathbf{k}$ is linear, so there exists a linear operator $\mathbf{A} : \mathcal{H} \to \mathcal{H}$ such that $\mathbf{s}_\mathbf{k} = \mathbf{A}\mathbf{k}$. Since $|\langle \mathbf{k}_1, \mathbf{A}\mathbf{k}_2 \rangle|^2 = |\langle \mathbf{k}_1, \mathbf{k}_2 \rangle_1|^2 \leq ||\mathbf{k}_1||^2 ||\mathbf{k}_2||^2$, we have in the case $\mathbf{k}_1 = \mathbf{A}\mathbf{k}_2$ the inequality $||\mathbf{A}\mathbf{k}_2||^4 \leq ||\mathbf{A}\mathbf{k}_2||^2 ||\mathbf{k}_2||^2$ or $||\mathbf{A}\mathbf{k}_2||^2 \leq ||\mathbf{k}_2||^2$. Thus, $\mathbf{A}$ is a bounded operator. Furthermore, $\langle \mathbf{k}_1, \mathbf{A}\mathbf{k}_2 \rangle = \langle \mathbf{k}_1, \mathbf{k}_2 \rangle_1 = \overline{\langle \mathbf{k}_2, \mathbf{k}_1 \rangle}_1 = \overline{\langle \mathbf{k}_2, \mathbf{A}\mathbf{k}_1 \rangle} = \langle \mathbf{A}\mathbf{k}_1, \mathbf{k}_2 \rangle$, so $\mathbf{A}$ is self-adjoint. Since $\langle \mathbf{k}, \mathbf{A}\mathbf{k} \rangle = \langle \mathbf{k}, \mathbf{k} \rangle_1 \geq 0$, $\mathbf{A}$ is nonnegative.)

By construction of $\langle \cdot, \cdot \rangle_1$ through the completion of L/N we have $\langle \mathbf{U}(g)\mathbf{k}_1, \mathbf{U}(g)\mathbf{k}_2 \rangle_1 = \langle \mathbf{k}_1, \mathbf{k}_2 \rangle$ for all $g \in G$. Hence, $\langle \mathbf{U}(g)\mathbf{k}_1, \mathbf{A}\mathbf{U}(g)\mathbf{k}_2 \rangle = \langle \mathbf{k}_1, \mathbf{A}\mathbf{k}_2 \rangle$ and $\mathbf{U}(g)^{-1}\mathbf{A}\mathbf{U}(g) = \mathbf{A}$, which implies $\mathbf{A}\mathbf{U}(g) = \mathbf{U}(g)\mathbf{A}$ for all $g \in G$. Since $\mathbf{U}$ is irreducible, we have $\mathbf{A} = a\mathbf{E}$ for some positive constant a. Thus $\rho_1(g) = \langle \mathbf{U}(g)\mathbf{f}, \mathbf{f} \rangle_1 = \langle \mathbf{U}(g)\mathbf{f}, \mathbf{A}\mathbf{f} \rangle = a\langle \mathbf{U}(g)\mathbf{f}, \mathbf{f} \rangle = a\rho(g)$, so ρ is indecomposable. $\square$

COROLLARY 5.2. *Let $\mathbf{U}$ be an irred unitary rep of G on $\mathcal{H}$, and suppose $\mathbf{f}_0, \mathbf{f}_1, \mathbf{f}_2$ are nonzero elements of $\mathcal{H}$ such that*

(5.59) $$\langle \mathbf{U}(g)\mathbf{f}_0, \mathbf{f}_0 \rangle = \langle \mathbf{U}(g)\mathbf{f}_1, \mathbf{f}_1 \rangle + \langle \mathbf{U}(g)\mathbf{f}_2, \mathbf{f}_2 \rangle$$

for all $g \in G$. Then $\mathbf{f}_1 = \kappa\mathbf{f}_2$, $\kappa \in \mathcal{C}$.

Proof. The functions of positive type $\rho_j(g) = \langle \mathbf{U}(g)\mathbf{f}_j, \mathbf{f}_j \rangle$ satisfy $\rho_0(g) = \rho_1(g) + \rho_2(g)$, so ρ_0 dominates ρ_1 and ρ_2. Since $\mathbf{U}$ is irred, ρ is indecomposable. Thus $\rho_0 = c_1^2 \rho_1 = c_2^2 \rho_2$ where c_1, c_2 are positive constants. Setting $\mathbf{f}_\ell' = c_\ell \mathbf{f}_\ell$, $\ell = 1, 2$ we have

$$\langle \mathbf{U}(g)\mathbf{f}_0, \mathbf{f}_0 \rangle = \langle \mathbf{U}(g)\mathbf{f}_1' \mathbf{f}_1' \rangle = \langle \mathbf{U}(g)\mathbf{f}_2', \mathbf{f}_2' \rangle.$$

By Corollary 5.1, $\mathbf{f}_1' = \lambda\mathbf{f}_2'$. Hence $\mathbf{f}_1 = \kappa\mathbf{f}_2$ for $\kappa = \lambda c_2/c_1$. $\square$

Note that Corollary 5.2 implies that the sum of two ambiguity functions corresponding to nonzero signals s_1, s_2 is again an ambiguity function if and only if $s_1 = \lambda s_2$.

5.9 Exercises.

5.1 Verify explicitly equation (5.10).

5.2 Show that if $\{f_n,\ n = 0, 1, 2, \cdots\}$ is an ON basis for $L_2(R)$ then the matrix elements $\{< \mathbf{T}^\lambda(x_1, x_2)f_n, f_m >\}$ form an ON basis for $L_2(R^2)$. Hint: From exercise 5.1, the matrix elements form an ON set. Hence to show that they form a basis it is enough to prove that if

$$\int\int_{-\infty}^{\infty} dx_1 dx_2 < \mathbf{T}^\lambda(x_1, x_2)f_n, f_m > \overline{g(x_1, x_2)} = 0$$

for $g \in L_2(R^2)$ and all n, m, then $g = 0$ almost everywhere. This shows that the ON set is also dense in $L_2(R^2)$, hence is a basis.

5.3 Set $\mathbf{B}(f,g) = < \mathbf{T}^\lambda(x_1, x_2)f, g >$ for $f, g \in L_2(R)$ and $\mathbf{B}(f) = \mathbf{B}(f, f)$. Show that

$$\mathbf{B}(f + g) = \mathbf{B}(f) + \mathbf{B}(f, g) + \mathbf{B}(g, f) + \mathbf{B}(g),$$
$$\mathbf{B}(f + ig) = \mathbf{B}(f) + i\mathbf{B}(g, f) - i\mathbf{B}(f, g) + \mathbf{B}(g).$$

5.4 Prove that the ambiguity functions $< \mathbf{T}^\lambda(x_1, x_2)f, f >$ for all $f \in L_2(R)$ span a dense subspace of $L_2(R^2)$.

5.5 Suppose $f \in L_2(R)$ such that $f_{\mathbf{P}} \neq 0$ almost everywhere. Prove that the set $\{e^{2\pi i(m_1 x_1 + m_2 x_2)} f_{\mathbf{P}}/|f_{\mathbf{P}}|,\ m_1, m_2 = \pm 1, \pm 2, \cdots\}$ is an ON basis for the lattice Hilbert space V'. Find an explicit expression for the corresponding ON basis of $L_2(R)$ obtained from the mapping $\mathbf{P}^{-1}$.

5.6 Show that the rep $\mathbf{T}^\lambda$ of H_R is *continuous* in the sense that

$$||\mathbf{T}^\lambda(x_1, x_2, x_3)f - f|| \to 0, \quad \text{as } (x_1, x_2, x_3) \to (0, 0, 0)$$

for each $f \in L_2(R)$.

5.7 Show that the ambiguity and cross-ambiguity functions $< \mathbf{T}^\lambda(x_1, x_2)f, g >$ are continuous functions of (x_1, x_2).

5.8 Construct the unitary rep $\tilde{\mathbf{T}}_n$ of H_R induced by the one-dimensional rep $\tilde{\mathbf{T}}_0^n(a_1, a_2, y_3) = e^{2\pi i n y_3}$ of the subgroup H^1, where n is an integer (not necessarily positive). Determine the action of H_R on the rep space, in analogy with (5.48). Under what conditions on n is $\tilde{\mathbf{T}}_n$ irred? Show that the rep space contains $|n|$ linearly independent ground state wave functions.

§6. REPRESENTATIONS OF THE AFFINE GROUP

6.1 Induced irreducible representations of G_A. Recall that the affine group G_A is the matrix group with elements

(6.1)
$$A(a,b) = \begin{pmatrix} a & b \\ 0 & 1 \end{pmatrix}, \quad a > 0$$

and multiplication rule

$$(a,b)(a',b') = (aa', ab' + b).$$

Even though this is a 2×2 matrix group with a very simple structure, there are some difficulties in developing its rep theory in accordance with the general results presented in Chapters 2-4. An indication of the complications appeared already in Chapter 4 where we showed that G_A is not unimodular, i.e., $d_\ell A \neq d_r A$. Indeed

(6.2)
$$d_\ell A = \frac{da\,db}{a^2}, \quad d_r A = \frac{da\,db}{a}.$$

We begin by deriving the irred unitary reps of G_A. One family of such reps is evident: $\chi_\rho[a,b] = a^{i\rho}$, ρ real. We use the method of induced reps to derive several forms of the remaining irred unitary reps.

Consider the subgroup $H_1 \cong R$ of elements of the form $A(1,b)$, $b \in R$. The unitary irred reps of H_1 take the form $\xi_\lambda(b) = e^{i\lambda b}$. We now construct the unitary rep of G_A induced by the rep ξ_λ of H_1. The rep is defined on a space of functions $\mathbf{f}(a,b)$ on G_A such that

$$\mathbf{f}(BA) = \xi_\lambda(B)\mathbf{f}(A), \quad B \in H_1, \ A \in G_A,$$

i.e.,

(6.3)
$$\mathbf{f}(a, b+b') = e^{i\lambda b'}\mathbf{f}(a,b).$$

Thus $\mathbf{f}(a,b) = e^{i\lambda b}\mathbf{f}(a,0) = e^{i\lambda b}\varphi(a)$ where φ is defined on the coset space $X_1 \cong H_1 \backslash G_A \cong H_2$ and H_2 is the subgroup of elements $A(a,0), a > 0$. The action of G_A in the induced rep is given by

$$[\mathbf{R}_\lambda(A)\mathbf{f}](A') = \mathbf{f}(A'A), \quad A, A' \in G_A$$

or, restricted to the functions φ:

(6.4)
$$(\mathbf{R}_\lambda[a,b]\varphi)(x) = e^{i\lambda x b}\varphi(ax)$$

where $A' = A'(x,y)$. The right-invariant measure on X_1 is $d\mu(x) = dx/x$, so the inner product $\langle \cdot, \cdot \rangle_1$ with respect to which the operators $\mathbf{R}_\lambda[a,b]$ are unitary, is

(6.5)
$$\langle \varphi_1, \varphi_2 \rangle_1 = \int_0^\infty \varphi_1(x)\bar{\varphi}_2(x)\frac{dx}{x}.$$

The elements of the Hilbert space $\mathcal{H}_1$ with the inner product are Lebesgue measurable functions $\varphi(x)$ such that $\|\varphi\|_1^2 = \langle \varphi, \varphi \rangle_1 < \infty$. The rep $\mathbf{R}_0$ is reducible. However, we have

THEOREM 6.1. *For* $\lambda \neq 0$ *the representation* $\mathbf{R}_\lambda$ *of* G_A *is irreducible.*

Proof. This demonstration is very similar to that of Theorem 5.2. For completeness we repeat the basic ideas of the proof, omitting some of the technical details. Our aim will be to show that if $\mathbf{L}$ is a bounded operator on $\mathcal{H}_1$ which commutes with the operators $\mathbf{R}_\lambda(A)$ for all $A \in G_A$ then $\mathbf{L} = \kappa \mathbf{E}$ for some constant κ, where $\mathbf{E}$ is the identity operator. (It follows from this that $\mathbf{R}_\lambda$ is irred, for if $\mathcal{M}$ were a proper closed subspace of $\mathcal{H}_1$, invariant under $\mathbf{R}_\lambda$, then the self-adjoint projection operator $\mathbf{P}$ on $\mathcal{M}$ would commute with the operators $\mathbf{R}_\lambda(A)$. This is impossible since $\mathbf{P}$ could not be a scalar multiple of $\mathbf{E}$.)

Suppose the bounded operator $\mathbf{L}$ satisfies

$$\mathbf{L}\mathbf{R}_\lambda[a, b] = \mathbf{R}_\lambda[a, b]\mathbf{L}$$

for all real a, b with $a > 0$. First consider the case $a = 1$: $\mathbf{L}$ commutes with the operation of multiplication by the function $e^{i\lambda b x}$. Clearly $\mathbf{L}$ must also commute with multiplication by finite sums of the form $\sum_{b_j} c_j e^{i\lambda b_j x}$ and, by using the well-known fact that trigonometric polynomials are dense in the space of measurable functions, $\mathbf{L}$ must commute with multiplication by any bounded function $f(x)$ on $(0, \infty)$. Now let Q be a bounded closed internal in $(0, \infty)$ and let $\chi_Q \in \mathcal{H}_1$ be the **characteristic function** of Q:

$$\chi_Q(x) = \begin{cases} 1 & \text{if } x \in Q \\ 0 & \text{if } x \notin Q \end{cases}.$$

Let $f_Q \in \mathcal{H}_1$ be the function $f_Q = \mathbf{L}\chi_Q$. Since $\chi_Q^2 = \chi_Q$ we have $f_Q(x) = \mathbf{L}\chi_Q(x) = \mathbf{L}\chi_Q^2(x) = \chi_Q(x)\mathbf{L}\chi_Q(x) = \chi_Q(x)f_Q(x)$ so f_Q is nonzero only for $x \in Q$. Furthermore, if Q' is a closed interval with $Q' \subseteq Q$ and $f_{Q'} = \mathbf{L}\chi_{Q'}$ then $f_{Q'}(x) = \mathbf{L}\chi_{Q'}\chi_q(x) = \chi_{Q'}(x)\mathbf{L}\chi_Q(x) = \chi_{Q'}(x)F_q(x)$ so $f_{Q'}(x) = f_Q(x)$ for $x \in Q'$ and $f_{Q'}(x) = 0$ for $x \notin Q'$. It follows that there is a unique function $f(x)$ such that $\chi_{\tilde{Q}}f \in \mathcal{H}_1$ and $\chi_{\tilde{Q}}(x)f(x) = \mathbf{L}\chi_{\tilde{Q}}(x)$ for any closed bounded interval $\tilde{Q}$ in $(0, \infty)$. Now let φ be a C^∞ function which is zero in the exterior of $\tilde{Q}$. Then $\mathbf{L}\varphi(x) = \mathbf{L}(\varphi\chi_{\tilde{Q}}(x)) = \varphi(x)\mathbf{L}\chi_{\tilde{Q}}(x) = \varphi(x)f(x)\chi_{\tilde{Q}}(x) = f(x)\varphi(x)$, so $\mathbf{L}$ acts on φ by multiplication by the function $f(x)$. Since as $\tilde{Q}$ runs over all finite subintervals of $(0, \infty)$ the functions φ are dense in $\mathcal{H}_1$, it follows that $\mathbf{L} = f(x)\mathbf{E}$.

Now we use the hypothesis that $\mathbf{L}\mathbf{R}_\lambda[a, 0]\varphi(x) = \mathbf{R}_\lambda[a, 0]\mathbf{L}\varphi(x)$ for all $a > 0$ and $\varphi \in \mathcal{H}_1$: $f(x)\varphi(ax) = f(ax)\varphi(ax)$. Thus $f(x) = f(ax)$ almost everywhere, which implies that $f(x)$ is a constant. $\square$

LEMMA 6.1. *For* $\tau > 0$ *the reps* $\mathbf{R}_\lambda$ *and* $\mathbf{R}_{\tau\lambda}$ *are equivalent.*

Proof. Let $\mathbf{S}_\tau : \mathcal{H}_0 \to \mathcal{H}_0$ be the linear unitary operator $\mathbf{S}_\tau\varphi(x) = \varphi(\tau x)$. Then $\mathbf{S}_\tau^{-1} = \mathbf{S}_{\tau^{-1}}$ and $\mathbf{R}_{\tau\lambda}[a, b]\mathbf{S}_\tau = \mathbf{S}_\tau\mathbf{R}_\lambda[a, b]$. $\square$

It follows that there are just two distinct irred reps in the family $\mathbf{R}_\lambda$. We choose these reps in the normalized form $\lambda = \pm 1$. Thus we have constructed the following irred unitary reps of $G_A : \chi_\rho, (-\infty < \rho < \infty)$; $\mathbf{R}_+$ and $\mathbf{R}_-$ where

(6.6)
$$\chi_\rho[a, b] = a^{i\rho}$$
$$\mathbf{R}_+[a, b] \, \boldsymbol{\varphi}(x) = e^{ixb}\boldsymbol{\varphi}(ax),$$
$$\mathbf{R}_-[a, b]\boldsymbol{\varphi}(x) = e^{-ixb}\boldsymbol{\varphi}(ax),$$

and $\boldsymbol{\varphi} \in \mathcal{H}_1$. It can be shown [K5] that these are the only unitary irred reps of G_A.

6.2 The wideband cross-ambiguity functions. Another important class of unitary reps of G_A can be induced from the subgroup $H_2 \cong R^+$ of elements of the form $A(a, 0), a > 0$. Consider the irred reps η_σ of H_2: $\eta_\sigma(a) = a^\sigma$, σ complex. We construct the rep $\mathbf{L}'_\sigma$ of G_A induced by η_σ. It is defined on a space of functions $\mathbf{f}(a, b)$ on G_A such that

$$\mathbf{f}(BA) = \eta_\sigma(B)\mathbf{f}(A), \quad B \in H_2, \ A \in G_A,$$

i.e.,

(6.7)
$$\mathbf{f}(a'a, a'b) = (a')^\sigma \mathbf{f}(a, b).$$

Thus $\mathbf{f}(a, b) = a^\sigma \mathbf{f}(1, b/a) = a^\sigma \boldsymbol{\varphi}(b/a)$ where $\boldsymbol{\varphi}$ is defined on the coset space $X_2 \cong H_2 \setminus G_A \cong H_1$. The action of G_A in the induced rep is given by

$$[\mathbf{L}'_\sigma(A)\mathbf{f}](A') = \mathbf{f}(A'A), \quad A, A' \in G_A,$$

or, restricted to the functions $\boldsymbol{\varphi}$:

(6.8)
$$(\mathbf{L}'_\sigma[a, b]\boldsymbol{\varphi})(t) = a^\sigma \boldsymbol{\varphi}\left(\frac{t+b}{a}\right).$$

There is no right-invariant measure on X_2. However, $d\rho(t) = dt$ goes to a multiple of itself, so an inner product $\langle \cdot, \cdot \rangle_2$ with respect to which the operators $\mathbf{L}'_\sigma[a, b]$ are unitary is

(6.9)
$$\langle \boldsymbol{\varphi}_1, \boldsymbol{\varphi}_2 \rangle_2 = \int_{-\infty}^{\infty} \boldsymbol{\varphi}_1(t)\overline{\boldsymbol{\varphi}_2(t)}dt,$$

provided $\sigma = i\mu - \frac{1}{2}$ where μ is real. The elements of the Hilbert space $\mathcal{H}_2 = L_2(R)$ with this inner product are Lebesgue measurable functions $\boldsymbol{\varphi}(t)$ such that $\|\boldsymbol{\varphi}\|_2^2 = \langle \boldsymbol{\varphi}, \boldsymbol{\varphi} \rangle_2 < \infty$. Since we will be restricting to the case of unitary reps, we introduce the notation $\mathbf{L}_\mu \equiv \mathbf{L}'_{i\mu - \frac{1}{2}}$:

(6.10)
$$(\mathbf{L}_\mu[a, b]\boldsymbol{\varphi})(t) = a^{i\mu - \frac{1}{2}}\boldsymbol{\varphi}\left(\frac{t+b}{a}\right), \quad \boldsymbol{\varphi} \in L_2(R).$$

The rep $\mathbf{L}_\mu$ is reducible. Indeed let us consider the Fourier transform $\mathcal{F}$ as a unitary map from $L_2(R)$ to $L_2(R)$:

(6.11)
$$\hat{\boldsymbol{\varphi}}(y) = \frac{1}{\sqrt{2\pi}}\int_{-\infty}^{\infty}\boldsymbol{\varphi}(t)e^{-ity}dt = \mathcal{F}\boldsymbol{\varphi}(y).$$

Then

$$\varphi(t) = \mathcal{F}^{-1}\hat{\varphi}(t) = \frac{1}{\sqrt{2\pi}} \int_{-\infty}^{\infty} \hat{\varphi}(y)e^{ity}\,dy$$

and the action $\hat{\mathbf{L}}_\mu$ of G_A on the function $\hat{\varphi}$ corresponding to the action $\mathbf{L}_\mu$ on φ is

(6.12)
$$\begin{aligned}
\hat{\mathbf{L}}_\mu[a,b]\hat{\varphi}(y) &= \mathcal{F}(\mathbf{L}_\mu[a,b]\varphi)(y) \\
&= \frac{1}{\sqrt{2\pi}} \int_{-\infty}^{\infty} a^{i\mu-\frac{1}{2}}\varphi\left(\frac{t+b}{a}\right)e^{-ity}\,dt \\
&= a^{i\mu+1/2}e^{iby}\hat{\varphi}(ay).
\end{aligned}$$

Now consider the map $\mathcal{S}_\mu : L_2(R) \to \mathcal{H}_1^+ \oplus \mathcal{H}_1^-$ defined by

$$\mathcal{S}_\mu\hat{\varphi}(y) = |y|^{i\mu+1/2}\hat{\varphi}(y); \quad y \neq 0.$$

and let

$$\hat{\varphi}^+(y) = \mathcal{S}_\mu\hat{\varphi}(y) \in \mathcal{H}_1^+ \text{ for } y > 0$$
$$\hat{\varphi}^-(z) = \mathcal{S}_\mu\hat{\varphi}(y) \in \mathcal{H}_1^- \text{ for } z = -y > 0.$$

Then the action induced on $\hat{\varphi}^+ \in \mathcal{H}_1^+$ by $\mathbf{L}_\mu[a,b]$ is

$$\mathbf{R}_+[a,b]\hat{\varphi}^+(y) = e^{iby}\hat{\varphi}^+(ay)$$

and the action induced on $\hat{\varphi}^- \in \mathcal{H}_1^-$ is

$$\mathbf{R}_-[a,b]\hat{\varphi}^-(z) = e^{-ibz}\hat{\varphi}^-(az).$$

(Here, the spaces $\mathcal{H}_1^+, \mathcal{H}_1^-$ consist of functions $\psi^+(y), \psi^-(z)$ on the positive real line with weight functions dy/y, dz/z, respectively. The operator $\mathcal{S}_\mu$ is responsible for the change in weight function.) As the reader can easily verify, we have the "Plancherel formula"

(6.13)
$$\langle \mathbf{L}_\mu[a,b]\varphi, \psi\rangle_2 = \langle \mathbf{R}_+[a,b]\varphi^+, \psi^+\rangle_1 + \langle \mathbf{R}_-[a,b]\varphi^-, \psi^-\rangle_1$$

where

(6.14)
$$\begin{aligned}
\psi^+(y) &= \mathcal{S}_\mu\mathcal{F}\psi(y), \quad y > 0, \\
\psi^-(z) &= \mathcal{S}_\mu\mathcal{F}\psi(y), \quad y = -z < 0.
\end{aligned}$$

Thus the rep $\mathbf{L}_\mu$ decomposes as the direct sum of the irred reps $\mathbf{R}_+$ and $\mathbf{R}_-$; the Fourier component $\hat{\varphi}(y)$ of φ corresponds to $\mathbf{R}_+$ for positive y and to $\mathbf{R}_-$ for negative y. (If, however, we restrict the rep $\mathbf{L}^\mu$ to those $\varphi \in L_2(R)$ whose Fourier transform $\hat{\varphi}(y)$ has support on the positive reals, then $\mathbf{L}_\mu$ is irreducible and equivalent to $\mathbf{R}_-$.)

Note that the matrix element $\langle \mathbf{L}_0[a,b]s_n, s_m\rangle_2$ coincides with the wideband cross-ambiguity function (2.19) with $y = a^{-1}, x = b/a$. Formula (6.13) suggests that, for computational simplicity, in selecting a basis for $L_2(R)$ in which to determine the cross-ambiguity function, one should choose the union of a basis on the subspace transforming according to $\mathbf{R}_+$ and a basis on the subspace transforming according to $\mathbf{R}_-$.

6.3 Decomposition of the regular representation. Even though G_A has a very simple structure, the decomposition of the regular rep of G_A and expansion formulas for functions on G_A in terms of the matrix elements of irred reps are not trivial consequences of the general theory worked out in the earlier chapters. In particular, G_A is not unimodular, i.e., $d_r A \neq d_\ell A$. (The machinery developed in Chapter 4 for averaging over a group assumed that the group is unimodular.) Choosing the measure $d_r A$ to be definite and assuming that the functions $\varphi_j(x) \in \mathcal{H}_1^+$ are C^∞ with compact support in $(0, \infty)$ to avoid convergence problems, we can verify directly that

$$(6.15) \qquad \int_0^\infty \int_{-\infty}^\infty \frac{dadb}{a} \langle \mathbf{R}_+[a, b]\varphi_1, \varphi_2 \rangle_1 \overline{\langle \mathbf{R}_-[a, b]\varphi_3, \varphi_4 \rangle_1} = 0,$$

in accordance with the earlier theory, but

$$(6.16) \qquad \begin{aligned} \int_0^\infty \int_{-\infty}^\infty \frac{dadb}{a} \langle \mathbf{R}_\pm[a, b]\varphi_1, \varphi_2 \rangle_1 \overline{\langle \mathbf{R}_\pm[a, b]\varphi_3, \varphi_4 \rangle_1} \\ = 2\pi \langle \varphi_1, \varphi_3 \rangle_1 \langle \varphi_4, \varphi_2' \rangle_1 \end{aligned}$$

where $\varphi_2'(t) = \varphi_2(t)/t$. To investigate the problem in more detail we will decompose explicitly the right regular rep of G_A into irred reps of G_A, [V].

Recall that the right regular rep of G_A is defined on the Hilbert space $\mathcal{H}_R$ of measurable functions $f(A(a, b)) \equiv f(a, b)$, square integrable with respect to the measure $d_r A = dadb/a$. The inner product is

$$(6.17) \qquad \langle f_1, f_2 \rangle = \int_0^\infty \int_{-\infty}^\infty \frac{dadb}{a} f_1(a, b)\overline{f_2(a, b)},$$

and G_A acts on this space in terms of the unitary operators $\mathbf{R}(A')$:

$$(6.18) \qquad \mathbf{R}(A')f(A) = f(AA').$$

To decompose $\mathcal{H}_R$ into irreducible components we project out subspaces of functions which transform irreducibly under the left action of the subgroup $H_2 = \{C(c) = A(c, 0)\}$. (Since the left action of G_A commutes with the right action, it follows that these subspaces will be invariant under the operators $\mathbf{R}(A')$.) The function $\chi_\mu(C(c)) = c^{i\mu+1/2}$, $\mu \in R$, defines a one-dimensional rep of H_2. Now consider the map $f \to f^\mu$ where $f \in \mathcal{H}_R$, given by

$$(6.19) \qquad f^\mu(A) = \int_{H_2} f(CA)\overline{\chi_\mu(C)}dC$$

where $dC(c) = dc/c$ is the two-sided invariant measure on H_2. Note that f^μ satisfies $f^\mu(C'A) = (c')^{i\mu-1/2}f^\mu(A)$. In terms of coordinates we have

$$(6.20) \qquad \begin{aligned} f^\mu(a, b) = \int_0^\infty f(\tau a, \tau b)\tau^{-i\mu-1/2}d\tau, \\ f^\mu(ca, cb) = c^{i\mu-1/2}f^\mu(a, b). \end{aligned}$$

Note also that the map $f \to f^\mu$ is invertible:

$$f(\tau a, \tau b) = \frac{1}{2\pi} \int_{-\infty}^{\infty} f^\mu(a, b) \tau^{i\mu - 1/2} d\mu.$$

(This is just the inverse formula for the Mellin transform, a variant of the Fourier transform, [EMOT2], [V].)

Now (6.20) agrees with (6.7) with $\sigma = i\mu - 1/2$, so the action of G_A on the functions f^μ induced by the operator $\mathbf{R}(A)$ is just $\mathbf{L}'_{i\mu-1/2}$, or in terms of the functions $\varphi(t)$ where

$$(6.21) \qquad f^\mu(a, b) = a^{i\mu - 1/2} \varphi^\mu \left(\frac{b}{a} \right),$$

it reads

$$(6.22) \qquad (\mathbf{L}_\mu [a, b] \varphi^\mu)(t) = a^{i\mu - 1/2} \varphi^\mu \left(\frac{t + b}{a} \right),$$

in agreement with (6.10). Furthermore, we have the decomposition formula

$$(6.23) \qquad \int_0^\infty \int_{-\infty}^\infty \frac{da\,db}{a} f_1(a, b) \overline{f_2(a, b)} = \frac{1}{2\pi} \int_{-\infty}^\infty d\mu \int_{-\infty}^\infty \varphi_1^\mu(t) \overline{\varphi_2^\mu(t)} dt$$

where the $\varphi_j^\mu \in L_2(R)$ are related to $f_j \in \mathcal{H}_R$ via (6.20) and (6.21). On each space of square integrable functions φ^μ, $-\infty < \mu < \infty$, the rep $\mathbf{L}_\mu$ decomposes into the direct sum of the irred reps $\mathbf{R}_+$ and $\mathbf{R}_-$, (6.12) and (6.13). Thus the right regular rep $\mathbf{R}$ decomposes into a **direct integral** (rather than a direct sum) of a continuous number of copies of the irred reps $\mathbf{R}_+$ and $\mathbf{R}_-$. (The one-dimensional unitary reps $\chi_\rho[a, b] = a^{i\rho}$ do not appear in the decomposition of $\mathbf{R}$.)

As we have shown

$$(6.24) \qquad \begin{aligned} f(a, b) = \frac{1}{(2\pi)^{3/2}} \int_{-\infty}^\infty d\mu \int_{-\infty}^\infty dy\ e^{iby} |y|^{-i\mu - 1/2} \times \\ \{\varphi^{\mu^+}(ay)\chi(y) + \varphi^{\mu^-}(-ay)\chi(-y)\} dy \end{aligned}$$

where

$$\chi(y) = \begin{cases} 1 & \text{if } y \geq 0 \\ 0 & \text{if } y < 0 \end{cases}.$$

We can express these results in another form by choosing an explicit ON basis $\{\overline{S_n^\mp}(y)\}$ for $\mathcal{H}_1^\pm$ in the reps $\mathbf{R}_\pm$. Then

$$(6.25) \qquad \varphi^{\mu\pm}(y) = \sum_{n=0}^\infty \kappa_n^{\mu\pm} \overline{S_n^\mp}(y)$$

where

$$(6.26) \qquad \kappa_n^{\mu\pm} = \int_0^\infty \varphi^{\mu\pm}(y) S_n^\mp(y) \frac{dy}{y}.$$

Substituting (6.25) and (6.26) into (6.24), using (6.20) and (6.14) to express the expansion in terms of $\{S_n^{\mp}\}$ and f alone, and making appropriate interchanges of integration and summation orders, we arrive at the formula

(6.27)
$$
\begin{aligned}
f(a,b) &= \frac{1}{2\pi}\sum_{n=0}^{\infty}(\langle F^-(f)t\circ S_n^-(t),\ \mathbf{R}_-[a,b]S_n^-(t)\rangle_1 \\
&\quad + \langle F^+(f)t\circ S_n^+(t),\mathbf{R}_+[a,b]S_n^+(t)\rangle_1) \\
&= \frac{1}{2\pi}\sum_{\pm}\operatorname{tr}(\mathbf{R}_{\pm}^*[a,b]F^{\pm}(f)\circ t).
\end{aligned}
$$

Here $[t\circ S_n](t) = tS_n(t)$,

$$
\langle S,T\rangle_1 = \int_0^{\infty} S(t)\bar{T}(t)\frac{dt}{t}
$$

and $F^{\pm}(f)$ is the operator

(6.28)
$$
F^{\pm}(f) = \int_{A\in G_A} f(A)\mathbf{R}_{\pm}(A)d_{\ell}A,\quad d_{\ell}A = \frac{dadb}{a^2}.
$$

(Since the bases $\{S_n^{\pm}\}$ are the same for all μ, dependence on this parameter disappears from the final result.)

Similarly, the Parseval formula

$$
\int_0^{\infty}\int_{-\infty}^{\infty}\frac{dadb}{a}|f(a,b)|^2 = \frac{1}{2\pi}\int_{-\infty}^{\infty}d\mu\sum_{n=0}^{\infty}(|\kappa_n^{\mu+}|^2 + |\kappa_n^{\mu-}|^2)
$$

yields

(6.29)
$$
\int_0^{\infty}\int_{-\infty}^{\infty}\frac{dadb}{a}|f(a,b)|^2 = \frac{1}{2\pi}\sum_{\pm}\operatorname{tr}([F^{\pm}(\delta f)\circ\sqrt{t}]^*[F^{\pm}(\delta f)\circ\sqrt{t}])
$$

where $\delta f(a,b) = a^{1/2}f(a,b)$. (The simple derivation of (6.27) and (6.29) given here is motivated by Vilenkin's treatment of the affine group, [V]. It is not our purpose here to give a rigorous derivation with precise convergence criteria. Rather, we want to demonstrate simply how group theory concepts lead to the correct expansion formulas. A rigorous, but much more complicated, derivation was given by Khalil [K4], and the relevance of these expansions to the radar ambiguity function was pointed out by Naparst [N3].)

The expansion (6.27) has been expressed in terms of the inner product on the Hilbert space $\mathcal{H}_1$ and the operators $\mathbf{R}_{\pm}$. It is enlightening to re-express it in terms of the operators $\mathbf{L}_0$, (6.10) and the Hilbert space $L_2(R)$. For this we set $\{S_n^{\pm}(t) = \sqrt{t}S_n^{'\pm}(t)\}$ where $\{S_n^{'+}\}$ and $\{S_n^{'-}\}$ are each ON bases for $L_2(R)$. Now let

$$
s_n^+(\tau) = \mathcal{F}^{-1}S_n^{'+}(\tau) = \frac{1}{\sqrt{2\pi}}\int_0^{\infty}e^{i t\tau}S_n^{'+}(t)dt
$$

and

$$s_n^-(\tau) = \mathcal{F}^{-1}S_n'^-(\tau) = \frac{1}{\sqrt{2\pi}}\int_{-\infty}^0 e^{it\tau}S_n'^-(-t)dt,$$

i.e., $s_n^\pm$ are the inverse Fourier transforms of $S_n'^\pm$. (Recall from the discussion following (6.12) that $S_n'^+$ corresponds to a Fourier transform with support on the positive t axis and $S_n'^-$ corresponds to a Fourier transform with support on the negative t axis.) Further, let $\tilde{s}_n^\pm(\tau) = \mathcal{F}^{-1}[t \circ S_n'^\pm(t)](\tau)$ with the same support conventions as for $s_n^\pm$. Then we can write (6.27) in the form

$$(6.30) \qquad f(a,b) = \sum_{n=0}^\infty (\langle F'(f)\tilde{s}_n^-,\ \mathbf{L}_0[a,b]s_n^-\rangle_2 + \langle F'(f)\tilde{s}_n^+,\ \mathbf{L}_0[a,b]s_n^+\rangle_2)$$

where $\langle\cdot,\cdot\rangle_2$ is the usual $L_2(R)$ inner product, and

$$\mathbf{L}_0[a,b]s_n^\pm(\tau) = a^{-1/2}s_n^\pm\left(\frac{\tau+b}{a}\right),$$

while

$$(6.31) \qquad \begin{aligned} F'(f)\tilde{s}_n(\tau) &= \int_{G_A} f(a',b')\mathbf{L}_0[a',b']\tilde{s}_n(\tau)d_\ell A \\ &= \int_{-\infty}^\infty\int_0^\infty f(a',b')\frac{1}{\sqrt{a'}}\tilde{s}_n\left(\frac{\tau+b'}{a'}\right)\frac{da'db'}{(a')^2} \\ &= \int_{-\infty}^\infty\int_0^\infty D(x,y)\sqrt{y}\tilde{s}_n(y[\tau+x])dydx \\ &= \tilde{e}_n(\tau), \end{aligned}$$

where $D(x,y) \equiv f(a',b')$ with $y = 1/a', x = b'$. Comparing (6.31) with (2.13) we see that $F'(f)\tilde{s}_n(\tau) = \tilde{e}_n(\tau)$ is just the echo generated at time τ from the signal $\tilde{s}_n(\tau)$ and the target position-velocity distribution $D(x,y)$. Each inner product on the right-hand side of (6.30) is the correlation function between the echo $\tilde{e}_n^\pm$ and a test signal $\mathbf{L}_0[a,b]s_n^\pm$:

$$(6.32) \qquad D(X,Y) = \sum_{n=0}^\infty (\langle\tilde{e}_n^-,\mathbf{L}_0[a,b]s_n^-\rangle_2 + \langle\tilde{e}_n^+,\mathbf{L}_0[a,b]s_n^+\rangle_2)$$

where $X = b, Y = a^{-1}$. Note that (6.32) provides a scheme for determining the distribution $D(X,Y)$ experimentally: we can send out signals $\tilde{s}_n^\pm$, measure the echos $\tilde{e}_n^\pm$ and then cross-correlate these echos with the test signals $\mathbf{L}_0[a,b]s_n^\pm$ to construct D.

6.4 Exercises.

6.1 Show that the representation $\mathbf{R}_0$ of the affine group G_A is reducible.

6.2 Verify directly equation (6.15).

6.3 Verify directly equation (6.16).

6.4 Show that the rep $\mathbf{R}_\lambda$ of G_A is *continuous* in the sense that

$$\|\mathbf{R}_\lambda(a,b)\varphi - \varphi\| \to 0, \quad \text{as } (a,b) \to (1,0)$$

for each $\varphi \in \mathcal{H}_1$.

6.5 Show that the ambiguity and cross-ambiguity functions $<\mathbf{L}_0[a,b]\varphi,\psi>$ are continuous functions of (a,b).

§7. WEYL-HEISENBERG FRAMES

7.1 Windowed Fourier transforms. In this and the next chapter we introduce and study two procedures for the analysis of time-dependent signals, locally in both frequency and time. The first procedure, the "windowed Fourier transform" is associated with the Heisenberg group while the second, the "wavelet transform" is associated with the affine group.

Let $g \in L_2(R)$ with $\|g\| = 1$ and define the time-frequency translation of g by

$$(7.1) \qquad g^{[x_1, x_2]}(t) = e^{2\pi i t x_2} g(t + x_1) = \mathbf{T}^1[x_1, x_2, 0]g(t)$$

where $\mathbf{T}^1$ is the unitary irred rep (5.5) of the Heisenberg group H_R with $\lambda = 1$. Now suppose g is centered about the point (t_0, ω_0) in phase (time-frequency) space, i.e., suppose

$$\int_{-\infty}^{\infty} t|g(t)|^2 dt = t_0, \qquad \int_{-\infty}^{\infty} \omega|\hat{g}(\omega)|^2 d\omega = \omega_0$$

where $\hat{g}(\omega) = \int_{-\infty}^{\infty} g(t)e^{-2\pi i \omega t}dt$ is the Fourier transform of $g(t)$. Then

$$\int_{-\infty}^{\infty} t|g^{[x_1, x_2]}(t)|^2 dt = t_0 - x_1, \qquad \int_{-\infty}^{\infty} \omega|\hat{g}^{[x_1, x_2]}(t)|^2 d\omega = \omega_0 + x_2$$

so $g^{[x_1, x_2]}$ is centered about $(t_0 - x_1, \omega_0 + x_2)$ in phase space. To analyze an arbitrary function $f(t)$ in $L_2(R)$ we compute the inner product

$$F(x_1, x_2) = \langle f, g^{[x_1, x_2]} \rangle = \int_{-\infty}^{\infty} f(t)\bar{g}^{[x_1, x_2]}(t)dt$$

with the idea that $F(x_1, x_2)$ is sampling the behavior of f in a neighborhood of the point $(t_0 - x_1, \omega_0 + x_2)$ in phase space. As x_1, x_2 range over all real numbers the samples $F(x_1, x_2)$ give us enough information to reconstruct $f(t)$. Indeed, since $\mathbf{T}^1$ is an irred rep of H_R the functions $\mathbf{T}^1[x_1, x_2, 0]g = g^{[x_1, x_2]}$ are dense in $L_2(R)$ as $[x_1, x_2]$ runs over R^2. Furthermore, $f \in L_2(R)$ is uniquely determined by the inner products $\langle f, g^{[x_1, x_2]} \rangle$, $-\infty < x_1, x_2 < \infty$. (Suppose $\langle f_1, g^{[x_1, x_2]} \rangle = \langle f_2, g^{[x_1, x_2]} \rangle$ for $f_1, f_2 \in L_2(R)$ and all x_1, x_2. Then with $f = f_1 - f_2$ we have $\langle f, g^{[x_1, x_2]} \rangle \equiv 0$, so f is orthogonal to the closed subspace of $L_2(R)$ generated by the $g^{[x_1, x_2]}$. Since $\mathbf{T}^1$ is irreducible this closed subspace is $L_2(R)$. Hence $f = 0$ and $f_1 = f_2$.)

However, the set of basis states $g^{[x_1, x_2]}$ is overcomplete: the coefficients $\langle f, g^{[x_1, x_2]} \rangle$ are not independent of one another, i.e., in general there is no $f \in L_2(R)$ such that $\langle f, g^{[x_1, x_2]} \rangle = F(x_1, x_2)$ for an arbitrary $F \in L_2(R^2)$. The $g^{[x_1, x_2]}$ are examples of **coherent states**, continuous overcomplete Hilbert space bases which are of interest in quantum optics, quantum field theory, group representation theory, etc., [KS].

As an important example we consider the case $g = \boldsymbol{\psi}_0(t) = \pi^{-1/4}e^{-t^2/2}$, (5.26), the ground state. (Recall that $\mathbf{a}\boldsymbol{\psi} = 0$ where $\mathbf{a}$ is the annihilation operator for bosons (5.18). This property uniquely determines the ground state.) Since $\boldsymbol{\psi}_0$ is essentially its own Fourier transform, (5.43), we see that $g = \boldsymbol{\psi}_0$ is centered about $(t_0, \omega_0) = (0, 0)$ in phase space. Thus

$$(7.2) \qquad g^{[x_1, x_2]}(t) = \pi^{-1/4}e^{2\pi i t x_2}e^{-(t+x_1)^2/2}$$

is centered about $(-x_1, x_2)$. It is very instructive to map these vectors in $L_2(R)$ with inner product $\langle \cdot, \cdot \rangle$ to the Bargmann-Segal Hilbert space $\mathcal{F}$ with inner product $(\cdot, \cdot)$, (5.31), via the unitary operator $\mathbf{A}$, (5.36). In $\mathcal{F}$ the ground state is $\mathbf{j}_0(z) = 1$. Thus the corresponding coherent states are

$$
\begin{aligned}
(7.3) \quad \mathbf{A}g^{[x_1, x_2]}(z) &= \mathbf{T}^1[x_1, x_2]\,\mathbf{j}_0(z) = \mathbf{j}_0^{[x_1, x_2]}(z) \\
&= \exp\left[-(x_1^2 + x_2^2)/4 + \frac{(x_1 - ix_2)}{\sqrt{2}}z - \frac{ix_1 x_2}{2}\right] \\
&= \exp[-(x_1^2 + x_2^2)/4 - ix_1 x_2/2]\mathbf{e}_{(x_1 + ix_2)/\sqrt{2}}(z)
\end{aligned}
$$

where $\mathbf{e}_b(z) = \exp(\bar{b}z) \in \mathcal{F}$ is the "delta function" with the property $(\mathbf{f}, \mathbf{e}_b) = \mathbf{f}(b)$ for each $\mathbf{f} \in \mathcal{F}$. Clearly

$$
\begin{aligned}
(7.4) \quad \langle g^{[x_1, x_2]}, g^{[y_1, y_2]} \rangle &= (\mathbf{j}_0^{[x_1, x_2]}, \mathbf{j}_0^{[y_1, y_2]}) \\
&= \exp[-(x_1^2 + x_2^2 + y_1^2 + y_2^2)/4 - ix_1 x_2 + iy_1 y_2] \times \\
&\quad \exp\left[\frac{(y_1 + iy_2)(x_1 - ix_2)}{2}\right]
\end{aligned}
$$

so the $g^{[x_1 x_2]}$ are not mutually orthogonal. Moreover, given $f \in L_2(R)$ with $\mathbf{f} = \mathbf{A}f \in \mathcal{F}$ we have

$$
(7.5) \qquad \langle f, g^{[x_1, x_2]} \rangle = (\mathbf{f}, \mathbf{j}_0^{[x_1, x_2]}) = \exp[-(x_1^2 + x_2^2)/4 + ix_1 x_2]\mathbf{f}\left(\frac{x_1 + ix_2}{\sqrt{2}}\right).
$$

Expression (7.5) displays clearly the overcompleteness of the coherent states. Since $\mathbf{f}$ is an entire function, it is uniquely determined by its values in an open set of the complex plane (or a line segment or even on a discrete set of points in $\mathcal{C}$ which have a limit point). Thus the values $\langle f, g^{[x_1 x_2]} \rangle$ cannot be prescribed arbitrarily. However, from the "delta function" property

$$
\mathbf{f}(b) = (\mathbf{f}, \mathbf{e}_b)
$$

we can easily expand $\mathbf{f} \in \mathcal{F}$ as a double integral over the coherent states $\mathbf{j}_0^{[x_1, x_2]}$, hence we can expand $f = \mathbf{A}^{-1}\mathbf{f} \in L_2(R)$ as the corresponding double integral over the coherent states $g^{[x_1, x_2]}$.

There are two features of the foregoing discussion that are worth special emphasis. First there is the great flexibility in the coherent function approach due to the fact that the function $g \in L_2(R)$ can be chosen to fit the problem at hand. Second is the fact that coherent states are always overcomplete. Thus it isn't necessary to compute the inner products $\langle f, g^{[x_1, x_2]} \rangle = F(x_1, x_2)$ for every point in phase space. In the windowed Fourier approach one typically samples F at the lattice points $(x_1, x_2) = (ma, nb)$ where a, b are fixed positive numbers and m, n range over the integers. Here, a, b and $g(t)$ must be chosen so that the map $f \to \{F(ma, nb)\}$ is one-to-one; then f can be recovered from the lattice point values $F(ma, nb)$.

7.2 The Weil-Brezin-Zak transform. The Weil-Brezin transform (earlier used in radar theory by Zak, so also called the Zak transform) (5.49) between the Schrödinger rep $\mathbf{T}^1$ of H_R on $L_2(R)$ and the lattice rep $\tilde{\mathbf{T}}$, (5.46)-(5.48), is very useful in studying the lattice sampling problem, particularly in the case $a = b = 1$. Restricting to this case for the time being, we let $\psi \in L_2(R)$. Then

$$(7.6) \qquad \psi_\mathbf{P}(x_1, x_2) = \mathbf{P}\psi(x_1, x_2, 0) = \sum_{k=-\infty}^{\infty} e^{2\pi i k x_2}\psi(x_1 + k)$$

satisfies

$$\psi_\mathbf{P}(k_1 + x_1, k_2 + x_2) = e^{-2\pi i k_1 x_2}\psi_\mathbf{P}(x_1, x_2)$$

for integers k_1, k_2. (Here (7.6) is meaningful if ψ belongs to, say, the Schwartz class. Otherwise $\mathbf{P}\psi = \lim_{n\to s} \mathbf{P}\psi_n$ where $\psi = \lim_{n=s} \psi_n$ and the ψ_n are Schwartz class functions. The limit is taken with respect to the Hilbert space norm.) Furthermore

$$\left[\mathbf{T}^1[y_1, y_2, 0]\psi\right]_\mathbf{P}(x_1, x_2, 0) = \tilde{\mathbf{T}}[y_1, y_2, 0]\psi_\mathbf{P}(x_1, x_2)$$
$$= \exp[2\pi i x_1 y_2]\psi_\mathbf{P}(x_1 + y_1, x_2 + y_2).$$

Hence if $\psi = g^{[m,n]} = \mathbf{T}^1[m, n]g$ we have

$$(7.7) \qquad g_\mathbf{P}^{[m,n]}(x_1, x_2) = \exp[2\pi i(x_1 n - x_2 m)]g_\mathbf{P}(x_1, x_2).$$

Thus in the lattice rep, the functions $g_\mathbf{P}^{[m,n]}$ differ from $g_\mathbf{P}$ simply by the multiplicative factor $e^{2\pi i(x_1 n - x_2 m)} = \mathbf{E}_{n,m}(x_1, x_2)$, and as n, m range over the integers the $\mathbf{E}_{n,m}$ form an ON basis for the Hilbert space of the lattice rep:

$$(7.8) \qquad (\boldsymbol{\varphi}_1, \boldsymbol{\varphi}_2) = \int_0^1 \int_0^1 \varphi_1(x_1, x_2)\overline{\varphi_2(x_1, x_2)}dx_1 dx_2.$$

THEOREM 7.1. For $(a, b) = (1, 1)$ and $g \in L_2(R)$ the transforms $\{g^{[m,n]} : m, n = 0 \pm 1, \pm 2, \cdots\}$ span $L_2(R)$ if and only if $\mathbf{P}g(x_1, x_2, 0) = g_\mathbf{P}(x_1, x_2) \neq 0$ a.e..

Proof. Let $\mathcal{M}$ be the closed linear subspace of $L_2(R)$ spanned by the $\{g^{[m,n]}\}$. Clearly $\mathcal{M} = L_2(R)$ iff $f = 0$ a.e. is the only solution of $\langle f, g^{[m,n]}\rangle = 0$ for all integers m and n. Applying the Weyl-Brezin -Zak isomorphism $\mathbf{P}$ we have

$$(7.9) \qquad \begin{aligned} \langle f, g^{[m,n]}\rangle &= (\mathbf{P}f, \mathbf{E}_{n,m}\mathbf{P}g) \\ &= ([\mathbf{P}f][\overline{\mathbf{P}g}], \mathbf{E}_{n,m}) = (f_\mathbf{P}\bar{g}_\mathbf{P}, \mathbf{E}_{n,m}). \end{aligned}$$

Since the functions $\mathbf{E}_{n,m}$ form an ON basis for the Hilbert space (7.8) it follows that $\langle f, g^{[m,n]}\rangle = 0$ for all integers m, n iff $f_\mathbf{P}(x_1, x_2)g_\mathbf{P}(x_1, x_2) = 0$, a.e.. If $g_\mathbf{P} \neq 0$, a.e. then $f_\mathbf{P} = f = 0$ and $\mathcal{M} = L_2(R)$. If $g_\mathbf{P} = 0$ on a set S of positive measure on the unit square, then the characteristic function $\chi_S = \mathbf{P}f = f_\mathbf{P}$ satisfies $f_\mathbf{P}g_\mathbf{P} = \chi_S g_\mathbf{P} = 0$ a.e., hence $\langle f, g^{[m,n]}\rangle = 0$ and $\mathcal{M} \neq L_2(R)$. □

In the case $g(t) = \pi^{-1/4} e^{-t^2/2}$ one finds that

$$(7.10) \qquad g_{\mathbf{P}}(x_1, x_2) = \pi^{-1/4} \sum_{k=-\infty}^{\infty} e^{2\pi i k x_2 - (x_1+k)^2/2}.$$

As is well-known, [EMOT1], [WW], the series (7.10) defines a Jacobi Theta function. Using complex variable techniques it can be shown that this function vanishes at the single point $(\frac{1}{2}, \frac{1}{2})$ in the square $0 \le x_1 < 1$, $0 \le x_2 < 1$, [WW]. Thus $g_{\mathbf{P}} \ne 0$ a.e. and the functions $\{g^{[m,n]}\}$ span $L_2(R)$. (However, the expansion of an $L_2(R)$ function in terms of this set is not unique and the $\{g^{[m,n]}\}$ do not form a frame in the sense of §7.4.)

COROLLARY 7.1. For $(a,b) = (1,1)$ and $g \in L_2(R)$ the transforms $\{g^{[m,n]} : m, n = 0, \pm 1, \cdots\}$ form an ON basis for $L_2(R)$ iff $|g_{\mathbf{P}}(x_1, x_2)| = 1$, a.e.

Proof. We have

$$\delta_{mm'} \delta_{nn'} = \langle g^{[m,n]}, g^{[m',n']} \rangle = (E_{n,m} g_{\mathbf{P}}, E_{n',m'} g_{\mathbf{P}})$$
$$= (|g_{\mathbf{P}}|^2, E_{n'-n, m'-m})$$

iff $|g_{\mathbf{P}}|^2 = 1$, a.e. □

As an example, let $g = \chi_{[0,1)}$ where

$$\chi_{[0,1)}(t) = \begin{cases} 1 & \text{if } 0 \le t < 1 \\ 0 & \text{otherwise} \end{cases}.$$

Then it is easy to see that $|g_{\mathbf{P}}(x_1, x_2)| \equiv 1$. Thus $\{g^{[m,n]}\}$ is an ON basis for $L_2(R)$.

THEOREM 7.2. For $(a,b) = (1,1)$ and $g \in L_2(R)$, suppose there are constants A, B such that

$$0 < A \le |g_{\mathbf{P}}(x_1, x_2)|^2 \le B < \infty$$

almost everywhere in the square $0 \le x_1, x_2 < 1$. Then $\{g^{[m,n]}\}$ is a basis for $L_2(R)$, i.e., each $f \in L_2(R)$ can be expanded **uniquely** in the form $f = \sum_{m,n} a_{mn} g^{[m,n]}$. Indeed,

$$a_{mn} = \left(f_{\mathbf{P}}, g_{\mathbf{P}}^{[m,n]} / |g_{\mathbf{P}}|^2 \right) = (f_{\mathbf{P}}/g_{\mathbf{P}}, E_{n,m}).$$

Proof. By hypothesis $|g_{\mathbf{P}}|^{-1}$ is a bounded function on the domain $0 \le x_1, x_2 < 1$. Hence $f_{\mathbf{P}}/g_{\mathbf{P}}$ is square integrable on this domain and, from the periodicity properties of elements in the lattice Hilbert space, $\frac{f_{\mathbf{P}}}{g_{\mathbf{P}}}(x_1 + n, x_2 + m) = \frac{f_{\mathbf{P}}}{g_{\mathbf{P}}}(x_1, x_2)$. It follows that

$$\frac{f_{\mathbf{P}}}{g_{\mathbf{P}}} = \sum a_{mn} E_{n,m}$$

where $a_{mn} = (f_{\mathbf{P}}/g_{\mathbf{P}}, E_{n,m})$, so $f_{\mathbf{P}} = \sum a_{mn} E_{n,m} g_{\mathbf{P}}$. This last expression implies $f = \sum a_{mn} g^{[m,n]}$. Conversely, given $f = \sum a_{mn} g^{[m,n]}$ we can reverse the steps in the preceding argument to obtain $a_{mn} = (f_{\mathbf{P}}/g_{\mathbf{P}}, E_{n,m})$. □

7.3 Windowed transforms and ambiguity functions. We can relate these expansions to radar cross-ambiguity functions as follows. The expansion $f = \sum a_{mn} g^{[n,n]}$ is equivalent to the lattice Hilbert space expansion $f_{\mathbf{P}} = \sum a_{mn} E_{n,m} g_{\mathbf{P}}$ or

$$(7.11) \qquad f_{\mathbf{P}} \bar{g}_{\mathbf{P}} = \sum (a_{mn} E_{n,m}) |g_{\mathbf{P}}|^2.$$

Now if $g_{\mathbf{P}}$ is a bounded function then $f_{\mathbf{P}} \bar{g}_{\mathbf{P}}(x_1, x_2)$ and $|g_{\mathbf{P}}|^2$ both belong to the lattice Hilbert space and are periodic functions in x_1 and x_2 with period 1. Hence,

$$f_{\mathbf{P}} \bar{g}_{\mathbf{P}} = \sum b_{mn} E_{n,m}$$
$$|g_{\mathbf{P}}|^2 = \sum c_{mn} E_{n,m}$$

with

$$b_{mn} = (f_{\mathbf{P}} \bar{g}_{\mathbf{P}}, E_{n,m}) = (f_{\mathbf{P}}, g_{\mathbf{P}} E_{n,m}) = \langle f, g^{[m,n]} \rangle = \langle f, \mathbf{T}^1[m,n]g \rangle,$$
$$c_{mn} = (g_{\mathbf{P}} \bar{g}_{\mathbf{P}}, E_{n,m}) = \langle g, g^{[m,n]} \rangle = \langle g, \mathbf{T}^1[m,n]g \rangle.$$

Thus (7.11) gives the Fourier series expansion for $f_{\mathbf{P}} \bar{g}_{\mathbf{P}}$ as the product of two other Fourier series expansions. (We consider the functions f, g, hence $f_{\mathbf{P}}, g_{\mathbf{P}}$ as known.) The Fourier coefficients in the expansions of $f_{\mathbf{P}} \bar{g}_{\mathbf{P}}$ and $|g_{\mathbf{P}}|^2$ are cross-ambiguity functions. If $|g_{\mathbf{P}}|^2$ never vanishes we can solve for the a_{mn} directly:

$$\sum a_{mn} E_{n,m} = (\sum b_{mn} E_{n,m})(\sum c'_{mn} E_{n,m})$$

where the c'_{mn} are the Fourier coefficients of $|g_{\mathbf{P}}|^{-2}$. However, if $|g_{\mathbf{P}}|^2$ vanishes at some point then the best we can do is obtain the convolution equations $b = a * c$, i.e.,

$$(7.12) \qquad b_{mn} = \sum_{\substack{k+k'=m \\ \ell+\ell'=n}} a_{k\ell} c_{\ell'_2 \ell'}.$$

(Auslander and Tolimieri [AT5] have shown how to approximate the coefficients $a_{k\ell}$ even in the cases where $|g_{\mathbf{P}}|^2$ vanishes at some points. The basic idea is to truncate $\sum a_{mn} E_{n,m}$ to a finite number of nonzero terms and to sample equation (7.11), making sure that $|g_{\mathbf{P}}|(x_1, x_2)$ is nonzero at each sample point. The a_{mn} can then be computed by using the inverse finite Fourier transform (3.37).)

The problem of $|g_{\mathbf{P}}|$ vanishing at a point is not confined to an isolated example, such as (7.10). Indeed it can be shown that if $g_{\mathbf{P}}$ is an everywhere continuous function in the lattice Hilbert space then it must vanish at at least one point, [HW].

7.4 Frames. To understand the nature of the complete sets $\{g^{[m,n]}\}$ it is useful to broaden our perspective and introduce the idea of a **frame** in an arbitrary Hilbert space $\mathcal{H}$. In this more general point of view we are given a sequence $\{\mathbf{f}_n\}$ of elements of $\mathcal{H}$ and we want to find conditions on $\{\mathbf{f}_n\}$ so that we can recover an arbitrary

$\mathbf{f} \in \mathcal{H}$ from the inner products $\langle \mathbf{f}, \mathbf{f}_n \rangle$ on $\mathcal{H}$. Let $L_2(Z)$ be the Hilbert space of countable sequences $\{\xi_n\}$ with inner product $(\xi, \eta) = \sum_n \xi_n \bar{\eta}_n$. (A sequence $\{\xi_n\}$ belongs to $L_2(Z)$ provided $\sum_n \xi_n \bar{\xi}_n < \infty$.) Now let $\mathbf{T} : \mathcal{H} \to L_2(Z)$ be the linear mapping defined by

$$(7.13) \qquad (\mathbf{Tf})_n = \langle \mathbf{f}, \mathbf{f}_n \rangle.$$

We require that $\mathbf{T}$ is a bounded operator from $\mathcal{H}$ to $L_2(Z)$, i.e., that there is a finite $B > 0$ such that $\sum_n |\langle \mathbf{f}, \mathbf{f}_n \rangle|^2 \le B\|\mathbf{f}\|^2$. In order to recover $\mathbf{f}$ from the $\langle \mathbf{f}, \mathbf{f}_n \rangle$ we want $\mathbf{T}$ to be invertible with $\mathbf{T}^{-1} : \mathcal{R}_\mathbf{T} \to \mathcal{H}$ where $\mathcal{R}_\mathbf{T}$ is the range $\mathbf{T}\mathcal{H}$ of $\mathbf{T}$ in $L_2(Z)$. Moreover, for numerical stability in the computation of $\mathbf{f}$ from the $\langle \mathbf{f}, \mathbf{f}_n \rangle$ we want $\mathbf{T}^{-1}$ to be bounded. (In other words we want to require that a "small" change in the data $\langle \mathbf{f}, \mathbf{f}_n \rangle$ leads to a "small" change in $\mathbf{f}$.) This means that there is a finite $A > 0$ such that $\sum_n |\langle \mathbf{f}, \mathbf{f}_n \rangle|^2 \ge A\|\mathbf{f}\|^2$. (Note that $\mathbf{T}^{-1}\xi = \mathbf{f}$ if $\xi_n = \langle \mathbf{f}, \mathbf{f}_n \rangle$.) If these conditions are satisfied, i.e., if there exist positive constants A, B such that

$$(7.14) \qquad A\|\mathbf{f}\|^2 \le \sum_n |\langle \mathbf{f}, \mathbf{f}_n \rangle|^2 \le B\|\mathbf{f}\|^2$$

for all $\mathbf{f} \in \mathcal{H}$, we say that the sequence $\{\mathbf{f}_n\}$ is a **frame** for $\mathcal{H}$ and that A and B are **frame bounds**.

The **adjoint** $\mathbf{T}^*$ of $\mathbf{T}$ is the linear mapping $\mathbf{T}^* : L_2(Z) \to \mathcal{H}$ defined by

$$\langle \mathbf{T}^*\xi, \mathbf{f} \rangle = (\xi, \mathbf{Tf})$$

for all $\xi \in L_2(Z)$, $\mathbf{f} \in \mathcal{H}$. A simple computation yields

$$(7.15) \qquad \mathbf{T}^*\xi = \sum_n \xi_n \mathbf{f}_n.$$

(Since $\mathbf{T}$ is bounded, so is $\mathbf{T}^*$ and the right-hand side of (7.15) is well-defined for all $\xi \in L_2(Z)$.) Now the bounded self-adjoint operator $\mathbf{S} = \mathbf{T}^*\mathbf{T} : \mathcal{H} \to \mathcal{H}$ is given by

$$(7.16) \qquad \mathbf{Sf} = \mathbf{T}^*\mathbf{Tf} = \sum_n \langle \mathbf{f}, \mathbf{f}_n \rangle \mathbf{f}_n,$$

and we can rewrite the defining inequality (7.14) for the frame as

$$(7.17) \qquad A\|\mathbf{f}\|^2 \le \langle \mathbf{T}^*\mathbf{Tf}, \mathbf{f} \rangle \le B\|\mathbf{f}\|^2.$$

Since $A > 0$, if $\mathbf{T}^*\mathbf{Tf} = \boldsymbol{\theta}$ then $\mathbf{f} = \boldsymbol{\theta}$, so $\mathbf{S}$ is one-to-one, hence invertible. Furthermore, the range $\mathbf{S}\mathcal{H}$ of $\mathbf{S}$ is $\mathcal{H}$. Indeed, if $\mathbf{S}\mathcal{H}$ is a proper subspace of $\mathcal{H}$ then we can find a nonzero vector $\mathbf{g}$ in $(\mathbf{S}\mathcal{H})^\perp : \langle \mathbf{Sf}, \mathbf{g} \rangle = 0$ for all $\mathbf{f} \in \mathcal{H}$. However, $\langle \mathbf{Sf}, \mathbf{g} \rangle = \langle \mathbf{T}^*\mathbf{Tf}, \mathbf{g} \rangle = (\mathbf{Tf}, \mathbf{Tg}) = \sum_n \langle \mathbf{f}, \mathbf{f}_n \rangle \langle \mathbf{f}_n, \mathbf{g} \rangle$. Setting $\mathbf{f} = \mathbf{g}$ we obtain

$$\sum_n |\langle \mathbf{g}, \mathbf{f}_n \rangle|^2 = 0.$$

By (7.14) we have $\mathbf{g} = \boldsymbol{\theta}$, a contradiction. thus $S\mathcal{H} = \mathcal{H}$ and the inverse operator S^{-1} exists and has domain $\mathcal{H}$.

Since $SS^{-1}\mathbf{f} = S^{-1}S\mathbf{f} = \mathbf{f}$ for all $\mathbf{f} \in \mathcal{H}$, we immediately obtain two expansions for $\mathbf{f}$ from (7.16):

(7.18)
$$a)\ \mathbf{f} = \sum_n \langle S^{-1}\mathbf{f}, \mathbf{f}_n \rangle \mathbf{f}_n = \sum_n \langle \mathbf{f}, S^{-1}\mathbf{f}_n \rangle \mathbf{f}_n$$
$$b)\ \mathbf{f} = \sum_n \langle \mathbf{f}, \mathbf{f}_n \rangle S^{-1}\mathbf{f}_n.$$

(The second equality in (7.18a) follows from the identity $\langle S^{-1}\mathbf{f}, \mathbf{f}_n \rangle = \langle \mathbf{f}, S^{-1}\mathbf{f}_n \rangle$, which holds since S^{-1} is self-adjoint.)

Recall that for a **positive** operator S, i.e., an operator such that $\langle S\mathbf{f}, \mathbf{f} \rangle \geq 0$ for all $\mathbf{f} \in \mathcal{H}$ the inequalities

$$A||\mathbf{f}||^2 \leq \langle S\mathbf{f}, \mathbf{f} \rangle \leq B||\mathbf{f}||^2$$

for $A, B > 0$ are equivalent to the inequalities

(7.19)
$$A||\mathbf{f}|| \leq ||S\mathbf{f}|| \leq B||\mathbf{f}||,$$

see [RN] or [DS2].

An examination of (7.18a) and (7.18b) suggests that if the $\{\mathbf{f}_n\}$ form a frame then so do the $\{S^{-1}\mathbf{f}_n\}$.

THEOREM 7.3. *Suppose $\{\mathbf{f}_n\}$ is a frame with frame bounds A, B and let $S = T^*T$. Then $\{S^{-1}\mathbf{f}_n\}$ is also a frame, called the **dual frame** of $\{\mathbf{f}_n\}$, with frame bounds B^{-1}, A^{-1}.*

Proof. Setting $\mathbf{f} = S^{-1}\mathbf{g}$ in (7.19) we have $B^{-1}||\mathbf{g}|| \leq ||S^{-1}\mathbf{g}|| \leq A^{-1}||\mathbf{g}||$. Since S^{-1} is self-adjoint, this implies $B^{-1}||\mathbf{g}||^2 \leq \langle S^{-1}\mathbf{g}, \mathbf{g} \rangle \leq A^{-1}||\mathbf{g}||^2$. From (7.18b) we have $S^{-1}\mathbf{g} = \sum_n \langle S^{-1}\mathbf{g}, \mathbf{f}_n \rangle S^{-1}\mathbf{f}_n$ so $\langle S^{-1}\mathbf{g}, \mathbf{g} \rangle = \sum_n \langle S^{-1}\mathbf{g}, \mathbf{f}_n \rangle \langle S^{-1}\mathbf{f}_n, \mathbf{g} \rangle = \sum_n |\langle \mathbf{g}, S^{-1}\mathbf{f}_n \rangle|^2$. Hence $\{S^{-1}\mathbf{f}_n\}$ is a frame with frame bounds B^{-1}, A^{-1}. $\square$

We say that $\{\mathbf{f}_n\}$ is a **tight frame** if $A = B$.

COROLLARY 7.2. *If $\{\mathbf{f}_n\}$ is a tight frame then every $\mathbf{f} \in \mathcal{H}$ can be expanded in the form*
$$\mathbf{f} = A^{-1} \sum_n \langle \mathbf{f}, \mathbf{f}_n \rangle \mathbf{f}_n.$$

Proof. Since $\{\mathbf{f}_n\}$ is a tight frame we have $A||\mathbf{f}||^2 = \langle S\mathbf{f}, \mathbf{f} \rangle$ or $\langle (S - AE)\mathbf{f}, \mathbf{f} \rangle = 0$ where E is the identity operator $E\mathbf{f} = \mathbf{f}$. Since $S - AE$ is a self-adjoint operator we have $||(S - AE)\mathbf{f}|| = 0$ for all $\mathbf{f} \in \mathcal{H}$. Thus $S = AE$. However, from (7.18), $S\mathbf{f} = \sum_n \langle \mathbf{f}, \mathbf{f}_n \rangle \mathbf{f}_n$. $\square$

7.5 Frames of $W - H$ type. We can now relate frames with the Heisenberg group lattice construction (7.6).

THEOREM 7.4. *For $(a, b) = (1, 1)$ and $g \in L_2(R)$, we have*

$$(7.20) \qquad\qquad 0 < A \le |g_{\mathbf{P}}(x_1, x_2)|^2 \le B < \infty$$

almost everywhere in the square $0 \le x_1, x_2 < 1$ iff $\{g^{[m,n]}\}$ is a frame for $L^2(R)$ with frame bounds A, B. (By Theorem 7.2 this frame is actually a basis for $L_2(R)$.).

Proof. If (7.20) holds then $g_{\mathbf{P}}$ is a bounded function on the square. Hence for any $f \in L_2(R)$, $f_{\mathbf{P}} \bar{g}_{\mathbf{P}}$ is a periodic function, in x_1, x_2 on the square. Thus

$$(7.21) \qquad \begin{aligned} \sum_{m,n=-\infty}^{\infty} |\langle f, g^{[m,n]} \rangle|^2 &= \sum_{m,n=-\infty}^{\infty} |(f_{\mathbf{P}}, E_{n,m} g_{\mathbf{P}})|^2 \\ &= \sum_{m,n=-\infty}^{\infty} |(f_{\mathbf{P}} \bar{g}_{\mathbf{P}}, E_{n,m})|^2 = ||f_{\mathbf{P}} \bar{g}_{\mathbf{P}}||^2 \\ &= \int_0^1 \int_0^1 |f_{\mathbf{P}}|^2 |g_{\mathbf{P}}|^2 dx_1 dx_2. \end{aligned}$$

(Here we have used the Plancherel theorem for the exponentials $E_{n,m}$) It follows from (7.20) that

$$(7.22) \qquad A||f||^2 \le \sum_{m,n=-\infty}^{\infty} |\langle f, g^{[m,n]} \rangle|^2 \le B||f||^2,$$

so $\{g^{[m,n]}\}$ is a frame.

Conversely, if $\{g^{[m,n]}\}$ is a frame with frame bounds A, B, it follows from (7.22) and the computation (7.21) that

$$A||f_{\mathbf{P}}||^2 \le \int_0^1 \int_0^1 |f_{\mathbf{P}}|^2 |g_{\mathbf{P}}|^2 dx_1 dx_2 \le B||f_{\mathbf{P}}||^2$$

for an **arbitrary** $f_{\mathbf{P}}$ in the lattice Hilbert space. (Here we have used the fact that $||f|| = ||f_{\mathbf{P}}||$, since $\mathbf{P}$ is a unitary transformation.) Thus the inequalities (7.20) hold almost everywhere. $\square$

Frames of the form $\{g^{[ma,nb]}\}$ are called **Weyl-Heisenberg** (or **W-H**) frames. The Weyl-Brezin-Zak transform is not so useful for the study of W-H frames with general frame parameters (a, b). (Note from (7.1) that it is only the product ab that is of significance for the W-H frame parameters. Indeed, the change of variable $t' = t/a$ in (7.1) converts the frame parameters (a, b) to $(a', b') = (1, ab)$.) An easy consequence of the general definition of frames is the following:

THEOREM 7.5. *Let $g \in L_2(R)$ and $a, b, A, B > 0$ such that*

1) $0 < A \le \sum_m |g(x + ma)|^2 \le B < \infty$, *a.e.,*
2) g *has support contained in an interval I where I has length b^{-1}.*

Then the $\{g^{[ma,nb]}\}$ are a W-H frame for $L_2(R)$ with frame bounds $b^{-1}A, b^{-1}B$.

Proof. For fixed m and arbitrary $f \in L_2(R)$ the function $F_m(t) = f(t)\overline{g(t+ma)}$ has support in the interval $I_m = \{t+ma : x \in I\}$ of length b^{-1}. Thus $F_m(t)$ can be expanded in a Fourier series with respect to the basis exponentials $E_{nb}(t) = e^{2\pi i bnt}$ on I_m. Using the Plancherel formula for this expansion we have

$$\sum_{m,n} |\langle f, g^{[ma,nb]}\rangle|^2 = \sum_{m,n} |\langle F_m, E_{nb}\rangle|^2$$

$$= \frac{1}{b}\sum_m |\langle F_m, F_m\rangle| = \frac{1}{b}\sum_m \int_{I_m} |f(t)|^2 |g(t+ma)|^2 dt$$

$$= \frac{1}{b}\int_{-\infty}^{\infty} |f(t)|^2 \sum_m |g(t+ma)|^2 dt.$$

From property 1) we have then

$$\frac{A}{b}||f||^2 \le \sum_{m,n} |\langle f, g^{[ma,nb]}\rangle|^2 \le \frac{B}{b}||f||^2,$$

so $\{g^{[ma,nb]}\}$ is a W-H frame. $\square$

There are no W-H frames with frame parameters (a,b) such that $ab > 1$, [BBGK], [R3]. For some insight into this case we consider the example $(a,b) = (N,1), N > 1, N$ an integer. Let $g \in L_2(R)$. There are two distinct possibilities:

1) There is a constant $A > 0$ such that $A \le |g_{\mathbf{P}}(x_1, x_2)|$ almost everywhere.

2) There is no such $A > 0$.

Let $\mathcal{M}$ be the closed subspace of $L_2(R)$ spanned by the functions $\{g^{[mN,n]}, m, n = 0 \pm 1, \pm 2, \cdots\}$ and suppose $f \in L_2(R)$. Then

$$\langle f, g^{[mN,n]}\rangle = (f_{\mathbf{P}}, E_{n,mN}g_{\mathbf{P}}) = (f_{\mathbf{P}}\bar{g}_{\mathbf{P}}, E_{n,mN}).$$

If possibility 1) holds, we set $f_{\mathbf{P}} = \bar{g}_{\mathbf{P}}^{-1}E_{n_0,1}$. Then $f_{\mathbf{P}}$ belongs to the lattice Hilbert space and $0 = (E_{n_0,1}, E_{n,mN}) = (f_{\mathbf{P}}\bar{g}_{\mathbf{P}}E_{n,mN}) = \langle f, g^{[mN,n]}\rangle$ so $f \in \mathcal{M}^\perp$ and $\{g^{[mN,n]}\}$ is not a frame. Now suppose possibility 2) holds. Then according to the proof of Theorem 7.4, g cannot generate a frame $\{g^{[m,n]}\}$ with frame parameters $(1,1)$ because there is no $A > 0$ such that $A||f||^2 < \sum_{m,n} |\langle f, g^{[m,n]}\rangle|^2$. Since the $\{g^{[mN,n]}\}$ corresponding to frame parameters $(1,N)$ is a proper subset of $\{g^{[m,n]}\}$, it follows that $\{g^{[mN,n]}\}$ cannot be a frame either.

For frame parameters (a,b) with $0 < ab < 1$ it is not difficult to construct W-H frames $\{g^{[ma,nb]}\}$ such that $g \in L_2(R)$ is a smooth function [DGM], [H3], [HW]. Taking the case $a = 1, b = \frac{1}{2}$, for example, let v be an infinitely differentiable function on R such that

$$v(x) = \begin{cases} 0 & \text{if } x \le 0 \\ 1 & \text{if } x \ge 1 \end{cases}$$

and $0 < v(x) < 1$ if $0 < x < 1$. Set

$$(7.23) \qquad g(x) = \begin{cases} 0, & x \leq 0 \\ v(x), & 0 < x < 1 \\ [1 - v^2(x-1)]^{\frac{1}{2}}, & 1 \leq x \leq 2 \\ 0, & 2 < x. \end{cases}$$

Then $g \in L_2(R)$ is infinitely differentiable and with support contained in the interval $[0,2]$. Moreover, $||g||^2 = 1$ and $\sum_n |g(x+m)|^2 \equiv 1$. It follows immediately from Theorem 7.5 that $\{g^{[m,n/2]}\}$ is a W-H frame with frame bounds $A = B = 2$.

We conclude this section by deriving some identities related to cross-ambiguity functions evaluated on lattices of the Heisenberg group.

THEOREM 7.6 [S1], [S2]. Let $f, g \in L_2(R)$ such that $|f_{\mathbf{P}}(x_1, x_2)|\, |g_{\mathbf{P}}(x_1, x_2)|$ are bounded almost everywhere. Then

$$\sum_{m,n} |\langle f, g^{[m,n]}\rangle|^2 = \sum_{m,n} \langle f, f^{[m,n]}\rangle \langle g^{[m,n]}, g\rangle.$$

Proof. Since $\langle f, g^{[m,n]}\rangle = (f_{\mathbf{P}}, E_{n,m} g_{\mathbf{P}}) = (f_{\mathbf{P}} \bar{g}_{\mathbf{P}}, E_{n,m})$ we have the Fourier series expansion

$$(7.24) \qquad f_P(x_1, x_2)\overline{g_P(x_1, x_2)} = \sum_{m,n} \langle f, g^{[m,n]}\rangle E_{n,m}(x_1, x_2).$$

Since $|f_P|, |g_P|$ are bounded, $f_P \bar{g}_P$ is square integrable with respect to the measure $dx_1 dx_2$ on the square $0 \leq x_1, x_2 < 1$. From the Plancherel formula for double Fourier series, we obtain the identity

$$(7.24) \qquad \int_0^1 \int_0^1 |f_{\mathbf{P}}|^2 |g_{\mathbf{P}}|^2 dx_1 dx_2 = \sum_{m,n} |\langle f, g^{[m,n]}\rangle|^2.$$

Similarly, we can obtain expansions of the form (7.24) for $f_{\mathbf{P}} \bar{f}_{\mathbf{P}}$ and $g_{\mathbf{P}} \bar{g}_{\mathbf{P}}$. Applying the Plancherel formula to these two functions we find

$$(7.25) \qquad \int_0^1 \int_0^1 |f_{\mathbf{P}}|^2 g_{\mathbf{P}}|^2 dx_1 dx_2 = \sum_{m,n} \langle f, f^{[m,n]}\rangle \langle g^{[m,n]}, g\rangle.$$

□

7.6 Exercises.

7.1 Verify that if $g \in L_2(R)$, $||g|| = 1$ and g is centered about (t_0, ω_0) in phase space, then $g^{[x_1, x_2]}$ is centered about $(t_0 - x_1, \omega_0 + x_2)$.

7.2 Given the function

$$g(t) = \begin{cases} 1, & |t| \leq \frac{1}{2} \\ 0, & |t| \geq \frac{1}{2}, \end{cases}$$

show that the set $\{g^{[m,n]}\}$ is an ON basis for $L_2(R)$.

§8. AFFINE FRAMES AND WAVELETS

8.1 Wavelets. Here we work out the analog for the affine group of the Weyl-Heisenberg frame for the Heisenberg group. Let $g \in L_2(R)$ with $||g|| = 1$ and define the affine translation of g by

$$(8.1) \qquad g^{(a,b)}(t) = a^{-1/2} g\left(\frac{t+b}{a}\right) = \mathbf{L}_0[a,b]g(t)$$

where $a > 0$ and $\mathbf{L}_0$ is the unitary rep (6.10) of the affine group. Recall that $\mathbf{L}_0 \approx \mathbf{R}_+ + \mathbf{R}_-$ is reducible. Indeed $L_2(R) = \mathcal{H}^+ \oplus \mathcal{H}^-$ where $\mathcal{H}^+$ consists of the functions f_+ such that the Fourier transform $\mathcal{F}f_+(y)$ has support on the positive y-axis and the functions f_- in $\mathcal{H}^-$ have Fourier transform with support on the negative y-axis. Thus the functions $\{g^{(a,b)}\}$ will not necessarily span $L_2(R)$. However, if we choose two functions $g_\pm \in \mathcal{H}^\pm$ with $||g_\pm|| = 1$ then the functions $\{g_+^{(a,b)}, g_-^{(a,b)} : a > 0\}$ **will** span $L_2(R)$. By translation in t if necessary, we can assume that $\int_{-\infty}^{\infty} t|g_\pm(t)|^2 dt = 0$. Let $k_+ = \int_0^\infty y|\mathcal{F}g_+(y)|^2 dy$, $k_- = \int_{-\infty}^0 y|\mathcal{F}g_-(y)|^2 dy$. Then $g_\pm$ are centered about the origin in position space and about $k_\pm$ in momentum space. It follows that

$$\int_{-\infty}^{\infty} t|g_\pm^{(a,b)}(t)|^2 dt = -b, \quad \pm \int_0^\infty y|\mathcal{F}g_\pm^{(a,b)}(\pm y)|^2 dy = a^{-1}k_\pm.$$

To define a lattice in the affine group space we choose two nonzero real numbers $a_0, b_0 > 0$ with $a_0 \neq 1$. Then the lattice points are $a = a_0^m, b = nb_0 a_0^m$, $m, n = 0, \pm 1, \cdots$, so

$$(8.2) \qquad g^{mn}(t) = g^{(a_0^m, nb_0 a_0^m)}(t) = a_0^{-m/2} g(a_0^{-m}t + nb_0).$$

Thus $g_\pm^{mn}$ is centered about $-nb_0 a_0^m$ in position space and about $a_0^{-m}k_\pm$ in momentum space. Note that if g has support contained in an interval of length ℓ then the support of g^{mn} is contained in an interval of length $a_0^{-m}\ell$. Similarly, if $\mathcal{F}g$ has support contained in an interval of length L then the support of $\mathcal{F}g^{mn}$ is contained in an interval of length $a_0^m L$. (Note that this behavior is very different from the behavior of the Heisenberg translates $g^{[ma,nb]}$. In the Heisenberg case the support of g in either position or momentum space is the same as the support of $g^{[ma,nb]}$. In the affine case the sampling of position-momentum space is on a logarithmic scale. There is the possibility, through the choice of m and n, of sampling in smaller and smaller neighborhoods of a fixed point in position space, [C], [D4].)

The affine translates $g_\pm^{(a,b)}$ are called **wavelets** and each of the functions $g_\pm$ is a **mother wavelet**. The map $\mathbf{T} : f \to \langle f, g_\pm^{mn} \rangle$ is the **wavelet transform**.

8.2 Affine frames. The general definitions and analysis of frames presented in Chapter 7 clearly apply to wavelets. However, there is no affine analog of the Weil-Brezin-Zak transform which was so useful for Weyl-Heisenberg frames. Nonetheless we can prove the following result directly.

LEMMA 8.1 [DGM]. *Let $g \in L_2(R)$ such that the support of $\mathcal{F}g$ is contained in the interval $[\ell, L]$ where $0 < \ell < L < \infty$, and let $a_0 > 1, b_0 > 0$ with $(L - \ell)b_0 \leq 1$. Suppose also that*

$$0 < A \leq \sum_m |\mathcal{F}g(a_0^m y)|^2 \leq B < \infty$$

for almost all $y \geq 0$. Then $\{g^{mn}\}$ is a frame for $\mathcal{H}^+$ with frame bounds $A/b_0, B/b_0$.

Proof. The demonstration is analogous to that of Theorem 7.5. Let $f \in \mathcal{H}^+$ and note that $g \in \mathcal{H}^+$. For fixed m the support of $\mathcal{F}f(a_0^m y)\overline{\mathcal{F}g}(y)$ is contained in the interval $\ell \leq y \leq \ell + 1/b_0$ (of length $1/b_0$). Then

$$\sum_{m,n} |\langle f, g^{mn}\rangle|^2 = \sum_{m,n} |\langle \mathcal{F}f, \mathcal{F}g^{mn}\rangle|^2$$

$$= \sum_{m,n} a_0^{-m} |\int_{-0}^{\infty} \mathcal{F}f(a_0^{-m}y)\overline{\mathcal{F}g}(y)e^{-inb_0 y} dy|^2$$

$$= (\text{ Plancherel theorem }) \sum_m \frac{a_0^{-m}}{b_0} \int_{\ell}^{\ell+1/b_0} |\mathcal{F}f(a_0^{-m}y)\mathcal{F}g(y)|^2 dy$$

$$= \frac{1}{b} \sum_m \int_0^{\infty} |\mathcal{F}f(y)\mathcal{F}g(a_0^m y)|^2 dy$$

$$= \frac{1}{b} \int_0^{\infty} |\mathcal{F}f(y)|^2 \left(\sum_m |\mathcal{F}g(a_0^m y)|^2 \right) dy.$$

Since $||f||^2 = \int_0^{\infty} |\mathcal{F}f(y)|^2 dy$ for $f \in \mathcal{H}^+$, the result

$$A||f||^2 \leq \sum_{m,n} |\langle f, g^{mn}\rangle|^2 \leq B||f||^2$$

follows. ∎

A very similar result characterizes a frame for $\mathcal{H}^-$. (Just let y run from $-\infty$ to 0.) Furthermore, if $\{g_+^{mn}\}, \{g_-^{mn}\}$ are frames for $\mathcal{H}^+, \mathcal{H}^-$, respectively, corresponding to lattice parameters a_0, b_0, then $\{g_+^{mn}, g_-^{mn}\}$ is a frame for $L_2(R)$.

Example 1. For lattice parameters $a_0 = 2, b_0 = 1$, choose $g_+ = \chi_{[1,2)}$ and $g_- = \chi_{(-2,-1]}$. Then g_+ generates a tight frame for $\mathcal{H}^+$ with $A = B = 1$ and g_- generates a tight frame for $\mathcal{H}^-$ with $A = B = 1$. Thus $\{g_+^{mn}, g_-^{mn}\}$ is a tight frame for $L_2(R)$. (Indeed, one can verify directly that $\{g_{\pm}^{mn}\}$ is an ON basis for $L_2(R)$.)

Example 2. Let g be the function such that

$$\mathcal{F}g(y) = \frac{1}{\sqrt{\ln a}} \begin{cases} 0 & \text{if } y \leq \ell \\ \sin \frac{\pi}{2} v\left(\frac{y-\ell}{\ell(a-1)}\right) & \text{if } \ell < y \leq a\ell \\ \cos \frac{\pi}{2} v\left(\frac{y-a\ell}{a\ell(a-1)}\right) & \text{if } a\ell < y \leq a^2\ell \\ 0 & \text{if } a^2\ell < y \end{cases}$$

where $v(x)$ is defined as in (7.23). Then $\{g^{mn}\}$ is a tight frame for $\mathcal{H}^+$ with $A = B = \frac{1}{b \ln a}$. Furthermore, if $g_+ = g$ and $g_- = \bar{g}$ then $\{g_{\pm}^{mn}\}$ is a tight frame for $L_2(R)$.

Suppose $g \in L_2(R)$ such that $\mathcal{F}g(y)$ is bounded almost everywhere and has support in the interval $\left[-\frac{1}{2b}, \frac{1}{2b}\right]$. Then for any $f \in L_2(R)$ the function

$$a_0^{-m/2}\mathcal{F}f(a_0^{-m}y)\overline{\mathcal{F}g(y)}$$

has support in this same interval and is square integrable. Thus

$$\sum_{m,n}|\langle f, g_{mn}\rangle|^2 = \sum_{m,n}|a_0^{-m/2}\int_{-\infty}^{\infty}\mathcal{F}f(a_0^{-m}y)\overline{\mathcal{F}g(y)}e^{-2\pi iyb_0}dy|^2$$

$$= b_0^{-1}\sum_{m}\int_{-\infty}^{\infty}a_0^{-m}|\mathcal{F}f(a_0^{-m}y)\mathcal{F}g(y)|^2dy$$

$$= \frac{1}{b_0}\int_{-\infty}^{0}|\mathcal{F}f(y)|^2\sum_{m}|\mathcal{F}g(a_0^m y)|^2dy$$

$$+ \frac{1}{b_0}\int_{0}^{\infty}|\mathcal{F}f(y)|^2\sum_{m}|\mathcal{F}g(a_0^m y)|^2dy.$$

It follows from the computation that if there exist constants $A, B > 0$ such that

$$A \leq \sum_{m}|\mathcal{F}g(a_0^m y)|^2 \leq B$$

for almost all y, then the single mother wavelet g generates an affine frame.

We conclude this section with two examples of wavelets whose properties do not follow directly from the preceding theory. The first is the Haar basis generated by the mother wavelet

$$g(t) = \begin{cases} 0 & \text{if } t < 0 \\ 1 & \text{if } 0 \leq t < \frac{1}{2} \\ -1 & \text{if } \frac{1}{2} \leq t \leq 1 \\ 0 & \text{if } 1 < t, \end{cases}$$

where $a = 2, b = 1$. One can check directly that $\{g^{mn}\}$ is not only a frame, it is an ON basis for $L_2(R)$.

The Haar wavelets have discontinuities. However, Y. Meyer discovered an ON basis for $L_2(R)$ whose mother wavelet g is an infinitely differential function such that $\mathcal{F}g$ has compact support. The lattice is $a = 2, b = 1$. The Meyer wavelet is defined by $\mathcal{F}g(y) = e^{iy/2}\omega(|y|)$ where

$$\omega(|y|) = \begin{cases} 0 & \text{if } y \leq \frac{1}{3} \\ \sin\frac{\pi}{2}v(3y - 1) & \text{if } \frac{1}{3} \leq y \leq \frac{2}{3} \\ \cos\frac{\pi}{2}v\left(\frac{3y}{2} - 1\right) & \text{if } \frac{2}{3} \leq y \leq \frac{4}{3} \\ 0 & \text{if } \frac{4}{3} \leq y \end{cases}$$

and v is defined as in (7.23), except that in addition we require $v(y)+v(1-y) = 1$ for $0 \leq y \leq 1$. One can check that $||g||^2 = 1$ and $\sum_m |\mathcal{F}g(2^m y)|^2 = 1$. Moreover, it can be shown that g generates a tight frame for $L_2(R)$ with frame bounds $A = B = 1$, and, indeed, that $\{g^{mn}\}$ is an ON basis for $L_2(R)$.

A theory which "explains" the orthogonality found in these last two examples is multiresolution analysis [HW], [LM], [D2]; it is beyond the scope of these notes.

8.3 Exercises.

8.1 Suppose $g \in L_2(R)$ with $||g|| = 1$ and g is centered about $(0, k)$ in the position-momentum space. Show that $g^{(a,b)}$ is centered about $(-b, a^{-1}k)$.

8.2 Prove directly that the Haar basis is an ON basis for $L_2(R)$.

8.3 For $g(t) = e^{-t^2}$, show that the functions $g^{(a,b)}$ are dense in $L_2(R)$.

§9. THE SCHRÖDINGER GROUP

9.1 Automorphisms of H_R. We have already seen that the infinite-dimensional irred unitary reps $\mathbf{T}^\lambda$ of the Heisenberg group H_R extend naturally to irred reps of the four-parameter oscillator group and that a study of the oscillator group reps provides insight into the behavior of the H_R reps, §5.7. In fact, the H_R reps extend to reps of the six-parameter Schrödinger group. An understanding of the action of the Schrödinger group provides an explanation for a number of the "deep" transformation properties of objects such as radar ambiguity functions and Jacobi Theta functions.

We start by searching for **automorphisms** of H_R, i.e. one-to-one maps ρ of H_R onto itself such that $\rho(AB) = \rho(A)\rho(B)$ for $A, B \in H_R$. Using the usual coordinate representation

(9.1)
$$A(x,y,z) = \begin{pmatrix} 1 & x & z \\ 0 & 1 & y \\ 0 & 0 & 1 \end{pmatrix}$$

for H_R, so that the group product is

$$A(x,y,z)A(x',y',z') = A(x+x', y+y', z+z'+xy'),$$

we can write

(9.2)
$$\rho(A)(x,y,z) = A(\rho_1(x,y,z), \rho_2(x,y,z), \rho_3(x,y,z))$$

where

(9.3)
$$
\begin{aligned}
a)\ & \rho_1(x+x', y+y', z+z'+xy') = \rho_1(x,y,z) + \rho_1(x',y',z') \\
b)\ & \rho_2(x+x', y+y', z+x'+xy') = \rho_2(x,y,z) + \rho_2(x',y',z') \\
c)\ & \rho_3(x+x', y+y', z+z'+xy') = \rho_3(x,y,z) + \rho_3(x',y',z') \\
& \qquad\qquad\qquad\qquad\qquad\qquad + \rho_1(x,y,z)\rho_2(z',y',z').
\end{aligned}
$$

Under the assumption that the ρ_j are continuously differentiable functions, we shall determine all such automorphisms.

Before proceeding with this task, let us see why it could be relevant to radar and sonar. The radar cross-ambiguity function takes the form $F(x,y,z) = \langle \mathbf{T}^1[x,y,z]f, g\rangle$ (up to a harmless exponential factor arising from the z coordinate) where $f, g \in L_2(R)$ and $\mathbf{T}^1[x,y,z] = \mathbf{T}^1(A(x,y,z))$ is the irred rep (5.5) of H_R. If ρ is an automorphism of H_R then $\mathbf{T}^1(\rho(A)(x,y,z)) = \mathbf{T}^1_\rho[x,y,z]$ also defines an irred unitary rep of H_R on $L_2(R)$. (A rep since ρ preserves the group multiplication property and irreducible since the operators $\mathbf{T}^1_\rho$ are just a reordering of the operators $\mathbf{T}^1$.) Thus $\mathbf{T}^1_\rho$ is equivalent, hence unitary equivalent, to one of the standard unitary irred reps $\mathbf{T}^\lambda$ of H_R that we have already studied. Suppose this rep is $\mathbf{T}^1$ itself, i.e., suppose there is a unitary operator $\mathbf{U}$ such that $\mathbf{T}^1_\rho[x,y,z] = \mathbf{U}^{-1}\mathbf{T}^1[x,y,z]\mathbf{U}$. Then $\langle \mathbf{T}^1(\rho(A))f, g\rangle = \langle \mathbf{T}^1_\rho[x,y,z]f, g\rangle = \langle \mathbf{U}^{-1}\mathbf{T}^1[x,y,z]\mathbf{U}f, g\rangle = \langle \mathbf{T}^1[x,y,z]\mathbf{U}f, \mathbf{U}g\rangle =$

$F'(x, y, z)$, which is again an ambiguity function. Thus if $F(x, y, z)$ is an ambiguity function, then so is $F(\rho_1(\mathbf{x}), \rho_2(\mathbf{x}), \rho_3(\mathbf{x}))$.

Now we compute the possible automorphisms ρ. Differentiating (9.3a) with respect to x' we find

$$\partial_1 \rho_1(x + x', y + y', z + z' + xy') = \partial_1 \rho_1(x', y', z').$$

Since the right-hand side of this equation is independent of x, y, z, we have $\partial_1 \rho_1(x, y, z)$ = α, a constant. Similarly, differentiating (9.3a) with respect to y we have $\partial_2 \rho_1(x, y, z) = \beta$. Differentiating (9.3a) with respect to x yields

$$\alpha + y' \partial_3 \rho_1(x + x', y + y', z + z' + xy') = \alpha.$$

Since this equation holds for general x', y', z', it follows that $\partial_3 \rho_1(x, y, z) = 0$. Thus $\rho_1(x, y, z) = \alpha x + \beta y + k$ where k is a constant. Substituting this expression back into (9.3a) we see that $k = 0$. Similarly, equation (9.3b) has only the solution $\rho_2(x, y, z) = \gamma x + \delta y$ where γ, δ are constants. The computation for equation (9.3c) is just as straightforward, although the details are a bit more complicated. The final result is

(9.4)
$$\rho_1(x, y, z) = \alpha x + \beta y, \quad \rho_2(x, y, z) = \gamma x + \delta y,$$
$$\rho_3(x, y, z) = ax + by + \frac{1}{2}(\alpha x + \beta y)(\gamma x + \delta y) + (\alpha \delta - \beta \gamma)(z - \frac{1}{2}xy).$$

Here, $\alpha, \beta, \gamma, \delta, a, b$ are real constants such that $\alpha \delta - \beta \gamma \neq 0$, so that ρ is 1-1.

Some of the automorphisms of H_R are **inner automorphisms**. These are the automorphisms of the form $\rho_B(A) = B^{-1}AB$ for $A \in H_R$, where B is a fixed member of H_R. (Clearly, ρ_B maps H_R onto itself and is one-to-one. Furthermore $\rho_B(A_1 A_2) = B^{-1} A_1 A_2 B = (B^{-1} A_1 B)(B^{-1} A_2 B) = \rho_B(A_1) \rho_B(A_2)$, so ρ_B is a group homomorphism.) If $B = B(a', b', c')$ then

$$\rho_B(A) = B^{-1} A(x, y, z) B = \begin{pmatrix} 1, & x, & z - a'y + b'x \\ 0, & 1, & y \\ 0, & 0, & 1 \end{pmatrix}$$
$$= A(x, y, z - a'y + b'x)$$

so the transformations

(9.5)
$$\rho_1(x, y, z) = z, \quad \rho_2(x, y, z) = y$$
$$\rho_3(x, y, z) = b'x - a'y + z$$

correspond to inner automorphisms. We are not very interested in inner automorphisms because they can easily be understood in terms of the Heisenberg group itself. Thus we set $a = b = 0$ in (9.4) and concentrate on the **outer automorphisms**

(9.6)
$$\rho_1(x, y, z) = \alpha x + \beta y, \quad \rho_2(x, y, z) = \gamma x + \delta y,$$
$$\rho_3(x, y, z) = \frac{1}{2}(\alpha x + \beta y)(\gamma x + \delta y) - (\alpha \delta + \beta \gamma)(z - \frac{1}{2}xy).$$

Recall that the infinite-dimensional irred unitary reps $\mathbf{T}^\lambda$ of H_R take the form, (5.5),

$$\mathbf{T}^\lambda[x, y, z]\mathbf{f}(t) = e^{2\pi i\lambda(z+ty)}\mathbf{f}(t+x)$$

for $\mathbf{f} \in L_2(R)$, where λ is a nonzero real constant. Note that the operators corresponding to the center C of H_R are just multiples of the identity operator: $\mathbf{T}^\lambda[0, 0, z] = e^{2\pi i\lambda z}\mathbf{E}$. Since $\mathbf{T}^\lambda_\rho[0, 0, z] = e^{2\pi i\lambda(\alpha\delta-\beta\gamma)z}\mathbf{E}$ the irred rep $\mathbf{T}^\lambda_\rho$ can possibly be equivalent to $\mathbf{T}^\lambda$ only if $\alpha\delta - \beta\gamma = 1$, so we now restrict our attention to this case.

With this restriction $\mathbf{T}^\lambda_\rho$ must be equivalent to $\mathbf{T}^\lambda$. Indeed, the reps $\mathbf{T}^\lambda, \mathbf{T}^\lambda_\rho$ coincide on the center C. Furthermore, the matrix elements $T_{\rho,jk}[x, y, 0]$ are square integrable with respect to the measure $dx dy$ in the plane. (In fact, these matrix elements differ from $T^\lambda_{jk}[x, y, 0]$ only by a factor of absolute value 1 and a change of variables with Jacobian $\alpha\delta - \beta\gamma = 1$.) Thus, if $\mathbf{T}^\lambda_\rho$ is not equivalent to $\mathbf{T}^\lambda$ we can use Corollary 3.1 and repeat the arguments leading to (5.8) for $\mathbf{T}^{(\mu)} \equiv \mathbf{T}^\lambda, \mathbf{T}^{(\nu)} \equiv \mathbf{T}^\lambda_\rho, \mu \neq \nu$, and measure $dx dy$ to obtain

$$\int_{-\infty}^{\infty}\int_{-\infty}^{\infty} T^\lambda_{\rho,j\ell}[x, y, 0]\overline{T^\lambda_{sk}[x, y, 0]}dx dy = 0$$

for all j, ℓ, s, k. Thus the matrix elements of $\mathbf{T}^\lambda_\rho$ are orthogonal to those of $\mathbf{T}^\lambda$ in $L_2(R^2)$. However, as we have shown in §5.6, the matrix elements $T^\lambda_{jk}(x, y, 0)$ form a basis for $L_2(R^2)$. This contradiction proves that $\mathbf{T}^\lambda_\rho \cong \mathbf{T}^\lambda$, hence that there exist unitary operators $\mathbf{U}(x, y)$ such that

$$\begin{aligned}
(9.7) \quad \mathbf{T}^\lambda_\rho[x, y, z] &= \mathbf{U}_D^{-1}\mathbf{T}^\lambda[x, y, z]\mathbf{U}_D \\
&= \mathbf{T}^\lambda\left[\alpha x + \beta y, \gamma x + \delta y, \frac{1}{2}(\alpha x + \beta y)(\gamma x + \delta y) + z - \frac{1}{2}xy\right] \\
&= \exp\left[\pi i\lambda\left((\alpha x + \beta y)(\gamma x + \delta y) - xy\right)\right]\mathbf{T}^\lambda[\alpha x + \beta y, \gamma x + \delta y, z]
\end{aligned}$$

where $D = \begin{pmatrix} \alpha & \beta \\ \gamma & \delta \end{pmatrix}$ and $\det D = \alpha\delta - \beta\gamma = 1$.

THEOREM 9.1. *Suppose* $F^\lambda(x, y, z)$ *is a matrix element of the irred unitary rep* $\mathbf{T}^\lambda$ *of* H_R. *Then*

$$\begin{aligned}
(9.8) \quad G^\lambda_D(x, y, z) &= \exp\left[\pi i\lambda\left((\alpha x + \beta y)(\gamma x + \delta y) - xy\right)\right] \times \\
&\quad F^\lambda(\alpha x + \beta y, \gamma x + \delta y, z)
\end{aligned}$$

is also a matrix element of $\mathbf{T}^\lambda$ *for any* $D = \begin{pmatrix} \alpha & \beta \\ \gamma & \delta \end{pmatrix}$ *with* $\det D = \alpha\delta - \beta\gamma = 1$.

Proof. Suppose $F^\lambda(x, y, z) = \langle\mathbf{T}^\lambda[x, y, z]\mathbf{f}_1, \mathbf{f}_2\rangle$ for $\mathbf{f}_1, \mathbf{f}_2 \in L_2(R)$. Setting $\mathbf{g}_j = \mathbf{U}\mathbf{f}_j, j = 1, 2$, we obtain (9.8) from (9.7) with $G^\lambda(x, y, z) = \langle\mathbf{T}^\lambda[x, y, z]\mathbf{g}_1, \mathbf{g}_2\rangle$. □

Note that (9.8) gives us information about the structure of the set of ambiguity and of cross-ambiguity functions.

9.2 The metaplectic representation. Next we turn to the problem of actually computing the operators $\mathbf{U}_D$. First of all, note from (9.7) that for any phase factor $e^{i\varphi(D)}$ (with $|e^{i\varphi}| = 1$) the unitary operators $\mathbf{U}'_D = e^{i\varphi}\mathbf{U}_D$ also satisfy (9.7). Indeed, a simple argument using Theorem 5.1 shows that the operators $\mathbf{U}_D$ are uniquely determined up to a phase factor. We shall find that it is possible to choose the operators $\mathbf{U}_D$ such that the mapping $\mathbf{f} \to \mathbf{U}_D\mathbf{f}$ is continuous in the norm as a function of the local parameters $\alpha, \beta, \gamma, \delta$ for every $\mathbf{f} \in L_2(R)$.

It is no accident that we have arranged the parameters $\alpha, \beta, \gamma, \delta$ in the form of the matrix $D \in SL(2, R)$, since $\det D = 1$. Indeed, it is straightforward to check that if

$$\mathbf{T}^\lambda_\rho[\mathbf{x}] = \mathbf{U}_D^{-1}\mathbf{T}^\lambda[\mathbf{x}]\mathbf{U}_D, \quad \mathbf{T}^\lambda_{\rho'}[\mathbf{x}] = \mathbf{U}_{D'}^{-1}T^\lambda[\mathbf{x}]\mathbf{U}_{D'}$$

for automorphisms ρ and ρ' of H_R, then the automorphism $\rho\rho' : \mathbf{x} \to \rho(\rho'(\mathbf{x}))$ corresponds to the matrix $DD' \in SL(2, R)$, (matrix product). However,

$$\begin{aligned}
\mathbf{T}^\lambda_{\rho\rho'}[\mathbf{x}] = \mathbf{T}^\lambda[\rho(\rho'(\mathbf{x}))] = \mathbf{T}^\lambda_\rho[\rho'(\mathbf{x})] &= \mathbf{U}_D^{-1}\mathbf{T}^\lambda[\rho'(\mathbf{x})]\mathbf{U}_D \\
&= \mathbf{U}_D^{-1}\mathbf{T}^\lambda_{\rho'}[\mathbf{x}]\mathbf{U}_D = \mathbf{U}_D^{-1}U_{D'}^{-1}\mathbf{T}^\lambda[\mathbf{x}]\mathbf{U}_{D'}\mathbf{U}_D \\
&= [\mathbf{U}_{D'}\mathbf{U}_D]^{-1}\mathbf{T}^\lambda[\mathbf{x}](\mathbf{U}_{D'}\mathbf{U}_D),
\end{aligned}$$

so

(9.9) $$\mathbf{U}_{DD'} = e^{e\psi(D',D)}\mathbf{U}_{D'}\mathbf{U}_D$$

for some phase factor $e^{i\psi(D',D)}$. (Note the reversal of order in (9.9).) It follows from (9.9) that the operators $\mathbf{U}_D$ determine a **projective representation** of $SL(2, R)$, i.e., a rep up to a phase factor.

It is easy to verify the operator identity

(9.10) $$\mathbf{R}(a)^{-1}\mathbf{T}^\lambda[x, y, z]\mathbf{R}(a) = \mathbf{T}^\lambda_{D_1}(a)[x, y, z]$$

where

$$D_1(a) = \begin{pmatrix} 1 & 0 \\ a & 1 \end{pmatrix}$$

and $\mathbf{R}(a)\mathbf{f}(t) = e^{i\pi\lambda at^2}\mathbf{f}(t)$.

Furthermore, defining the unitary operator $\mathbf{V}(b)$ by

$$\mathbf{V}(b)\mathbf{f}(t) = b^{1/2}\mathbf{f}(t), \quad \mathbf{f} \in L_2(R), \quad b > 0,$$

we find

(9.11) $$\mathbf{V}^{-1}(b)\mathbf{T}^\lambda[x, y, z]\mathbf{V}(b) = \mathbf{T}^\lambda[by, b^{-1}y, z] = \mathbf{T}^\lambda_{D_2(b)}[\mathbf{x}]$$

where

$$D_2(b) = \begin{pmatrix} b & 0 \\ 0 & b^{-1} \end{pmatrix}.$$

Matrices of the form $D_1(a)$, $D_2(b)$ generate a two-dimensional subgroup of $SL(2, R)$. To generate the full group we need a third one-parameter subgroup. We have already derived such operators, the $\mathbf{U}(\alpha)$ in (5.43). However, from the form (5.43) it is not easy to verify relations (9.7). It is much easier to use the unitary transformation $\mathbf{A} : L_2(R) \to \mathcal{F}$ to realize the rep $\mathbf{T}^\lambda$ on the Bargmann-Segal Hilbert space $\mathcal{F}$. Recall that $\mathbf{U}'(\alpha) = \mathbf{A}\mathbf{U}(\alpha)\mathbf{A}^{-1}$ takes the form (5.41):

$$\mathbf{U}'(\alpha)\mathbf{f}(w) = \mathbf{f}(e^{i\alpha}w), \quad \mathbf{f} \in \mathcal{F}.$$

The action of $\mathbf{T}^\lambda$ on $\mathcal{F}$ is given by (5.35):

$$\mathbf{T}'^\lambda(\mathbf{x})\mathbf{f}(w) = \exp\left[-\frac{1}{4}(x^2 + 4\pi^2\lambda^2 y^2) + 2^{-1/2}(x - 2\pi\lambda iy)w\right.$$
$$\left. -\pi\lambda ixy + 2\pi\lambda iz\right]\mathbf{f}(w - 2^{-1/2}[x + 2\pi\lambda iy]).$$

Now it is easy to verify the identity

$$\mathbf{U}'(-\alpha)\mathbf{T}'^\lambda[x, y, z]\mathbf{U}'(\alpha)$$

(9.12)
$$= \mathbf{T}'^\lambda\left[x\cos\alpha + 2\pi\lambda y\sin\alpha, \frac{-x}{2\pi\lambda}\sin\alpha + y\cos\alpha,\right.$$
$$\left. z + \frac{1}{2}(x\cos\alpha + 2\pi\lambda y\sin\alpha)\left(\frac{-x}{2\pi\lambda}\sin\alpha + y\cos\alpha\right) - \frac{1}{2}xy\right]$$
$$= \mathbf{T}'^\lambda_{D_3(\lambda,\alpha)}[\mathbf{x}]$$

where

$$D_3(\lambda, \alpha) = \begin{pmatrix} \cos\alpha & 2\pi\lambda\sin\alpha \\ -\sin\alpha/2\pi\lambda & \cos\alpha \end{pmatrix}.$$

Transforming back to $L_2(R)$ we see that the operators $\mathbf{U}(\alpha)$ in (5.43) must satisfy

$$\mathbf{U}(\alpha)^{-1}\mathbf{T}^\lambda[\mathbf{x}]\mathbf{U}(\alpha) = \mathbf{T}^\lambda_{D_3(\lambda,\alpha)}[\mathbf{x}].$$

Since

$$\begin{pmatrix} b & 0 \\ 0 & b^{-1} \end{pmatrix}\begin{pmatrix} \cos\alpha & \sin\alpha \\ -\sin\alpha & \cos\alpha \end{pmatrix}\begin{pmatrix} b^{-1} & 0 \\ 0 & b \end{pmatrix} = \begin{pmatrix} \cos\alpha & b^2\sin\alpha \\ -b^{-2}\sin\alpha & \cos\alpha \end{pmatrix},$$

setting $b = \sqrt{2\pi\lambda}$ in the case where $\lambda > 0$ we find the operator

$$\mathbf{W}(\alpha) = \mathbf{V}\left(\sqrt{2\pi\lambda}\right)\mathbf{U}(\alpha)\mathbf{V}\left(\frac{1}{\sqrt{2\pi\lambda}}\right)$$

or

(9.13)
$$\mathbf{W}(\alpha)\mathbf{f}(t) = \sqrt{\frac{\lambda}{2}}\lim_{n\to\infty}\int_{-n}^{n}\frac{e^{-i\epsilon\left(\frac{\pi}{4} - \frac{\theta}{2}\right)}}{(|\sin\alpha|)^{1/2}}\exp\left[\pi\lambda i(\cot\alpha)\left(\frac{t^2 + \tau^2}{2}\right) - \frac{\pi\lambda it\tau}{\sin\alpha}\right]\mathbf{f}(\tau)d\tau,$$

where $\alpha = 2k\pi + \epsilon\beta$, k an integer, $\epsilon = \pm 1$, $0 < \beta < \pi$ and $\mathbf{f} \in L_2(R)$. Here $\mathbf{W}(\alpha)$ satisfies

$$\mathbf{W}(-\alpha)\mathbf{T}^\lambda[\mathbf{x}]\mathbf{W}(\alpha) = \mathbf{T}^\lambda_{D_3(\alpha)}[\mathbf{x}]$$

where

$$D_3(\alpha) = \begin{pmatrix} \cos\alpha & \sin\alpha \\ -\sin\alpha & \cos\alpha \end{pmatrix}.$$

NOTE: The operators $e^{2\pi\lambda iat^2}$, $\mathbf{V}(b)$ and $\mathbf{W}(\alpha)$ do not generate a rep of $SL(2, R)$ but the first two types of operators and the operators $\mathbf{W}'(\alpha) = \mathbf{W}(\alpha)e^{-i\alpha/2}$ do generate a rep of a two-fold covering group $\widetilde{SL(2, R)}$ of $SL(2, R)$. This rep is called the **metaplectic representation**. See [M6] and [S2] for more details. For $2\pi\lambda = 1$ the rep $\mathbf{T}^\lambda$ of H_R together with the metaplectic rep of $\widetilde{SL(2, R)}$ extends uniquely to an irred unitary rep of the 6-parameter **Schrödinger group**, the semi-direct product of H_R and $\widetilde{SL(2, R)}$. The Schrödinger group is the symmetry group of the time-dependent Schrödinger equations for each of the free-particle, the harmonic oscillator and the linear potential in two-dimensional space time. See [M6] for a detailed analysis.

The formula

$$\begin{pmatrix} 1 & 0 \\ \tan\theta & 1 \end{pmatrix} \begin{pmatrix} \cos\theta & \sin\theta \\ -\sin\theta & \cos\theta \end{pmatrix} \begin{pmatrix} 1/\cos\theta & 0 \\ 0 & \cos\theta \end{pmatrix} = \begin{pmatrix} 1 & \cos\theta\sin\theta \\ 0 & 1 \end{pmatrix}$$

shows that the unitary operator $\mathbf{Z}(\tau)$ corresponding to the matrix $D = \begin{pmatrix} 1 & \tau \\ 0 & 1 \end{pmatrix}$ can be defined (unique to within a phase factor) by

$$\mathbf{Z}(\tau)\mathbf{f}(x) = \mathbf{V}\left(\frac{1}{\cos\theta}\right)\mathbf{W}'(\theta)\mathbf{R}(\tan\theta)\mathbf{f}(x)$$

$$= \sqrt{\frac{\lambda}{i\tau}} \lim_{n\to\infty} \int_{-n}^{n} \exp\left[-\frac{\pi\lambda(t-y)^2}{2i\tau}\right] \mathbf{f}(y)dy$$

where $\tau = \sin\theta\cos\theta$. Clearly this operator is well defined and unitary for $|\tau| \leq 1$, since it is a product of unitary operators. Indeed $\mathbf{Z}(\tau)$ is well defined and unitary for all real τ. To show this we use the fact that the family of all functions of the form $\mathbf{f}(x) = e^{-b(x-a)^2}$, for $b > 0$ and a real, spans $L_2(R)$, i.e., the set of all dilations and translations of e^{-x^2} spans $L_2(R)$. (See [K1, page 494] and Exercise 8.3.) Now the integral

$$(9.14) \qquad \mathbf{Z}(\tau)\mathbf{f}(x) = \sqrt{\frac{\lambda}{2i\tau}} \int_{-\infty}^{\infty} e^{-\pi\lambda(x-y)^2/2i\tau}\mathbf{f}(y)dy$$

is well-defined for all τ and agrees with the preceding integral for $|\tau| \leq 1$. An explicit evaluation of the integral yields

$$(9.14') \qquad \mathbf{Z}(\tau)\mathbf{f}(x) = \frac{1}{\sqrt{1 + \frac{2ib\tau}{\pi\lambda}}} e^{-b(x-a)^2/(1+\frac{2ib\tau}{\pi\lambda})},$$

so

$$\langle \mathbf{Z}(\tau)\mathbf{f}_1, \mathbf{Z}(\tau)\mathbf{f}_2 \rangle = \langle \mathbf{f}_1, \mathbf{f}_2 \rangle = \sqrt{\frac{\pi}{b_1 + b_2}} e^{-b_1 b_2 (a_1 - a_2)^2 / (b_1 + b_2)}.$$

(Here the parameters a_i, b_i correspond to the functions f_i.) Furthermore,

$$\mathbf{Z}(\tau_1)[\mathbf{Z}(\tau_2)f](x) = \left[1 + \frac{2ib(\tau_1 + \tau_2)}{\pi\lambda}\right]^{-1/2} \exp\left[-b(x-a)^2 / \left(1 + \frac{2ib(\tau_1 + \tau_2)}{\pi\lambda}\right)\right]$$
$$= \mathbf{Z}(\tau_1 + \tau_2)f(x),$$

so

$$\mathbf{Z}(\tau_1)\mathbf{Z}(\tau_2) = \mathbf{Z}(\tau_1 + \tau_2)$$

for all τ_1, τ_2. Since $\mathbf{Z}(\tau)$ is unitary for $|\tau| \leq 1$ it follows easily that $\mathbf{Z}(\tau)$ is unitary for all τ. Note also that $\mathbf{Z}(0) = \mathbf{E}$.

By explicit differentiation in (9.14′) we see that for $\mathbf{g}(x, \tau) = \mathbf{Z}(\tau)\mathbf{f}(x)$, $\partial_\tau \mathbf{g} = \frac{i}{2\pi\lambda}\partial_{xx}\mathbf{g}$, $\mathbf{g}(x, 0) = \mathbf{f}(x)$. Thus, $\mathbf{Z}(\tau)\mathbf{f}(x)$ gives the unique solution of the Cauchy problem for the time dependent free particle Schrödinger equation. In particular

$$\mathbf{Z}(\tau)\mathbf{f}(x) = e^{\frac{i\tau}{2\pi\lambda}\partial_{xx}}\mathbf{f}(x)$$

where $e^{i\tau\mathbf{H}}$ is the unitary operator generated by the self-adjoint operator $\mathbf{H}$ via the spectral theorem, [K1]. In [M6] it is shown that the Schrödinger group acts as the symmetry group of the time dependent Schrödinger equation, i.e., it maps solutions into solutions of this equation, and that the possible solutions which are obtainable by separation of variables can be characterized by the group action. Similarly, the operator $\mathbf{W}'(\tau)$ satisfies the equation

$$\partial_\tau \mathbf{g} = \frac{i}{2\pi\lambda}\left(\partial_{xx} + \frac{x^2}{4}\right)\mathbf{g}, \quad \mathbf{g}(x, 0) = \mathbf{f}(x)$$

where $\mathbf{g}(x, \tau) = \mathbf{W}'(\tau)\mathbf{f}(x)$. Thus, $\mathbf{g}$ is the unique solution of the Cauchy problem for the time dependent Schrödinger equation equation with a harmonic oscillator potential, [M6]. Such considerations are beyond the scope of these notes.

9.3 Theta functions and the lattice Hilbert space. For another application of the use of the metaplectic formula (9.6) let us reconsider our construction of the lattice representation of H_R. According to (5.45) this rep is defined on functions $f[\mathbf{x}] = f(A(\mathbf{x}))$ on H_R such that

(9.15) $\qquad \mathbf{f}(a_1 + x_1, a_2 + x_2, y_3 + x_3 + a_1 x_2) = e^{2\pi i y_3}\mathbf{f}(x_1, x_2, x_3)$

where a_1, a_2 are integers. For ρ an automorphism (9.6) of H_R, with $\alpha\delta - \beta\gamma = 1$, it is natural to look for the conditions such that $\mathbf{f}_\rho(\mathbf{x}) \equiv \mathbf{f}(\rho(\mathbf{x}))$ belongs to the lattice Hilbert space for every $\mathbf{f}$ belonging to this Hilbert space. If ρ corresponds to the matrix

$$\begin{pmatrix} \alpha & \beta \\ \gamma & \delta \end{pmatrix}, \quad \alpha\delta - \beta\gamma = 1,$$

the conditions are

(9.16)

$$\mathbf{f}\Big(\alpha x_1 + \beta x_2 + \alpha a_1 + \beta a_2, \ \gamma x_1 + \delta x_2 + \gamma a_1 + \delta a_2,$$

$$y_3 + x_3 + a_1 x_2 + \frac{1}{2}(\alpha x_1 + \beta x_2 + \alpha a_1 + \beta a_2)(\gamma x_1 + \delta x_2 + \gamma a_1 + \delta a_2)$$

$$-\frac{1}{2}(a_1 + x_1)(a_2 + x_2)\Big)$$

$$= e^{2\pi i y_3}\mathbf{f}\left(\alpha x_1 + \beta x_2, \gamma x_1 + \delta x_2, x_3 + \frac{1}{2}(\alpha x_1 + \beta x_2)(\gamma x_1 + \delta x_2) - \frac{1}{2}x_1 x_2\right).$$

Since $\mathbf{f}$ satisfies only (9.15) we see that $\alpha, \beta, \gamma, \delta$ must be integers. Then (9.15) implies

(9.17)
$$\mathbf{f}([\alpha x_1 + \beta x_2] + [\alpha a_1 + \beta a_2], \ [\gamma x_1 + \delta x_2] + [\gamma a_1 + \delta a_2], \ \tilde{y}_3 + \tilde{x}_3$$
$$+ [\alpha a_1 + \beta a_2][\gamma x_1 + \delta x_2]) = e^{2\pi i \tilde{y}_3}\mathbf{f}(\alpha x_1 + \beta x_2, \gamma x_1 + \delta x_2, \tilde{x}_3).$$

Setting $\tilde{x}_3 = x_3 + \frac{1}{2}[\alpha x_1 + \beta x_2][\gamma x_1 + \delta x_2] - \frac{1}{2}x_1 x_2, \tilde{y}_3 = x_3 + \frac{1}{2}(\alpha a_1 + \beta a_2)(\gamma a_1 + \delta a_2) - \frac{1}{2}a_1 a_2$ in (9.17), we recover (9.16) provided $(\alpha a_1 + \beta a_2)(\gamma a_1 + \delta a_2) - a_1 a_2 = \alpha\gamma a_1^2 + 2\beta\gamma a_1 a_2 + \beta\delta a_2^2$ is an even integer for all integers a_1, a_2. This will be the case if and only if

(9.18)
$$\alpha\gamma \equiv \beta\delta \equiv 0 \mod 2,$$

i.e., $\alpha\gamma$ and $\beta\delta$ must be even integers.

Thus we see that if $D = \begin{pmatrix} \alpha & \beta \\ \gamma & \delta \end{pmatrix} \in SL(2, Z)$, i.e., if $D \in SL(2, R)$ and the matrix elements of D are integers, and if conditions (9.18) are satisfied, then $\mathbf{f}_{\rho_D}$ belongs to the lattice Hilbert space whenever $\mathbf{f}$ so belongs. Reduced to the space of functions $\varphi(x_1, x_2)$ where $\mathbf{f}(x_1, x_2, x_3) = \varphi(x_1, x_2)e^{2\pi i x_3}$, so that

(9.19)
$$\varphi(x_1 + a_1, x_2 + a_2) = e^{-2\pi i a_1 x_2}\varphi(x_1, x_2)$$

for $a_1, a_2 \in Z$, the action is

(9.20)
$$\varphi_{\rho_D}(x_1, x_2) = \varphi(\alpha x_1 + \beta x_2, \gamma x_1 + \delta x_2)e^{\pi i[(\alpha x_1 + \beta x_2)(\gamma x_1 + \delta x_2) - x_1 x_2]}.$$

The action of $SL(2, Z)$ on the lattice Hilbert space will lead us to a number of interesting transformation formulas for Theta functions.

As we showed earlier, (7.10), the ground state wave function $\psi_0(t) = \pi^{-1/4}e^{-x^2/2} \in L_2(R)$ is mapped by the Weil-Brezin-Zak transform to

(9.21)
$$\mathbf{P}\psi(x_1, x_2) = \pi^{-1/4}e^{-x_1^2/2}\theta_3(x_2 + \frac{ix_1}{2\pi} \mid \frac{i}{2\pi})$$

where θ_3 is the Jacobi theta function, [EMOT1], [WW],

(9.22)
$$\theta_3(z \mid \tau) = \sum_{n=-\infty}^{\infty} \exp[\pi i \tau n^2 + 2\pi i n z].$$

Here for τ such that $\text{Im}\tau > 0$, θ_3 is an entire function of z. Moreover, the function $\Theta^\tau(t) = e^{\pi i r t^2} \in L_2(R)$, with $\text{Im}\tau > 0$ is mapped to

(9.23) $$\hat{\Theta}^\tau(x_1, x_2) = \mathbf{P}\Theta^\tau(x_1, x_2) = e^{\pi i \tau x_1^2}\theta_3(x_1\tau + x_2 \mid \tau)$$

in the lattice Hilbert space. An elementary complex variable argument [WW] shows that $\hat{\Theta}^\tau(x_1, x_2)$ vanishes precisely once in the square $0 \le x_1 < 1$, $0 \le x_2 < 1$, with a simple zero at the point $(\frac{1}{2}, \frac{1}{2})$. Thus by Theorem 7.1 the functions $\hat{\Theta}^\tau(x_1, x_2)e^{2\pi i(m_1 x_1 + m_2 x_2)}$, $m_1, m_2 \in Z$, span the lattice Hilbert space. Another way to state this is to say that every element in the lattice Hilbert space can be written in the form $\hat{\Theta}^\tau(x_1, x_2)h(x_1, x_2)$ where h is a periodic function in x_1 and x_2 : $h(x_1 + a_1, x_2 + a_2) = h(x_1, x_2)$ for integers a_1, a_2. Since $\hat{\Theta}^\tau$ belongs to the lattice Hilbert space, so does $\hat{\Theta}^\tau_{\rho_D}$ where $D \in SL(2, Z)$, and satisfies (9.18). Thus

(9.24) $$\hat{\Theta}^\tau_{\rho_D}(\mathbf{x}) = \hat{\Theta}^{\tau'}(\mathbf{x})h^{\tau'}(\mathbf{x})$$

for some periodic function $h^{\tau'}$. Expression (9.24) describes the framework for a family of transformation formulas obeyed by the Theta functions. Note that τ' need not be the same as τ. In the derivations to follow we will choose τ' so that the expressions for $h^{\tau'}$ are as simple as possible.

As a nontrivial example we take the case $D = \begin{pmatrix} 0 & 1 \\ -1 & 0 \end{pmatrix}$, see Exercise 9.3. The result is

$$e^{-2\pi i x_1 x_2}e^{\pi i \tau x_2^2}\theta_3(x_2\tau - x_1 \mid \tau) = e^{-i\frac{\pi x_1^2}{\tau}}\sqrt{\frac{i}{\tau}}\,\theta_3\left(-\frac{x_1}{\tau} + x_2 \mid -\frac{1}{\tau}\right).$$

(We have chosen $\tau' = -1/\tau$.) This is equivalent to the transformation formula

(9.25) $$\theta_3(z \mid \tau) = \sqrt{\frac{i}{\tau}}e^{-\pi i z^2/\tau}\theta_3\left(\frac{z}{\tau} \mid \frac{-1}{\tau}\right).$$

As a second example we take $D = \begin{pmatrix} 1 & 0 \\ 2 & 1 \end{pmatrix}$. Then

$$\begin{aligned}
\hat{\Theta}^\tau_{\rho_D}(\mathbf{x}) &= e^{2\pi i x_1^2}e^{\pi i \tau x_1^2}\theta_3(x_1\tau + 2x_1 + x_2 \mid \tau) \\
&= e^{2\pi i x_1^2}e^{\pi i \tau x_1^2}\sum_n e^{i\pi\tau n^2 + 2\pi i n(x_1\tau + 2x_1 + x_2)} \\
&= e^{2\pi i x_1^2}\sum_n e^{i\pi\tau(n+x_1)^2 + 2\pi i n(2x_1 + 1/2)} = \sum_n e^{i\pi(\tau+2)(n+x_1)^2}e^{2\pi i n x_2} \\
&= e^{i\pi\tau' x_1^2}\theta_3(\tau' x_1 + x_2 \mid \tau'), \quad \tau' = \tau + 2.
\end{aligned}$$

Thus,

(9.26) $$\theta_3(z \mid \tau) = \theta_3(z \mid \tau + 2).$$

Even in the cases where the parity conditions (9.18) don't hold, we get useful information. For example, consider the case $D = \begin{pmatrix} 1 & 0 \\ 1 & 1 \end{pmatrix}$. With this automorphism we are replacing the function $\varphi(x_1, x_2)$

$$(9.27) \qquad \varphi(x_1, x_2 + 1) = \varphi(x_1, x_2), \quad \varphi(x_1 + 1, x_2) = e^{-2\pi i x_2} \varphi(x_1, x_2)$$

in the lattice Hilbert space by the function

$$\eta(x_1, x_2) = \varphi(x_1, x_1 + x_2) e^{\pi i x_1^2}.$$

Now it is easy to check that

$$(9.28) \qquad \eta(x_1, x_2 + 1) = \eta(x_1, x_2), \quad \eta(x_1 + 1, x_2) = -e^{-2\pi i x_2} \eta(x_1, x_2),$$

so η doesn't belong to the lattice Hilbert space. However it is straightforward to show that $\tilde{\Theta}^\tau(x_1, x_2) = \hat{\Theta}^\tau \left(x_1, x_2 + \frac{1}{2}\right)$ transforms according to (9.28). It follows from this remark that any square integrable (on the unit square) function η satisfying (9.28) can be written in the form $\hat{\Theta}^{\tau'} \left(x_1, x_2 + \frac{1}{2}\right) h^{\tau'}(x_1, x_2)$ where $h^{\tau'}$ is periodic in x_1, x_2. Indeed we find

$$\hat{\Theta}^\tau_{\rho_D}(\mathbf{x}) = e^{\pi i x_1^2} e^{\pi i \tau x_1^2} \theta_3(x_1 \tau + x_1 + x_2 \mid \tau) = e^{\pi i x_1^2} e^{\pi i \tau x_1^2} \sum_2 e^{i\pi \tau n^2 + 2\pi i n (x_1 \tau + x_1 + x_2)}$$

$$= e^{\pi i x_1^2} \sum_n e^{i\pi \tau (n+x_1)^2 + 2\pi i n (x_1 + x_2)} = \sum_n e^{i\pi (\tau+1)(n+x_1)^2} e^{2\pi i n \left(x_2 + \frac{1}{2}\right)}$$

$$= e^{i\pi \tau' x_1^2} \theta_3 \left(x_1 \tau' + x_2 + \frac{1}{2} \mid \tau'\right), \quad \tau' = \tau + 1.$$

(Here we have used the fact that $e^{-i\pi n^2} = e^{i\pi n}$ for any integer n.) Thus we have the transformation formula

$$(9.29) \qquad \theta_3(z \mid \tau) = \theta_3 \left(z + \frac{1}{2} \mid \tau + 1\right).$$

Note that by using (9.29) twice we get (9.26), in accordance with the fact that $\begin{pmatrix} 1 & 0 \\ 1 & 1 \end{pmatrix}^2 = \begin{pmatrix} 1 & 0 \\ 2 & 0 \end{pmatrix}$. Note: The other three basic Jacobi Theta functions θ_1, θ_2, and θ_4 (or θ_0) can easily be expressed in terms of θ_3, [EMOT1], [WW].

Since the modular group elements $\begin{pmatrix} 1 & 0 \\ 1 & 0 \end{pmatrix}$ and $\begin{pmatrix} 0 & 1 \\ -1 & 0 \end{pmatrix}$ generate $SL(2, Z)$, see for example [H3, pages 168-171], it follows that all the $SL(2, Z)$ transformation formulas can be derived by repeated use of (9.25) and (9.29). See [AT1] and [EMOT1] for details. It is worth remarking that the appropriate τ' corresponding to each $\begin{pmatrix} \alpha & \beta \\ \gamma & \delta \end{pmatrix} \in SL(2, Z)$ is

$$\tau' = \frac{\delta \tau + \gamma}{\beta \tau + \alpha}.$$

In the preceding discussion we have been concerned with nonorthogonal bases for the lattice Hilbert space. For completeness we also compute the ON basis

$$\varphi_n(x_1, x_2) = \mathbf{P}\boldsymbol{\psi}_n(x_1, x_2), \quad n = 0, 1, 2, \cdots$$

corresponding, via the Weyl-Brezin-Zak transform, to the ON basis (5.26) for $L_2(R)$:

$$\boldsymbol{\psi}_n(t) = \pi^{-1/4}(n!)^{-1/2}(-1)^n 2^{-n/2} e^{-t^2/2} H_n(t),$$

where $H_n(t)$ is a Hermite polynomial. We have already seen that the ground state wave function $\boldsymbol{\psi}_0(t) = \pi^{-1/4} e^{-t^2/2}$ maps to, (9.21),

$$\varphi_0(x_1, x_2) = \pi^{-1/4} e^{-x_1^2/2} \theta_3\left(x_2 + \frac{ix_1}{2\pi} \mid \frac{i}{2\pi}\right).$$

Applying the transform $\mathbf{P}$ to both sides of the generating function (5.25) for the $\boldsymbol{\psi}_n(t)$ and using the fact that

$$\mathbf{P}\mathbf{f}(x_1, x_2) = \pi^{-1/4} e^{-\beta^2 - 2\beta x_1 - \frac{1}{2}x_1^2} \theta_3\left(x_2 + \frac{i}{2\pi}[x_1 + 2\beta] \mid \frac{i}{2\pi}\right)$$

for $\mathbf{f}(t) = \pi^{-1/4}\exp(-\beta^2 - 2\beta t - \frac{1}{2}t^2)$, we obtain

$$\pi^{-1/4}\exp(-\beta^2 - 2\beta x_1 - \frac{1}{2}x_1^2)\theta_3(x_2 + \frac{i}{2\pi}[x_1 + 2\beta] \mid \frac{i}{2\pi})$$
$$= \sum_{n=0}^{\infty} \frac{2^{n/2}\beta^n}{(n!)^{1/2}} \varphi_n(x_1, x_2).$$

The left-hand side of this expression is an entire function of β.

With this brief look at the Schrödinger group, an interesting group for future study which contains both H_R and the affine group as subgroups, we conclude these notes.

9.4 Exercises.

9.1 Compute the automorphism group of G_A. Does G_A have any outer automorphisms?

9.2 For $\gamma > 0$ verify the identity

$$D = \begin{pmatrix} \alpha & \beta \\ \gamma & \delta \end{pmatrix} = D_1(\frac{\alpha}{\gamma}) D_3(\frac{1}{2\pi}, -\frac{\pi}{2}) D_2(\gamma) D_1(\frac{\delta}{\gamma}).$$

Find a similar factorization for $\gamma < 0$ and $\gamma = 0$. Show that the automorphisms of H_R are generated by $D_1(\alpha)$, $D_3(\frac{1}{2\pi}, -\frac{\pi}{2})$, and $D_2(\gamma)$.

9.3 Apply the automorphism $D = \begin{pmatrix} 0 & 1 \\ -1 & 0 \end{pmatrix}$ to the Theta function (9.23) in the lattice Hilbert space to derive the formula

$$e^{-2\pi i x_1 x_2} e^{\pi i \tau x_2^2} \theta_3(x_2 \tau - x_1 \mid \tau) = e^{-i\frac{\pi x_1^2}{\tau}} \sqrt{\frac{i}{\tau}}\, \theta_3\left(-\frac{x_1}{\tau} + x_2 \mid -\frac{1}{\tau}\right).$$

9.4 Show that the functions $e^{\pi i \tau x_1^2}\theta_3(x_1\tau + x_2 \mid \tau)e^{2\pi i(m_1 x_1 + m_2 x_2)}$ form an ON basis for the lattice Hilbert space. What is the corresponding ON basis for $L_2(R)$ under the inverse Weil-Brezin-Zak transform?

9.5 Express the relation, Exercise 4.7,

$$\frac{1}{\sqrt{2\pi t}} \sum_{n=-\infty}^{\infty} e^{-\frac{n^2}{2t}} = \sum_{n=-\infty}^{\infty} e^{-2\pi^2 n^2 t}$$

as a Theta function identity. Compare with equation (9.25).

REFERENCES

[AG1]   M. I. AKHIEZER AND I. M. GLAZMAN, *The Theory of Linear Operators in Hilbert Space Vol I*, Frederick Ungar, New York, 1961.

[AG2]   M. I. AKHIEZER AND I. M. GLAZMAN, *The Theory of Linear Operators in Hilbert Space Vol II*, Frederick Ungar, New York, 1963.

[AK]   E. W. ASLAKSEN AND J. R. KLAUDER, *Unitary representation of the affine group*, J. Math. Physics, 9 (1968), pp. 208–211.

[AB]   L. AUSLANDER AND T. BREZIN, *Fiber bundle structures and harmonic analysis of compact Heisenberg manifolds*, Conference on Harmonic Analysis, Lecture Notes in Mathematics 266, Springer Verlag, New York (1971).

[AG3]   L. AUSLANDER AND I. GERTNER, *Wide-band ambiguity functions and a · x + b group*, Signal Processing, Part I: Signal Processing Theory, Vol 22. IMA Volumes in Mathematics and its Applications, L. Auslander, T. Kailath and S. Mitter, eds., Springer Verlag, New York (1990), pp. 1–12.

[AT1]   L. AUSLANDER AND R. TOLIMIERI, *Abelian harmonic analysis, theta functions and function algebras on a nilmanifold*, Lecture Notes in Mathematics 436, Springer-Verlag, New York (1975).

[AT2]   L. AUSLANDER AND R. TOLIMIERI, *Characterizing the radar ambiguity functions*, IEEE Transactions on Information Theory, IT-30 (6), November (1984), pp. 832–836.

[AT3]   L. AUSLANDER AND R. TOLIMIERI, *Radar ambiguity function and group theory*, SIAM J. Math. Anal., 16, No. 3 (1985), pp. 577–601.

[AT4]   L. AUSLANDER AND R. TOLIMIERI, *Computing decimated finite cross-ambiguity functions*, IEEE Transactions on Acoustics, Speech, and Signal Processing, vol 36, No (March 1988), pp. 359–363.

[AT5]   L. AUSLANDER AND R. TOLIMIERI, *On finite Gabor expansions of signals*, Signal Processing, Part I: Signal Processing Theory, Vol 22. IMA Volumes in Mathematics and its Applications, L. Auslander, T. Kailath and S. Mitter, eds., Springer Verlag, New York (1990), pp. 13–23.

[BGZ]   H. BACRY, A. GROSSMAN, AND J. ZAK, *Proof of completeness of lattice states in the kq representation*, Phys. Rev. B., 12 (1975), pp. 1118–1120.

[B1]   R. BALIAN, *Un principe d'incertitude fort en théorie du signal on en mécanique quantique*, C.R. Acad. Sci. Paris, 292 (1981), pp. 1357–1362.

[B2]   V. BARGMANN, *On a Hilbert space of analytic functions and an associated integral transform, I*, Comm. Pure Appl. Math., 14 (1961), pp. 187–214.

[BBGK] V. BARGMANN, P. BUTERA, L. GIRARDELLO AND J.R. KLAUDER, *On the completeness of coherent states*, Rep. Math. Phys., 2 (1971), pp. 221–228.

[B3]   M.J. BASTIANS, *The expansion of an optical signal into a discrete set of Gaussian beams*, Optik, 57, No. 1 (1980), pp. 95–102.

[B4]   M.J. BASTIANS, *Gabor's expansion of a signal onto Gaussian elementary signals*, IEEE Proc., 68 (1980).

[B5]   J. BENEDETTO, *Gabor representations and wavelets*, Commutative Harmonic Analysis, D. Colella, Ed., Contemp. Math. 19, American Mathematical Society, Providence (1989), pp. 9–27.

[BJ]   J.W.M. BERGMANS AND A.J.E.M. JANSSEN, *Robust data equalization, fractional tap spacing and the Zak transform*, Phillips J. Res., 42 (1987), pp. 351–398.

[B6]   M. BERNFELD, *Chirp Doppler radar*, Proceedings of the IEEE, 72(4), April (1984), pp. 540–541.

[B7]   H. BOERNER, *Representations of Groups*, North-Holland, Amsterdam, 1969.

[BZ]   M. BOON AND J. ZAK, *Amplitudes on von Neumann lattices*, J. Math. Phys., 22 (1981), pp. 1090–1099.

[BZZ]   M. BOON, J. ZAK, AND I.J. ZUCKER, *Rational von Neumann lattices*, J. Math. Phys., 24 (1983), pp. 316–323.

[B8]   J. BREZIN, *Function theory on metabelian solvmanifolds*, J. Func. Anal., 10 (1972), pp. 33–51.

[B9]   N.G. DE BRUIJN, *Uncertainty principles in Fourier analysis*, in Inequalities (ed. O. Shisha) Academic Press, New York (1967), pp. 55–71.

[CM]   T.A.C.M. CLAASSEN AND W.F.G. MECKLENBRÄUKER, *The Wigner distributions — A tool for time-frequency signal analysis*, Phillips J. Res., 35 (1980), pp. 217–250, 276–300, 372–389.

[C]   R.R. COIFMAN, *Wavelet analysis and signal processing*, Signal Processing, Part I: Signal Processing Theory, Vol 22. IMA Volumes in Mathematics and its Applications, L. Auslander, T. Kailath and S. Mitter, eds., Springer Verlag, New York (1990), pp. 59–68.

[CR] R.R. COIFMAN AND R. ROCHBERG, *Representation theorems for holomorphic and harmonic functions in L^p*, Astérique, 77 (1980), pp. 11–66.

[CB] C. E. COOK AND M. BERNFELD, *Radar Signals*, Academic Press, New York (1967).

[D1] I. DAUBECHIES, *Discrete sets of coherent states and their use in signal analysis*, In International Conferences on Differential Equations and Mathematical Physics, Birmingham, Alabama (1986).

[D2] I. DAUBECHIES, *Orthonormal bases of compactly supported wavelets*, Comm. Pure Appl. Math., 41 (1988), pp. 909–996.

[D3] I. DAUBECHIES, *Time-frequency localization operators: a geometric phase space approach*, IEEE Trans. Inform. Theory, 34 (1988), pp. 605–612.

[D4] I. DAUBECHIES, *The wavelet transform, time-frequency localization and signal analysis*, IEEE Trans. Inform. Theory, 36 (1990), pp. 961–1005.

[DGM] I. DAUBECHIES, A. GROSSMAN, AND Y. MEYER, *Painless nonorthogonal expansions*, J. Math. Phys., 27 (1986), pp. 1271–1283.

[DP] I. DAUBECHIES AND T. PAUL, *Time-frequency localization operators-a geometric phase space approach: II The use of dilations*, Inverse Problems, 4 (1988), pp. 661–680.

[DG] M.E. DAVISON AND F.A. GRÜNBAUM, *Tomographic reconstruction with arbitrary directions*, Communications on Pure and Applied Math, 34 (1981), pp. 77–120.

[DS1] R.J. DUFFIN AND A.C. SCHAEFFER, *A class of nonharmonic Fourier series*, Trans. Amer. Math. Soc..

[DM1] M. DUFLO AND C.C. MOORE, *On the regular representation of a nonunimodular locally compact group*, Journal of Functional Analysis, 21 (1976), pp. 209–243.

[DS2] N. DUNFORD AND J. T. SCHWARTZ, *Linear Operators*, Interscience Publishers, New York (1958).

[DM2] H. DYM AND H.P. MCKEAN, *Fourier Series and Integrals*, Academic Press, New York (1972).

[EMOT1] A. ERDÉLYI, W. MAGNUS, F. OBERHETTINGER AND F. TRICOMI, *Higher Transcendental Functions, Vol. II*, McGraw-Hill, New York, 1953.

[EMOT2] A. ERDÉLYI, W. MAGNUS, F. OBERHETTINGER AND F. TRICOMI, *Tables of Integral Transforms, Vol. I*, McGraw-Hill, New York, 1954.

[FG3] E. FEIG AND F. A. GRÜNBAUM, *tomographic methods in range-Doppler radar*, Inverse Problems, 2(2) (1986), pp. 185–195.

[G1] S. GAAL, *Linear Analysis and Representation Theory*, Springer-Verlag, New York, 1973.

[G2] D. GABOR, *Theory of communication*, J. Inst. Electr. Engin. (London), 93 (III) (1946), pp. 429–457.

[GMS] I.M. GEL'FAND, R.A. MINLOS AND Z.YA. SHAPIRO, *Representations of the Rotation and Lorentz Groups and their Applications*, Pergamon, New York, 1963.

[G3] R.J. GLEISER, *Doppler shift for a radar echo*, American Journal of Physics, 47(8), August (1979), p. 735.

[G4] A. GROSSMAN, *Wavelet transforms and edge detection*, Stochastic Processing in Physics and Engineering, S. Albeverio et al., eds., D. Reidel, Dordrecht, the Netherlands (1988), pp. 149–157.

[GM1] A. GROSSMAN AND J. MORLET, *Decomposition of Hardy functions into square integrable wavelets of constant shape*, SIAM Journal of Mathematical Analysis, 15 (1984), pp. 723–726.

[GM2] A. GROSSMAN AND J. MORLET, *Decomposition of functions into wavelets of constant shape, and related transforms*, in Mathematics and Physics, Lectures on recent results, World Scientific (Singapore) (1985).

[GMP1] A. GROSSMAN AND J. MORLET, AND T. PAUL, *Transforms associated to square integrable group representations. Part 1: General results*, Journal of Mathematical Physics, 26 (10), October (1985), pp. 2473–2479.

[GMP2] A. GROSSMAN AND J. MORLET, AND T. PAUL, *Transforms associated to square integrable group representations. Part 2: examples*, Ann. Inst. Henri Poincaré, 45(3) (1986), pp. 293–309.

[HS]   M. HAUSNER AND J. T. SCHWARTZ, Gordon and Breach, London (1968), *Lie Groups, Lie Algebras*,.

[H1]   C. E. HEIL, *Generalized harmonic analysis in higher dimensions; Weyl-Heisenberg frames and the Zak transform*, Ph.D. thesis, University of Maryland, College Park, MD (1990).

[H2]   C. E. HEIL, *Wavelets and frames*, Signal Processing, Part I: Signal Processing Theory, Vol 22. IMA Volumes in Mathematics and its Applications, L. Auslander, T. Kailath and S. Mitter, eds., Springer Verlag, New York (1990), pp. 147–160.

[HW]   C. E. HEIL AND D. F. WALNUT, *Continuous and discrete wavelet transforms*, SIAM Review, Vol 31, No 4 (December 1989), pp. 628–666.

[H3]   E. HILLE, *Analytic Function Theory, Volume II*, Ginn and Company, Boston, 1962.

[J1]   J.D. JACKSON, *Classical Electrodynamics*, John Wiley & Sons, New York (1975).

[J2]   A.J.E.M. JANSSEN, *Weighted Wigner distributions vanishing on lattices*, J. of Mathematical Analysis and Applications, 80, No. 1 (1981), pp. 156–167.

[J3]   A.J.E.M. JANSSEN, *Gabor representation of generalized functions*, J. of Mathematical Analysis and Applications, 83 (1981), pp. 377–394.

[J4]   A.J.E.M. JANSSEN, *Bargmann transform, Zak transform, and coherent states*, J. Math. Phys., 23 (1982), pp. 720–731.

[J5]   A.J.E.M. JANSSEN, *The Zak transform: a signal transform for sampled time-continuous signals*, Philips J. Res., Vol. 43 (1988), pp. 23–69.

[K1]   T. KATO, *Perturbation Theory for Linear Operators*, Springer-Verlag, New York, 1966.

[K2]   Y. KATZNELSON, *An Introduction to Harmonic Analysis*, John Wiley, New York, pp. 1968.

[KK]   J.B. KELLER AND H.B. KELLER, *Determination of reflected and transmitted fields by geometrical optics*, Journal of the Optical Society of America, 40 (1949), pp. 48–52.

[K3]   E.J. KELLY, *The radar measurement of range, velocity and acceleration*, IRE Transactions on Military Electronics, MIL-5, April (1961), pp. 51–57.

[K4]   I. KHALIL, *Sur l'analyse harmonique du groupe affine de la droite*, Studia Mathematica, LI(2) (1974), pp. 139–167.

[K5]   A.A. KIRILLOV, *Elements of the Theory of Representations*, Springer-Verlag, New York, pp. 1976.

[K6]   J.R. KLAUDER, *The design of radar signals having both high range resolution and high velocity resolution*, The Bell System Technical Journal, July (1960), pp. 808–819.

[KS]   J.R. KLAUDER AND B.S. SKAGERSTAM, *Coherent states*, World Scientific (Singapore) (1985).

[K7]   A. KOHARI, *Harmonic analysis on the group of linear transformations of the straight line*, Proceedings of the Japanese Academy, 37 (1961), pp. 250–254.

[K8]   J. KOREVAAR, *Mathematical Methods, Vol. 1*, Academic Press, New York, 1968.

[KMG]   R. KRONLAND-MARTINET, J. MORLET, AND A. GROSSMANN, *Analysis of sound patterns through wavelet transforms*, Internet. J. Pattern Recog. Artif. Int., 1 (1987), pp. 273–302.

[LM]   P. LEMARIÉ AND Y. MEYER, *Ondelettes et bases hilbertiennes*, Rev. Mat. Iberoamericana, 2 (1986), pp. 1–18.

[L]   F. LOW, *Complete sets of wave-packets*, in A passion for physics - Essays in honor of Geoffrey Chew, World Scientific (Singapore) (1985), pp. 17–22.

[M1]   Y. MEYER, *Principe d'incertitude, bases hilbertiennes et algèbres d'opérateurs*, Séminaire Bourbaki.

[M2]   Y. MEYER, *Ondelettes, function splines, et analyses graduées*, Univ. of Torino (1986).

[M3]   W. MILLER, JR., *On the special function theory of occupation number space*, Comm. Pure Appl. Math. (1965), pp. 679–696.

[M4]   W. MILLER, JR., *Lie Theory and Special Functions*, Academic Press, New York, 1968.

[M5]   W. MILLER, JR., *Symmetry Groups and their Applications*, Academic Press, New York, 1972.

[M6]   W. MILLER, JR., *Symmetry and Separation of Variables*, Addison-Wesley, Reading, Massachusetts, 1977.

[MAFG] J. MORLET, G. AREHS, I. FORUGEAU AND D. GIARD, *Wave propagation and sampling theory*, Geophysics, 47 (1982), pp. 203–236.

[N1] L. NACHBIN, *Haar Integral*, Van Nostrand, Princeton (1965).

[N2] M. NAIMARK, *Normed Rings*, English Transl. Noordhoff, Groningen, The Netherlands, 1959.

[N3] H. NAPARST, *Radar signal choice and processing for a dense target environment*, Signal Processing, Part II: Control Theory and Applications, Vol 23, IMA Volumes in Mathematics and its Applications, F.A. Grunbaum, J.W. Helton and P. Khargonekar, eds., Springer Verlag, New York (1990), pp. 293–319.

[N4] F. NATTERER, *On the inversion of the attenuated Radon transform*, Numerische Mathematik, 32 (1979), pp. 431–438.

[NS] A. NAYLOR AND G. SELL, *Linear Operator Theory in Engineering and Science*, Holt, New York, 1971.

[P1] A. PAPOULIS, *Ambiguity function in Fourier optics*, Journal of the Optical Society of America, June (1974), pp. 779–788.

[P2] A. PAPOULIS, *Signal Analysis*, McGraw-Hill, New York (1977).

[P3] T. PAUL, *Functions analytic on the half-plane at quantum mechanical states*, J. Math. Phys., 25 (1984), pp. 3252–3263.

[P4] A.M. PERELOMOV, *Note on the completeness of systems of coherent states*, Teor. I. Matem. Fis., 6 (1971), pp. 213–224.

[RR] T. RADO AND P. REICHELDERFER, *Continuous Transformations in Analysis* , Springer-Verlag, Berlin, New York (1955).

[R1] H. REITER, *Classical harmonic Analysis and Locally Compact Groups*, Oxford University Press, Oxford, 1968.

[R2] A. W. RIBACZEK, *Radar resolution of moving targets*, IEEE Transactions on Information Theory, IT-13(1) (1967), pp. 51–56.

[R3] M. RIEFFEL, *Von Neumann algebras associated with pairs of lattices in Lie groups*, Math. Ann., 257 (1981), pp. 403–418.

[RN] F. RIESZ AND B. SZ.-NAGY, *Functional Analysis*, English Transl. Ungar, New York, 1955.

[R4] W. RUDIN, *Principles of Mathematical Analysis*, McGraw-Hill, New York, 1964.

[S1] W. SCHEMPP, *Radar ambiguity functions, the Heisenberg group, and holomorphic theta series*, Proc. Amer. Math. Soc., 92 (1984), pp. 103–110.

[S2] W. SCHEMPP, *Harmonic analysis on the Heisenberg nilpotent Lie group with applications to signal theory*, Longman Scientific and technical, Pitman Research Notes in Mathematical Sciences, 147, Harlow, Essex, UK (1986).

[S2'] W. SCHEMPP, *Neurocomputer architectures*, Resultate der Mathematik – Results in Mathematics ((To appear)).

[S3] J. M. SPEISER, *Wideband ambiguity functions*, IEEE Transactions on information Theory, IT-13, January (1967), pp. 122–123.

[S4] M. SPIVAK, *Calculus on Manifolds*, Benjamin, New York, 1965.

[S5] S. SUSSMAN, *Least square synthesis of radar ambiguity functions*, IRE Trans. Info. Theory (April 1962), pp. 246–254.

[S6] D. A. SWICK, *An ambiguity function independent of assumption about bandwidth and carrier frequency*, NRL Report 6471, Naval Research Laboratory, Washington, D.C., December (1966).

[S7] D. A. SWICK, *A Review of Wideband Ambiguity Functions*, NRL Report 6994, Naval Research Laboratory, Washington, D.C., December (1969).

[T1] N. TATSUUMA, *Plancherel formula for nonunimodular locally compact groups*, J. Math. Kyoto University, 12 (1972), pp. 179–261.

[T2] P. TCHAMITCHIAN, *Calcul symbolique sur les opérateurs de Caldéron-Zygmund et bases inconditionnelles de $L^2(\mathbf{R}^n)$*, C.R. Acad. Sc. Paris, 303, série 1 (1986), pp. 215–218.

[V] N. JA. VILENKIN, *Special Functions and the Theory of Group Representations*, American Mathematical Society, Providence, Rhode Island (1966).

[W1] D. WALNUT, *Weyl-Heisenberg wavelet expansions: existence and stability in weighed spaces*, Ph.D. thesis, University of Maryland, College Park, MD (1989).

[W2] A. WEIL, *Sur certains groupes d'operatours unitaires*, Acta Math., 111 (1964), pp. 143–211.

[WW] E.T. WHITTAKER AND G.N. WATSON, *A Course in Modern Analysis*, Cambridge University Press, Cambridge.

[W3] N. WIENER, *The Fourier integral and certain of its applications*, MIT Press, Cambridge (1933).

[W4] E.P. WIGNER, *On the Quantum correction for thermodynamic equilibrium*, Phys. Rev. (1932), pp. 749–759.

[W5] C. H. WILCOX, *The Synthesis Problem for Radar Ambiguity Functions*, MRC Technical Summary Report 157, Mathematics Research Center, United States Army, University of Wisconsin, Madison, Wisconsin, April (1960).

[W6] P.M. WOODWARD, *Probability and Information theory, with Applications to Radar*, Pergamon Press, New York (1953).

[Y] R. YOUNG, *An Introduction to Nonharmonic Fourier Series*, Academic Press, New York (1980).

[Z1] J. ZAK, *Finite translations in solid state physics*, Phys. Rev. Lett., 19 (1967), pp. 1385–1397.

[Z2] J. ZAK, *Dynamics of electrons in solids in external fields*, Phys. rev., 168 (1968), pp. 686–695.

[Z3] J. ZAK, *The kq-representation in the dynamics of electrons in solids*, Solid State Physics, 27 (1972), pp. 1–62.

[Z4] J. ZAK, *Lattice operators in crystals for Bravais and reciprocal vectors*, Phys. Rev. B, 12 (1975), pp. 3023–3026.

SONAR AND RADAR ECHO STRUCTURE

CALVIN H. WILCOX†

Abstract. The structure of pulse mode sonar and radar echoes is derived from the underlying field equations of acoustics and electromagnetics, respectively. The scattering object Γ is assumed to lie in the far field of both the transmitter and the receiver. In this approximation, the sonar or radar pulse mode signals are shown to be represented by plane waves $s(x \cdot \theta_0 - t, \theta_0)$ at all points x near Γ. In the derivation of this result the signal speed is normalized to unity and θ_0 denotes a unit vector which is directed from the transmitter toward Γ. The echoes are also represented by plane waves $e(x \cdot \theta - t, \theta, \theta_0)$ at all points x near the receiver, where θ is a unit vector which is directed from Γ toward the receiver. For a stationary scatterer Γ the principal result of these lectures is the relation

$$e(\tau, \theta, \theta_0) = Re \left\{ \int_0^\infty e^{i\tau\omega} T_+(\omega\theta, \omega\theta_0)\, \hat{s}(\omega, \theta_0)\, d\omega \right\}$$

where $\hat{s}(\omega, \theta_0)$ is the Fourier transform of $s(\tau, \theta_0)$ and $T_+(\omega\,\theta, \omega\,\theta_0)$ is the scattering amplitude for Γ. Thus $T_+(\omega\,\theta, \omega\,\theta_0)$ is the amplitude of the scattered field in the direction θ due to the scattering by Γ of a plane wave $e^{i\omega\,\theta_0 \cdot x}$ with frequency ω and propagation direction θ_0. For scatterers Γ that move with velocity v such that $|v| \ll 1$ it is shown that the echo waveform is given by

$$e(\tau, \theta, \theta_0) = Re \left\{ \int_0^\infty e^{i\left(\frac{\gamma_0}{\gamma}\tau\right)\omega} T_+(\omega\gamma_0\theta', \omega\gamma_0\theta_0')\hat{s}(\omega, \theta_0)\, d\omega \right\} .$$

where

$$\gamma = \frac{1 - v \cdot \theta}{\sqrt{1 - v^2}}, \qquad \gamma_0 = \frac{1 - v \cdot \theta_0}{\sqrt{1 - v^2}}$$

and θ', θ_0' are related to θ, θ_0 by a Lorentz transformation based on v. Finally, it is known that the high frequency limit

$$\lim_{\omega \to \infty} T_+(\omega\,\theta, \omega\,\theta_0) = T_+^\infty(\theta, \theta_0).$$

exists and is real. Hence if $\hat{s}(\omega, \theta_0)$ is concentrated in a high frequency band where $T_+(\omega\,\theta, \omega\,\theta_0)$ is essentially constant then one has the approximation

$$e(\tau, \theta, \theta_0) = \left(\frac{\pi}{2}\right)^{1/2} T_+^\infty(\theta', \theta_0')\; s(\frac{\gamma_0}{\gamma}\tau, \theta_0).$$

AMS(MOS) subject classifications. 35P25, 45B05, 76Q05, 78A45

---

†Department of Mathematics, University of Utah, Salt Lake City, Utah 84112.

TABLE OF CONTENTS

INTRODUCTION. A sonar or radar system consists of a signal source or "transmitter"and a signal detector or "receiver". The transmitter radiates an acoustic or electromagnetic signal which is scattered by objects in its environment. The resulting echoes are recorded by the receiver. In the simplest sonar and radar systems the directions of arrival, time delays and Doppler shifts of the echoes are used to estimate the directions, ranges and speeds of the scattering objects. The goal of a sophisticated system is to classify objects into identifiable classes by means of their echoes. This is an inverse scattering problem. An associated problem is that of waveform design; that is, the choice of a signal waveform which optimizes the information obtainable from the echoes. Before either of these problems can be attacked, the direct problem of echo prediction when the signal and scattering object are known, must be understood. The purpose of these lectures is to present solutions to the sonar and radar echo prediction problems that are based directly on the field equations of acoustics and electromagnetics, respectively. The presentation is based of the author's publications [24] and [25].

Part 1. SONAR ECHO STRUCTURE

1.1 The Sonar Echo Prediction Problem. Sonar systems are normally operated in the pulse mode in which a sequence of equally spaced short pulses is emitted. A second mode of operation is the CW (continuous wave) mode in which a steady tone of fixed frequency is emitted. The construction of pulse mode echoes given below is derived from the theory of CW mode echoes. The analysis of each of these modes is based on a boundary value problem for an acoustic potential function.

Sonar echo prediction is analyzed below under the following physical assumptions.

- The sonar system (transmitter and receiver) operates in a stationary homogeneous unlimited fluid medium.
- The system is stationary with respect to the medium.
- The sonar signals are generated by conservative force fields.
- The scattering objects are rigid bodies.
- The transmitter and receiver are in the far field of the scattering objects.
- The speeds of the scattering objects are less than the speed of sound and are essentially constant during the interval required for the sonar pulse to sweep over the objects.
- Secondary echoes due to the sonar system components are negligible.
- Noise in the medium is negligible.

The case of a scattering object which is stationary with respect to the sonar system is analyzed first. The analysis is based on the usual linear theory of acoustics [11,26]. The following notation is used. $x = (x_1, x_2, x_3) \in R^3$ denotes spatial coordinates of a Cartesian system which is fixed in the medium. $t \in R$ denotes a time coordinate. $\Omega \subset R^3$ denotes the domain exterior to the scattering object and $\Gamma = R^3 - \Omega$ denotes the object. The common frontier of Γ and Ω, which describes the surface of the object, is denoted by $\partial \Gamma$ or $\partial \Omega$. The equilibrium state of the medium is characterized by its constant density ρ_0, sound speed c_0 and pressure p_0. It will be assumed for notational simplicity that $\rho_0 = 1$ and $c_0 = 1$ since this can be achieved by a suitable choice of units.

The acoustic field generated by the transmitter is characterized by a real-valued acoustic potential function

$$(1.1.1) \qquad u = u(t, x), \quad t \in R, \quad x \in \Omega,$$

which satisfies the inhomogeneous d'Alembert equation

$$(1.1.2) \qquad \partial_t^2 u - \Delta u = f(t, x) \quad \text{for} \quad t \in R, \quad x \in \Omega.$$

Here $\partial_t = \partial/\partial t$, Δ is the Laplacian and $f(t, x)$ is a function, characteristic of the transmitter, which will be called the source function. It has the structure

$$(1.1.3) \qquad f(t, x) = \partial_t V(t, x) + q(t, x)$$

where $V(t, x)$ is a potential for the conservative forces acting in the transmitter and $q(t, x)$ is the volume flow in the transmitter [11,p.280]. It will be assumed, tentatively, that $f(t, x)$ is known. It will be seen below that the true task of the sonar design engineer is to design a transmitter with a prescribed pulse waveform and radiation pattern.

Pulse Mode Scattering. It will suffice to analyze the scattering of a single short pulse of duration T which is emitted by a transmitter that is localized near a point x_0. Hence the support of $f(t, x)$ will be assumed to satisfy

$$(1.1.4) \qquad \text{supp } f \subset \{(t, x) : t_0 \leq t \leq t_0 + T \quad \text{and} \quad |x - x_0| \leq \delta_0\}$$

where δ_0 and t_0 are suitable constants and $|x - x_0|$ is the distance between x and x_0. The acoustic field generated by $f(t, x)$ is characterized by a potential $u(t, x)$ that satisfies (1.1.2), together with the Neumann boundary condition,

$$(1.1.5) \qquad \partial_\nu u \equiv \nabla u \cdot \vec{\nu} = 0 \quad \text{for} \quad t \in R, \quad x \in \partial\Omega,$$

where $\vec{\nu} = \vec{\nu}(x)$ is a normal vector to $\partial\Omega$, and the initial condition,

$$(1.1.6) \qquad u(t, x) = 0 \quad \text{for} \quad t < t_0, \quad x \in \Omega.$$

The Neumann boundary condition characterizes a rigid object while the initial condition implies that there is no signal before the sources begin to act.

The primary field of the transmitter is the field $u_0(t, x)$ that is generated by the sources when no scattering object is present. It is given by the retarded potential formula

$$(1.1.7) \quad u_0(t, x) = \frac{1}{4\pi} \int_{|x' - x_0| \leq \delta_0} \frac{f(t - |x - x'|, x')}{|x - x'|} \, dx' \quad \text{for} \quad t \in R, \quad x \in R^3,$$

where $dx' = dx'_1 dx'_2 dx'_3$. The variation of $u_0(t, x)$ near the scatterer Γ takes a simpler form when the transmitter lies in the far field of Γ. To demonstrate this it will be convenient to assume that the origin of coordinates lies in Γ and to let $\delta > 0$ be the smallest constant such that

$$(1.1.8) \qquad \Gamma \subset \{x : |x| \leq \delta\}.$$

Then the far field assumption takes the form

$$(1.1.9) \qquad |x_0| >> \delta_0 + \delta.$$

If θ_0 is the unit vector defined by

$$(1.1.10) \qquad x_0 = -|x_0|\theta_0,$$

and if $|x - x'|$ is developed in inverse powers of $|x_0|$, it is found that

$$
\begin{aligned}
|x - x'| &= |x_0 + (x' - x_0 - x)| \\
&= (|x_0|^2 + 2x_0 \cdot (x' - x_0 - x) + |x' - x_0 - x|^2)^{1/2} \\
&= |x_0|(1 - 2\theta_0 \cdot (x' - x_0 - x)/|x_0| + |x' - x_0 - x|^2/|x_0|^2)^{1/2} \\
&= |x_0|(1 - \theta_0 \cdot (x' - x_0 - x)/|x_0| + \mathcal{O}(1/|x_0|^2)) \\
&= |x_0| + \theta_0 \cdot x - \theta_0 \cdot (x' - x_0) + \mathcal{O}(1/|x_0|) \quad \text{for} \quad |x_0| \to \infty,
\end{aligned}
$$

(1.1.11)

and the error term is uniformly small for all $|x' - x_0| \leq \delta_0$ and $|x| \leq \delta$. Hence, (1.1.11) may be substituted in (1.1.7) and gives

(1.1.12)
$$
u_0(t, x) = \frac{s(x \cdot \theta_0 - t + |x_0|, \theta_0)}{|x_0|} + \mathcal{O}(\frac{1}{|x_0|^2}), \quad |x_0| \to \infty,
$$

uniformly for $t \in R$, $|x| \leq \delta$, where

(1.1.13)
$$
s(\tau, \theta_0) = \frac{1}{4\pi} \int_{|x' - x_0| \leq \delta_0} f(\theta_0 \cdot (x' - x_0) - \tau, x') \, dx' \quad \text{for} \quad \tau \in R.
$$

If the error term in (1.1.12) is dropped then the primary field becomes a plane wave pulse propagating in the direction of the unit vector θ_0. The profile $s(\tau, \theta_0)$ will be called the signal waveform. One of the primary tasks of the sonar design engineer is to design the transmitter in such a way that the signal waveform is a prescribed function. Note that (1.1.4) and (1.1.13) imply that supp $s(\cdot, \theta_0) \subset [-t_0 - \delta_0 - T, -t_0 + \delta_0]$.

The acoustic field produced when a plane wave

(1.1.14)
$$
u_0(t, x) = s(x \cdot \theta_0 - t, \theta_0), \quad \text{supp} \quad s(\cdot, \theta_0) \subset [a, b]
$$

is scattered by Γ is the solution $u(t, x)$ of the boundary value problem

(1.1.15) $\qquad \partial_t^2 u - \Delta u = 0 \quad \text{for} \quad t \in R, \quad x \in \Omega$

(1.1.16) $\qquad \partial_\nu u = 0 \quad \text{for} \quad t \in R, \quad x \in \partial\Omega$

(1.1.17) $\qquad u(t, x) \equiv u_0(t, x) \quad \text{for} \quad t + b + \delta < 0.$

The secondary field, or echo, is defined by

(1.1.18) $\qquad u_{sc}(t, x) = u(t, x) - u_0(t, x) \quad \text{for} \quad t \in R, x \in \Omega.$

It is shown below that in the far field of the scatterer Γ the echo is a diverging spherical wave:

(1.1.19)
$$
u_{sc}(t, x) \approx \frac{e(|x| - t, \theta, \theta_0)}{|x|}, \quad x = |x|\theta.
$$

The profile $e(\tau, \theta, \theta_0)$ will be called the echo waveform. It depends of the direction of incidence θ_0 of the plane wave (1.1.14) and the direction of observation θ.

The principal goal of this article is to calculate the relationship between the echo waveform and the signal waveform. The calculation will be based on the theory of

CW Mode Scattering. CW mode fields are generated by source functions of the form

$$(1.1.20) \qquad f(t, x) = g_1(x)\cos\omega t + g_2(x)\sin\omega t = Re\{g(x)e^{-i\omega t}\}$$

where $\omega > 0$ is a fixed frequency and $g = g_1 + ig_2$. The corresponding CW mode field has the same time-dependence:

$$(1.1.21) \qquad u(t, x) = w_1(x)\cos\omega t + w_2(x)\sin\omega t = Re\{w(x)e^{-i\omega t}\}$$

where $w = w_1 + iw_2$. $u(t, x)$ must satisfy the d'Alembert equation (1.1.12), with $f(t, x)$ defined by (1.1.20), and the Neumann boundary condition (1.1.5). The initial condition (1.1.6) is not appropriate for CW mode fields and is replaced by the Sommerfeld radiation condition. The corresponding boundary value problem for the complex-valued wave function $w(x)$ is

$$(1.1.22) \qquad \Delta w + \omega^2 w = -g(x) \quad \text{for} \quad x \in \Omega$$

$$(1.1.23) \qquad \partial_\nu w = 0 \quad \text{for} \quad x \in \partial\Omega$$

$$(1.1.24) \qquad \partial w/\partial|x| - i\omega w = \mathcal{O}(1/|x|^2) \quad \text{for} \quad |x| \to \infty,$$

where it is assumed that

$$(1.1.25) \qquad \text{supp} \quad g \subset \{x : |x - x_0| \le \delta_0\}$$

in agreement with (1.1.4). The necessity of the Sommerfeld radiation condition was proved in [17]. Physically, it ensures that u is a pure outgoing wave in the far field. Mathematically, it guarantees the uniqueness of the CW mode field w; see [16] and [21].

The primary CW mode field is the CW mode field $w_0(x)$ generated by $g(x)$ when no scattering object is present. It is given by

$$(1.1.26) \qquad w_0(x) = \frac{1}{4\pi} \int_{|x'-x_0|\le\delta_0} \frac{e^{i\omega|x-x'|}}{|x-x'|} g(x')\, dx' \quad \text{for} \quad x \in R^3.$$

The variation of $w_0(x)$ near Γ takes a simpler form, when Γ lies in the far field of the transmitter, just as in the case of the pulse mode fields. On substituting (1.1.11) into (1.1.26) one finds that

$$(1.1.27) \qquad w_0(x) = \frac{T(\omega\theta_0)}{|x_0|} e^{i\omega\theta_0 \cdot x} + \mathcal{O}(1/|x_0|^2), \quad |x_0| \to \infty$$

uniformly for $|x| \le \delta$ where

$$(1.1.28) \qquad T(\omega\theta_0) = \frac{1}{4\pi} \int_{|x'-x_0|\le\delta_0} e^{-i\omega\theta_0 \cdot x'} g(x')\, dx'.$$

If the error term in (1.1.27) is dropped the primary field becomes a CW mode plane wave of frequency ω propagating in the direction θ_0. It will be convenient to renormalize the primary field to

$$(1.1.29) \qquad w_0(x, \omega\theta_0) = (2\pi)^{-3/2} e^{i\omega\theta_0 \cdot x}.$$

The CW mode field which is produced when $w_0(x, \omega\theta_0)$ is scattered by Γ will be denoted by $w^+(x, \omega\theta_0)$. It is the solution

$$(1.1.30) \qquad w^+(x, \omega\theta_0) = w_0(x, \omega\theta_0) + w_{sc}^+(x, \omega\theta_0), \quad x \in \Omega,$$

of the boundary value problem

$$(1.1.31) \qquad \Delta w^+ + \omega^2 w^+ = 0 \quad \text{for} \quad x \in \Omega,$$
$$(1.1.32) \qquad \partial_\nu w^+ = 0 \quad \text{for} \quad x \in \partial\Omega,$$
$$(1.1.33) \qquad \partial w_{sc}^+ / \partial |x| - i\omega w_{sc}^+ = \mathcal{O}(1/|x|^2), \quad |x| \to \infty.$$

It is shown below that in the far field of Γ the secondary field $w_{sc}^+(x, \omega\theta_0)$ is a diverging spherical wave

$$(1.1.34) \qquad w_{sc}^+(x, \omega\theta_0) \approx \frac{e^{i\omega|x|}}{4\pi|x|} T_+(\omega\theta, \omega\theta_0), \quad x = |x|\theta.$$

The coefficient $T_+(\omega\theta, \omega\theta_0)$ is called the scattering amplitude of Γ. It determines the amplitude and phase of the CW mode echo in the direction θ due to the primary wave (1.1.29) with frequency ω and propagation direction θ_0.

The solution of the echo prediction problem for stationary scatterers can now be formulated. It is given by the integral relation

$$(1.1.35) \qquad e(\tau, \theta, \theta_0) = Re\{ \int_0^\infty e^{i\tau\omega} T_+(\omega\theta, \omega\theta_0) \hat{s}(\omega, \theta_0) \, d\omega \}$$

where $\hat{s}(\omega, \theta_0)$ is the Fourier transform of the signal waveform:

$$(1.1.36) \qquad \hat{s}(\omega, \theta_0) = \frac{1}{(2\pi)^{1/2}} \int_{-\infty}^\infty e^{-i\omega\tau} s(\tau, \theta_0) \, d\tau.$$

The derivation of this relation is the principal result of Part 1 of these lectures.

It is clear that (1.1.35) is not valid for moving scatterers because it predicts that a signal and its echo contain the same frequencies, and thereby fails to predict the well known Doppler shift. However, (1.1.35) can be generalized to the case of a moving scatterer by first applying it in a coordinate frame that is fixed with respect to the scatterer and then passing by means of a Lorentz transformation to a frame which is fixed with respect to the sonar transmitter and receiver. This calculation is carried out in §1.5 below.

Relation (1.1.35) reveals the fundamental importance for sonar echo analysis of the scattering amplitude $T_+(\omega\theta, \omega\theta_0)$. One simple consequence of (1.1.35) is the result that echoes of high frequency pulses are undistorted. Indeed, it was shown by A. Majda [6] that

(1.1.37)
$$\lim_{\omega \to \infty} T_+(\omega\theta, \omega\theta_0) = T_+^\infty(\theta, \theta_0)$$

exists and is real-valued when Γ is a smooth convex object. It follows from (1.1.35) that if $\hat{s}(\omega, \theta_0)$ is concentrated in a high frequency band $\omega \geq \omega_0$ where $T_+(\omega\theta, \omega\theta_0)$ is nearly constant then

(1.1.38)
$$e(\tau, \theta, \theta_0) \approx T_+^\infty(\theta, \theta_0) Re\{ \int_0^\infty e^{i\tau\omega} \hat{s}(\omega, \theta_0)\, d\omega \}$$
$$= \left(\frac{\pi}{2}\right)^{1/2} T_+^\infty(\theta, \theta_0) s(\tau, \theta_0),$$

by Fourier's theorem. More interesting, however, is the possibility of applying (1.1.35) to the study of pulse distortion by scattering and its dependence on the geometry of the scatterer. In particular, (1.1.35) shows that resonances in $T_+(\omega\theta, \omega\theta_0)$ will produce selective enhancement of sonar echoes at the resonance frequencies. Another possible application of (1.1.35) is to use Fourier analysis of observed sonar echoes produced by known signals to estimate the function $T_+(\omega\theta, \omega\theta_0)$ over a range of frequencies. These studies are applicable to the analysis and design of sonar systems.

The theory presented in the remainder of Part 1 is based of the author's monograph [21] to which reference is made for most of the analytical details. The theory developed in [21] is applicable to a large class of objects Γ with irregular, non-smooth boundaries. The theory was developed in this generality not just for mathematical completeness but because most applications require it. The class of allowable objects Γ includes all the simple objects that arise in applications, such as polyhedra, finite sections of cylinders, cones, spheres, disks and, more generally, all objects with piecewise smooth surfaces that have a finite number of smooth edges and vertices.

The remainder of Part 1 is organized as follows. §1.2 reviews briefly the definition and structure of CW mode wave fields. §1.3 reviews the definition of pulse mode sonar wave fields and their representation by means of CW mode fields. §1.4 describes the theory of asymptotic wave fields, as developed in [21], and applies it to the derivation of the far field form (1.1.19) of sonar echoes and the integral relation (1.1.35). §1.5 contains the generalization of (1.1.35) to the case of moving scatterers.

Exercises for §1.1.

1. Derive the retarded potential representation (1.1.7). What hypotheses on $f(t, x)$ are needed for your derivation?

2. Derive the approximation (1.1.11) for $|x - x'|$ and show that it is uniform for all $|x' - x_0| \leq \delta_0$ and $|x| \leq \delta$.

3. Use the estimate (1.1.11) and the retarded potential formula (1.1.7) to derive the far field representation (1.1.12), (1.1.13). Verify that (1.1.4) and (1.1.13) imply that supp $s \subset [-t_0 - \delta_0 - T, -t_0 + \delta_0]$.

4. Derive the primary CW mode field (1.1.26) from the retarded potential formula (1.1.7) by the limiting amplitude principle; i.e., put $f(t, x) = e^{-i\omega t} g(x) H(t - t_0)$ in (1.1.7), where $H(t)$ is the Heaviside function, and show that $u_0(t, x) \sim e^{-i\omega t} w_0(x)$ for $t \to \infty$.

5. Derive the far field formula (1.1.27), (1.1.28) from the representation (1.1.26) and the approximation (1.1.11).

1.2 The Structure of CW Mode Sonar Echoes. The definition and basic properties of the CW mode fields $w^+(x,p)$ are reviewed in this section. The definition is based of the boundary value problem (1.1.30)-(1.1.33), but in a formulation appropriate for scatterers Γ with non-smooth boundaries. In the classical theory due to Kupradze [5] and Weyl [15] a solution is a function in the class $C^2(\Omega) \cap C^1(\bar\Omega)$ where $\bar\Omega = \Omega \cup \partial\Omega$. It is known that if $\partial\Omega$ is a smooth surface then a unique classical solution $w^+(x,p)$ exists for each $p = \omega\theta_0 \in R^3 - \{0\}$. However, if $\partial\Omega$ has edges and/or vertices then at such points the normal vector $\vec{\nu}$ is undefined and the boundary condition (1.1.32) is not meaningful. In this case it was discovered that if (1.1.32) is enforced only at points where $\partial\Omega$ is smooth then all solutions have first derivatives that are singular at the edges and vertices and no classical solution exists [3,7]. Moreover, if the strength of these singularities is not controlled then the uniqueness of the solution is lost [8]. Physically, non-uniqueness can occur because line or point sources can be situated in the edges or vertices. This type of non-uniqueness is eliminated by the "edge condition" [3] which requires that the energy in each bounded portion of Ω should be finite. It may be formulated mathematically as the condition that $u(x) = w^+(x, \omega\theta_0)$ should satisfy [21]

$$(1.2.1) \qquad \int_{K \cap \Omega} \{|\nabla u(x)|^2 + |u(x)|^2\}\, dx < \infty \quad \text{for every cube} \quad K \subset R^3.$$

In the theory developed in [21] the edge condition (1.2.1) is formulated by means of the function classes

$$(1.2.2) \qquad L_2^{loc}(\bar\Omega) = \{u : u \in L_2(K \cap \Omega) \quad \text{for every cube} \quad K \subset R^3\},$$

and

$$(1.2.3) \qquad L_2^{1,loc}(\bar\Omega) = L_2^{loc}(\bar\Omega) \cap \{u : \partial_j u \in L_2^{loc}(\bar\Omega) \quad \text{for} \quad j = 1,2,3\},$$

where $L_2(M)$ denotes the usual Lebesgue class of a set $M \subset R^3$ and $\partial_j = \partial/\partial x_j$. Condition (1.2.1) is equivalent to the condition $u \in L_2^{1,loc}(\bar\Omega)$. Moreover, if $u = w^+ \in L_2^{1,loc}(\bar\Omega)$ is a weak solution of (1.1.31), in the sense of the theory of distributions, then u is also in the class

$$(1.2.4) \qquad L_2^{1,loc}(\Delta, \bar\Omega) = L_2^{1,loc}(\bar\Omega) \cap \{u : \Delta u \in L_2^{loc}(\bar\Omega)\}.$$

For general domains $\Omega \subset R^3$ and functions $u \in L_2^{1,loc}(\Delta, \bar\Omega)$, the Neumann condition (1.1.32) is replaced by the generalized Neumann condition of [21]:

$$(1.2.5) \qquad \int_\Omega \{(\Delta u)v + \nabla u \cdot \nabla v\}\, dx = 0$$

for all $v \in L_2(\Omega)$ such that $\nabla v \in L_2(\Omega)$ and $v(x) = 0$ outside of a bounded set. The terminology is justified by the fact that (1.2.5) implies that the classical Neumann condition $\partial_\nu u(x) = 0$ is satisfied at every boundary point x where $\partial\Omega$ is smooth [21,p.41].

The CW mode field $w^+(x, p)$ that corresponds to the primary field

(1.2.6) $$w_0(x, p) = (2\pi)^{-3/2} e^{ip\cdot x}, \quad p \in R^3 - \{0\},$$

and the exterior domain $\Omega \subset R^3$ is defined in [21] by the conditions

(1.2.7) $$w^+(x, p) \in L_2^{N, loc}(\Delta, \bar{\Omega}) = L_2^{1, loc}(\Delta, \bar{\Omega}) \cap \{u : u \quad \text{satisfies} \quad (1.2.5)\}$$

(1.2.8) $$(\Delta + |p|^2) w^+(x, p) = 0 \quad \text{for} \quad x \in \Omega$$

(1.2.9) $$\partial w_{sc}^+ / \partial |x| - i|p| w_{sc}^+ = \mathcal{O}(1/|x|^2), \quad |x| \to \infty$$

where $w_{sc}^+ = w^+ - w_0$. In this formulation the edge condition and the generalized Neumann condition are incorporated into condition (1.2.7). Moreover, since it is known that weak solutions of (1.2.8) are necessarily analytic functions of $x \in \Omega$, (1.2.9) is meaningful for functions which satisfy (1.2.7) and (1.2.8). The uniqueness of the functions $w^+(x, p)$ was proved in [21] for arbitrary exterior domains and their existence was proved for allowable objects in the sense of [21].

The remainder of this section presents a calculation of the far field form (1.1.34) of the CW echo w_{sc}^+. It is based on the integral representation [21, p.90]

$$w_{sc}^+(x, p)$$
(1.2.10)
$$= \int_{|x'|=r_0} \left\{ w^+(x', p) \frac{\partial G_0(x - x', |p|)}{\partial |x'|} - \frac{\partial w^+(x', p)}{\partial |x'|} G_0(x - x', |p|) \right\} dS',$$

valid for all $|x| > r_0 > \delta$, where

(1.2.11) $$G_0(x - x', \omega) = \frac{e^{i\omega|x-x'|}}{4\pi|x - x'|}$$

is the free space Green's function for the Helmholtz equation (1.2.8). The asymptotic form of $w_{sc}^+(x, p)$ for $x = |x|\theta$, $|x| \to \infty$ may be calculated from (1.2.10) by using the approximation $|x - x'| = |x| - x' \cdot \theta + \mathcal{O}(1/|x|)$, as in §1.1. The result is

(1.2.12) $$w_{sc}^+(x, p) = \frac{e^{i|p||x|}}{4\pi|x|} T_+(|p|\theta, p) + \mathcal{O}(1/|x|^2), \quad |x| \to \infty$$

uniformly for all directions θ and p in any bounded set, where the scattering amplitude $T_+(p, p')$ may be written

$$T_+(p, p')$$
(1.2.13)
$$= \int_{|x|=r_0} \left\{ w^+(x, p') \frac{\partial e^{-ip\cdot x}}{\partial |x|} - e^{-ip\cdot x} \frac{\partial w^+(x, p')}{\partial |x|} \right\} dS.$$

It follows from Green's theorem that the integral in (1.2.13) is independent of r_0 provided that $r_0 > \delta$ and $|p'| = |p|$.

The well-known reciprocity theorem for $T_+(p, p')$ states that

$$(1.2.14) \qquad\qquad T_+(p, p') = T_+(-p', -p)$$

for all p and p' such that $|p| = |p'|$. This relation may be verified in the present setting by applying Green's theorem to $w^+(x, p)$ and $w^+(x, p')$ in the region $\Omega(r_0) = \Omega \cap \{x : |x| < r_0\}$ and using the representation (1.2.13). If $\partial\Omega$ is not smooth the calculation must be based on the version of Green's theorem given by the generalized Neumann condition (1.2.5). The technique for doing this is described in [21,pp.57-8].

It can be shown that if the scatterer Γ has a piecewise smooth surface $\partial\Gamma = \partial\Omega$ then $w^+(x, p)$ is continuous and $\partial w^+(x, p)/\partial\nu = 0$ at all boundary points except edges and vertices. In this case Green's theorem can be applied to $w_0(x, -p)$ and $w^+(x, p')$ in $\Omega(r_0)$. This calculation, together with (1.2.13), implies the representation

$$(1.2.15) \qquad\qquad T_+(p, p') = \int_{\partial\Omega} w^+(x, p') \frac{\partial e^{-ip\cdot x}}{\partial\nu}\, dS.$$

Moreover, under the same hypothesis the application of Green's theorem to $G_0(x - x', p)$ and $w^+(x', p)$ gives the representation

$$(1.2.16) \qquad w^+(x, p) = w_0(x, p) + \int_{\partial\Omega} \frac{\partial G_0(x - x', |p|)}{\partial\nu'} w^+(x', p)\, dS',$$

for all $x \in \Omega$, and the unknown boundary values of $w^+(x, p)$ satisfy the Fredholm integral equation [2]

$$(1.2.17) \qquad w^+(x, p) = 2w_0(x, p) + 2\int_{\partial\Omega} \frac{\partial G_0(x - x', |p|)}{\partial\nu'} w^+(x', p)\, dS'$$

for all $x \in \partial\Omega$. These relations provide a basis for both exact and approximate calculations of $w^+(x, p)$ and $T_+(p, p')$.

Exercises for §1.2.

1. Derive the integral representation (1.2.10), (1.2.11) for the CW mode scattered field $w_{sc}^+(x, p)$. Assume that Green's theorem may be applied in the domain $\Omega(r_0) = \Omega \cap \{x : |x| < r_0\}$. (True if $\partial\Omega$ is smooth.)

2. Show how the generalized boundary condition (1.2.5) can be used to extend the representation (1.2.10) to non-smooth boundaries $\partial\Omega$. (See [21].)

3. Derive the CW mode far field representation (1.2.12), (1.2.13) for $w_{sc}^+(x, p)$ from the integral representation (1.2.10) and show that it holds uniformly for all unit vectors θ and all $p \in B \subset R^3$ where B is bounded.

4. Derive the reciprocity relation (1.2.14) for the scattering amplitude $T_+(p, p')$.

5. Derive the integral representation (1.2.15) for $T_+(p, p')$.

6. Use Green's theorem to derive the integral representation (1.2.16) for the CW mode field $w^+(x, p)$.

7. Use the representation (1.2.16) and the standard jump relations of potential theory [2] to derive the Fredholm integral equation (1.2.17) for the boundary value of $w^+(x, p)$.

1.3 The Structure of Pulse Mode Sonar Echoes. In this section the Hilbert space method of [21] is used to construct the pulse mode sonar echoes produced by arbitrary pulse mode signals. This method has several advantages. First, the spectral theorem yields an immediate construction of the scattered wave field for essentially arbitrary scatterer geometry and pulse structure. Second, the generalized eigenfunction expansion of [21] provides an integral representation of the pulse mode echoes that makes clear the fundamental role of the scattering amplitude $T_+(p, p')$.

Some readers may find the use of functional analysis excessively abstract or inappropriate in a work intended for applied mathematicians, engineers and applied scientists. Such readers may nevertheless follow the presentation given below by disregarding the underlying functional analysis and treating our operator-theoretic equations as heuristic calculations of the kind often used by physicists. Of course, if this approach is followed then it should be verified by direct calculation that the final integral representation for the pulse mode echoes does indeed satisfy the wave equation and boundary and initial conditions.

The construction of the pulse mode echoes given below is based on the fact that the negative Laplacian $A = -\Delta$, acting on functions that satisfy the edge condition and the generalized Neumann condition, defines a selfadjoint operator in the Hilbert space $L_2(\Omega)$. It was shown in [21] that the CW mode fields $w^+(x, p)$ form a complete set of generalized eigenfunctions for A. This result is used below to derive an integral representation of pulse mode sonar echoes.

The edge condition and generalized Neumann condition for functions in $L_2(\Omega)$ were formulated in [21] by means of the function classes

(1.3.1)
$$L_2^1(\Omega) = L_2(\Omega) \cap \{u : \partial_j u \in L_2(\Omega) \quad \text{for} \quad j = 1, 2, 3\},$$

(1.3.2)
$$L_2^1(\Delta, \Omega) = L_2^1(\Omega) \cap \{u : \Delta u \in L_2(\Omega)\},$$

(1.3.3)
$$L_2^N(\Delta, \Omega) = L_2^1(\Delta, \Omega) \cap \{u : \quad (1.2.5) \text{ holds for all} \quad v \in L_2^1(\Omega)\}.$$

The operator A in $L_2(\Omega)$ was defined by

(1.3.4)
$$D(A) = L_2^N(\Delta, \Omega),$$

(1.3.5)
$$Au = -\Delta u \quad \text{for all} \quad u \in D(A).$$

It was shown in [21,p.41] that A is a selfadjoint non-negative operator in $L_2(\Omega)$:

(1.3.6)
$$A = A^* \geq 0.$$

The non-negative square root of A, denoted by $A^{1/2}$, is also needed below. It was shown in [21] that

(1.3.7)
$$D(A^{1/2}) = L_2^1(\Omega),$$

(1.3.8)
$$\|A^{1/2}u\| = \|\nabla u\| \quad \text{for all} \quad u \in D(A^{1/2})$$

where $\| \cdot \|$ denotes the $L_2(\Omega)$-norm.

The d'Alembert equation (1.1.2) is interpreted below as an ordinary differential equation

$$(1.3.9) \qquad \partial_t^2 u + Au = f(t, \cdot), \quad t \in R,$$

for a function $t \to u(t, \cdot) \in L_2(\Omega)$. It is assumed that

$$(1.3.10) \qquad \mathrm{supp} \quad f \subset \{t : t_0 \le t \le t_0 + T\}$$

and a solution of (1.3.9) is sought which satisfies

$$(1.3.11) \qquad u(t, \cdot) = 0 \quad \text{for} \quad t < t_0$$

in agreement with (1.1.4) and (1.1.6).

A formal solution of (1.3.9) and (1.3.11) is defined by the Duhamel integral

$$(1.3.12) \qquad u(t, \cdot) = \int_{t_0}^{t} \{A^{-1/2} \sin(t - \tau)A^{1/2}\} f(\tau, \cdot) \, d\tau,$$

for all $t \ge t_0$, where $A^{-1/2} \sin t A^{1/2}$ is defined by means of the spectral theorem. Integration by parts in (1.3.12) gives the alternative representation

$$(1.3.13) \qquad u(t, \cdot) = \int_{t_0}^{t} \{\cos(t - \tau)A^{1/2}\} \int_{t_0}^{\tau} f(t', \cdot) \, dt' \, d\tau.$$

It will be assumed that the source function f satisfies

$$(1.3.14) \qquad f \in C(R, L_2^1(\Omega)) = C(R, D(A^{1/2})).$$

It can then be verified that (1.3.12) defines a function

$$(1.3.15) \qquad u \in C^2(R, L_2(\Omega)) \cap C^1(R, D(A^{1/2})) \cap C(R, D(A))$$

whose first and second t-derivatives satisfy

$$(1.3.16) \qquad \partial_t u = \int_{t_0}^{t} \{\cos(t - \tau)A^{1/2}\} f(\tau, \cdot) \, d\tau$$

and

$$(1.3.17) \qquad \begin{aligned} \partial_t^2 u &= f(t, \cdot) - \int_{t_0}^{t} \{A^{1/2} \sin(t - \tau)A^{1/2}\} f(\tau, \cdot) \, d\tau \\ &= f(t, \cdot) - Au. \end{aligned}$$

In particular, $u(t, \cdot) \in D(A)$ for every $t \in R$ and hence $u(t, x)$ satisfies the edge condition and generalized Neumann condition for every t. The uniqueness of this kind of solution was proved in [19].

The remainder of this article deals with acoustic fields of the form (1.3.12) at times $t \geq t_0 = T$; i.e., after the primary field has been established and the sources have ceased to act. It will be assumed that

$$(1.3.18) \qquad \int_{t_0}^{t_0+T} f(t, x)\, dt = 0 \quad \text{for all} \quad x \in \Omega.$$

This hypothesis is made to simplify the analysis and is not essential to the method. The physical meaning of (1.3.18) is clear from (1.1.3).

Equations (1.3.13) and (1.3.18) imply that

$$(1.3.19) \qquad u(t, \cdot) = \int_{t_0}^{t_0+T} \{\cos(t-\tau)A^{1/2}\} \int_{t_0}^{\tau} f(t', \cdot)\, dt'\, d\tau.$$

for all $t \geq t_0 + T$. Moreover, since $f(t, x)$ is real-valued, (1.3.19) implies that

$$(1.3.20) \qquad u(t, x) = Re\{v(t, x)\} \quad \text{for} \quad t \geq t_0 + T,$$

where

$$(1.3.21) \qquad v(t, \cdot) = \int_{t_0}^{t_0+T} e^{-i(t-\tau)A^{1/2}} \int_{t_0}^{\tau} f(t', \cdot)\, dt'\, d\tau = e^{-itA^{1/2}} h$$

and

$$(1.3.22) \qquad h = \int_{t_0}^{t_0+T} e^{i\tau A^{1/2}} \int_{t_0}^{\tau} f(t', \cdot)\, dt'\, d\tau.$$

These equations contain, in abstract form, the solution of the sonar echo prediction problem. A more concrete representation is provided by the eigenfunction expansion theorem of [21] which will be reviewed here briefly and applied to (1.3.21).

The expansion theorem states that every $h \in L_2(\Omega)$ has a generalized Fourier transform

$$(1.3.23) \qquad \hat{h}_+(p) = L_2(\mathbf{R}^3)\text{-}\lim_{M \to \infty} \int_{\Omega(M)} \overline{w^+(x, p)}\; h(x)\, dx,$$

where $\Omega(M) = \Omega \cap \{x : |x| < M\}$, and a corresponding eigenfunction representation

$$(1.3.24) \qquad h(x) = L_2(\Omega)\text{-}\lim_{M \to \infty} \int_{|p| \leq M} w^+(x, p)\; \hat{h}_+(p)\, dp.$$

Moreover, the transformation $\Phi_+ : L_2(\Omega) \to L_2(R^3)$ defined by

$$(1.3.25) \qquad \Phi_+ h = \hat{h}_+, \quad h \in L_2(\Omega),$$

is unitary and diagonalizes A in the sense that

$$(1.3.26) \qquad (\Phi_+ F(A)h)(p) = F(|p|^2)\Phi_+ h(p)$$

for every bounded measurable function $F(\lambda)$ which is defined for $\lambda \geq 0$. In what follows the relations (1.3.23), (1.3.24) will be written in the symbolic form

(1.3.27)
$$\hat{h}_+(p) = \int_\Omega \overline{w^+(x,p)} \ h(x) \, dx$$

(1.3.28)
$$h(x) = \int_{R^3} w^+(x,p) \ \hat{h}_+(p) \, dp.$$

However, these integrals are not in general convergent and must be interpreted in the sense of (1.3.23), (1.3.24).

A second complete family of generalized eigenfunctions for A is defined by

(1.3.29)
$$w^-(x,p) = \overline{w^+(x,-p)}.$$

This family can also be characterized by the boundary value problem (1.2.5)- (1.2.9), but with the outgoing radiation condition (1.2.9) replaced by the incoming radiation condition

(1.3.30)
$$\partial w_{sc}^-/\partial |x| + i|p| w_{sc}^- = \mathcal{O}(1/|x|^2), \quad |x| \to \infty$$

where $w_{sc}^- = w^- - w_0$. The relations (1.3.23)-(1.3.26) hold with $w^+(x,p)$, $\hat{h}_+(p)$ and Φ_+ replaced by $w^-(x,p)$, $\hat{h}_-(p)$ and Φ_-, respectively. This was proved in [21]. It also follows directly from (1.3.23)-(1.3.26) and (1.3.29). The far field behavior of $w^-(x,p)$ follows from (1.2.12) and (1.3.29):

(1.3.31)
$$w_{sc}^-(x,p) = \frac{e^{-i|p||x|}}{4\pi|x|} T_-(|p|\theta,p) + \mathcal{O}(1/|x|^2), \quad |x| \to \infty$$

where

(1.3.32)
$$T_-(p,p') = \overline{T_+(p,-p')}.$$

Application of (1.3.26) to (1.3.21) gives

(1.3.33)
$$(\Phi_+ \ e^{-itA^{1/2}} h)(p) = e^{-it|p|} \ \hat{h}_+(p),$$

and hence, by (1.3.24),

(1.3.34)
$$v(t,x) = \int_{R^3} w^+(x,p) \ e^{-it|p|} \ \hat{h}_+(p) \, dp.$$

The family $\{w^-(x,p) : p \in R^3\}$ provides a second representation

(1.3.35)
$$v(t,x) = \int_{R^3} w^-(x,p) \ e^{-it|p|} \ \hat{h}_-(p) \, dp.$$

where

(1.3.36)
$$\hat{h}_-(p) = \int_\Omega \overline{w^-(x,p)} \ h(x) \, dx.$$

If h is defined by (1.3.22) where f satisfies (1.3.10), (1.3.14), (1.3.18) then the continuity of Φ_+ and Φ_-, together with (1.3.26), imply that

$$(1.3.37) \qquad \hat{h}_\pm(p) = \int_{t_0}^{t_0+T} e^{i\tau|p|} \int_{t_0}^{\tau} \Phi_\pm f(t,p)\,dt d\tau.$$

Interchange of the t and τ integrations and use of (1.3.18) gives

$$(1.3.38) \qquad \hat{h}_\pm(p) = i|p|^{-1} \int_{t_0}^{t_0+T} \int_{\Omega} \overline{e^{-i|p|t}\,w^\pm(x,p)}f(t,x)\,dx dt,$$

or

$$(1.3.39) \qquad \hat{h}_\pm(p) = (2\pi)^{1/2}i|p|^{-1}\hat{f}_\pm(-|p|,p),$$

where

$$(1.3.40) \qquad \hat{f}_\pm(\omega,p) = \frac{1}{(2\pi)^{1/2}} \int_{t_0}^{t_0+T} \int_{\Omega} \overline{e^{i\omega t}w^\pm(x,p)}f(t,x)\,dx dt.$$

The total field produced by the scattering of acoustic waves with a source function $f(t,x)$ is represented by (1.3.34) and (1.3.35) where $\hat{h}_+(p)$ and $\hat{h}_-(p)$ are defined by (1.3.39) and (1.3.40). These equations are not directly applicable to the boundary value problem (1.1.14)-(1.1.17) for the scattering of plane waves. However, it will be shown that this problem can be reduced to the preceding one. To this end let r_0 and R be radii such that $\delta < r_0 < R$, where $\Gamma \subset \{x : |x| \le \delta\}$, and let $j(x) \in C^\infty(R^3)$ be a function with the properties

$$(1.3.41) \qquad j(x) = 0 \quad \text{for} \quad |x| \le r_0, \quad j(x) = 1 \quad \text{for} \quad |x| \ge R.$$

The solution of the boundary value problem (1.1.14)-(1.1.17) will be constructed as a sum

$$(1.3.42) \qquad u(t,x) = j(x)u_0(t,x) + \tilde{u}_{sc}(t,x).$$

Note that $\tilde{u}_{sc}(t,x)$ coincides with the echo (1.1.18) for $|x| \ge R$ and all $t \in R$, by (1.3.41).

The solution of (1.1.14)-(1.1.17) is given by (1.3.42) if $\tilde{u}_{sc}(t,x)$ satisfies

$$(1.3.43) \qquad \partial_t^2 \tilde{u}_{sc} - \Delta\tilde{u}_{sc} = f(t,x) \quad \text{for} \quad t \in R, x \in \Omega,$$
$$(1.3.44) \qquad \partial_\nu \tilde{u}_{sc} = 0 \quad \text{for} \quad t \in R, x \in \partial\Omega,$$
$$(1.3.45) \qquad \tilde{u}_{sc}(t,x) = 0 \quad \text{for} \quad t + b + R < 0, x \in \Omega,$$

where

$$(1.3.46) \qquad f(t,x) = -\left(\partial_t^2 - \Delta\right)j(x)u_0(t,x).$$

Note that supp $f \subset \{(t,x) : -b - R \le t \le -a + R$ and $r_0 \le |x| \le R\}$, by (1.1.14) and (1.3.41). In particular, (1.3.10) holds with $t_0 = -b - R$ and $T = 2R + b - a$. Moreover, (1.3.14) holds if

$$(1.3.47) \qquad s(\cdot, \theta_0) \in L_2^2(R) = L_2(R) \cap \{s(\tau, \theta_0) : \partial_\tau^j s \in L_2(R), j = 1, 2\},$$

and (1.3.18) holds if

$$(1.3.48)) \qquad \int_a^b s(\tau)\, d\tau = 0.$$

Of course, if $\partial\Gamma$ is not smooth then $\tilde{u}_{sc}$ must satisfy the edge condition and generalized Neumann condition; i.e., $\tilde{u}_{sc}(t, \cdot) \in D(A) = L_2^N(\Delta, \Omega)$ for all $t \in R$. The solution is given by

$$(1.3.49) \qquad \tilde{u}_{sc}(t, x) = Re\{\tilde{v}_{sc}(t, x)\},$$

where $\tilde{v}_{sc}$ has two representations,

$$(1.3.50) \qquad \tilde{v}_{sc}(t, x) = \int_{R^3} w^\pm(x, p)\, e^{-it|p|}\, \hat{h}_\pm(p)\, dp,$$

corresponding to w^+ and w^-, and $\hat{h}_+(p)$ and $\hat{h}_-(p)$ are defined by (1.3.39), (1.3.40) and (1.3.46). This section is concluded with a calculation of $\hat{h}_-(p)$.

The support of f in space-time is bounded, by (1.3.47), and a simple calculation gives

$$(1.3.51) \qquad \frac{1}{(2\pi)^{1/2}} \int_{-\infty}^{\infty} e^{-i\omega t}\, f(t, x)\, dt = \hat{s}(-\omega, \theta_0)(\Delta + \omega^2)\, j(x)\, e^{-i\omega\theta_0 \cdot x}$$

where $\hat{s}(\omega, \theta_0)$ is defined by (1.1.36). Thus, by (1.3.29), (1.3.40) and (1.3.51), one has

$$(1.3.52) \qquad \hat{f}_-(\omega, p) = \hat{s}(-\omega, \theta_0) \int_\Omega w^+(x, -p)\, (\Delta + \omega^2)\, j(x)\, e^{-i\omega\theta_0 \cdot x}\, dx.$$

Now by (1.3.41), one has

$$(1.3.53) \qquad \text{supp } (\Delta + \omega^2)\, j(x)\, e^{-i\omega\theta_0 \cdot x} \subset \{x : r_0 \le |x| \le R\} = \Omega(R) - \Omega(r_0).$$

Hence, replacing Ω by $\Omega(R) - \Omega(r_0)$ in (1.3.52) and applying Green's theorem to $w^+(x, -p)$ and $j(x)\, e^{i|p|\theta_0 \cdot x}$ in that region, gives

$$(1.3.54) \qquad \begin{aligned} &\hat{f}_-(-|p|, p) \\ &= \hat{s}(|p|, \theta_0) \int_{|x|=R} \left\{ w^+(x, -p)\frac{\partial e^{i|p|\theta_0 \cdot x}}{\partial |x|} - e^{i|p|\theta_0 \cdot x}\frac{\partial w^+(x, -p)}{\partial |x|} \right\} dS \\ &= \hat{s}(|p|, \theta_0)\, T_+(p, |p|\theta_0) \end{aligned}$$

by (1.2.13) and (1.2.14). Thus, by (1.3.39), one has

$$(1.3.55) \qquad \hat{h}_-(p) = (2\pi)^{1/2} i \, |p|^{-1} \, T_+(p, |p|\theta_0) \hat{s}(|p|, \theta_0)$$

Combining (1.3.49), (1.3.50) and (1.3.55) gives the representation

(1.3.56)
$$u_{sc}(t,x) = Re \left\{ (2\pi)^{1/2} i \int_{R^3} w^-(x,p) \, e^{-it|p|} \, |p|^{-1} \, T_+(p, |p|\theta_0) \hat{s}(|p|, \theta_0) \, dp \right\},$$

for $|x| \geq R$. Actually, (1.3.56) is valid for all $x \in \Omega$. This can be seen by noting that (1.3.56) does not depend on $j(x)$ and that condition (1.3.41) may be replaced by the condition that supp $\nabla j(x)$ is contained in an arbitrarily small neighborhood of Γ.

Exercises for §1.3.

1. Verify by formal differentiation that (1.3.12) defines a solution of (1.3.9) that satisfies the initial conditions $u(t_0, \cdot) = 0$ and $\partial_t u(t_0, \cdot) = 0$. This may be done by verifying equations (1.3.16) and (1.3.17).

2. Use condition (1.3.18) to derive the representation (1.3.19) for the total wave field $u(t, \cdot)$.

3. Use the representation (1.3.19) to derive the representation (1.3.20)- (1.3.22) for $u(t, \cdot)$.

4. Use equations (1.3.29) and (1.2.12) to calculate the far field form (1.3.31), (1.3.32) of $w_{sc}^-(x, p)$.

5. Show that if h is defined by (1.3.22) then $\hat{h}_\pm(p)$ is given by (1.3.39), (1.3.40).

6. Show that if the scattered field $\tilde{u}_{sc}$ is defined by (1.3.42) then equations (1.3.43)-(1.3.46) hold.

7. Starting with the definition (1.3.46), derive the representation (1.3.55) of $\hat{h}_-$ by carrying out the steps (1.3.51)-(1.3.54), as indicated in the text.

8. Show that the representation (1.3.56) for $u_{sc}(t, x)$ actually holds for all $t \geq t_0 + T$ and all $x \in \Omega$.

1.4 Sonar Echoes in the Far Field. The sonar echo $u_{sc}(t, x)$ of the primary field $u_0(t, x) = s(x \cdot \theta_0 - t, \theta_0)$ originates at Γ and reaches points x in the far field, characterized by $|x| >> \delta$, after a time interval of magnitude comparable to $|x|$. Hence the far field form of $u_{sc}(t, x)$ coincides with its asymptotic form for large t. The latter is provided by the theory of asymptotic wave functions developed in [21]. The relevant parts of the theory are summarized here and applied to $u_{sc}(t, x)$.

The simplest theorem concerning asymptotic wave functions states that if

$$(1.4.1) \qquad u(t, x) = Re\{v(t, x)\}, \quad v(t, \cdot) = e^{-itA^{1/2}} h, \ h \in L_2(\Omega),$$

then

$$(1.4.2) \qquad \lim_{t \to \infty} \|u(t, \cdot) - u^\infty(t, \cdot)\| = 0$$

where $u^\infty(t, x)$, the asymptotic wave function for $u(t, x)$, is defined by

$$(1.4.3) \qquad u^\infty(t, x) = \frac{F(|x| - t, \theta)}{|x|}, \quad x = |x|\theta,$$

and

$$(1.4.4) \qquad F(\tau, \theta) = Re\left\{ \frac{1}{(2\pi)^{1/2}} \int_0^\infty e^{i\tau\omega} \hat{h}_-(\omega\theta) (-i\omega) \, d\omega \right\},$$

where $\hat{h}_-(p)$ is defined by (1.3.36). This result is proved in [21], Corollary 8.3.

Equations (1.4.1)-(1.4.4) can be applied to the echo $u_{sc}(t, x)$. Indeed, (1.3.56) implies that

$$(1.4.5) \qquad u_{sc}(t, x) = Re\{v_{sc}(t, x)\}, \quad v_{sc}(t, \cdot) = e^{-itA^{1/2}} h,$$

where h is defined by (1.3.55). Equation (1.3.22) states that

$$(1.4.6) \qquad h(x) = \int_{t_0}^{t_0+T} e^{i\tau A^{1/2}} \int_{t_0}^\tau f(t, x) \, dt d\tau.$$

A simple calculation based on (1.3.46) and (1.4.6) gives the alternative representation

(1.4.7)
$$h(x) = \int_{t_0}^{t_0+T} e^{i\tau A^{1/2}} \left\{ \Delta j(x) \int_{x \cdot \theta_0 - \tau}^\infty s(t, \theta_0) \, dt - (\theta_0 \cdot \nabla j(x)) s(x \cdot \theta_0 - \tau, \theta_0) \right\} d\tau.$$

It follows that if

$$(1.4.8) \qquad s(\cdot, \theta_0) \in L_2^2(R), \quad \text{supp } s(\cdot, \theta_0) \subset [a, b], \quad \int_a^b s(\tau, \theta_0) \, d\tau = 0,$$

as was assumed above, then $h \in D(A)$. In particular, $h \in L_2(\Omega)$ and $u_{sc}(t, x)$ has the asymptotic wave function

$$(1.4.9) \qquad u_{sc}^\infty(t, x) = \frac{e(|x| - t, \theta, \theta_0)}{|x|}, \quad x = |x|\theta,$$

where $e(\tau, \theta, \theta_0) = F(\tau, \theta)$ is given by (1.4.4) with $\hat{h}_-$ defined by (1.3.55). Substitution gives equation (1.1.35) of §1.1; i.e.,

$$(1.4.10) \qquad e(\tau, \theta, \theta_0) = Re \left\{ \int_0^\infty e^{i\tau\omega} T_+(\omega\theta, \omega\theta_0) \, \hat{s}(\omega, \theta_0) \, d\omega \right\}.$$

Moreover, (1.1.19) of §1.1 now has the interpretation

$$(1.4.11) \qquad \lim_{t \to \infty} \|u_{sc}(t, \cdot) - u_{sc}^\infty(t, \cdot)\| = 0.$$

Stronger convergence results follow from the hypotheses (1.4.8). Indeed, (1.4.8) implies that $h \in D(A^{1/2}) = L_2^1(\Omega)$ and hence Theorem 8.5 of [21] implies that

$$(1.4.12) \qquad \lim_{t \to \infty} \|\partial_j u_{sc}(t, \cdot) - u_{sc,j}^\infty(t, \cdot)\| = 0, j = 0, 1, 2, 3,$$

where $\partial_j = \partial/\partial x_j$ and $x_0 = t$,

$$(1.4.13) \qquad u_{sc,j}^\infty(t, x)) = \frac{e_j(|x| - t, \theta, \theta_0)}{|x|}, \quad j = 0, 1, 2, 3,$$

$$(1.4.14) \qquad \begin{aligned} e_0(\tau, \theta, \theta_0) &= -\partial_\tau e(\tau, \theta, \theta_0) \\ &= Re \left\{ \int_0^\infty e^{i\tau\omega} (-i\omega) T_+(\omega\theta, \omega\theta_0) \, \hat{s}(\omega, \theta_0) \, d\omega \right\} \end{aligned}$$

and

$$(1.4.15) \qquad e_j(\tau, \theta, \theta_0) = -\theta_j \, e_0(\tau, \theta, \theta_0) \quad \text{for} \quad j = 1, 2, 3.$$

Note that

$$(1.4.16) \qquad u_{sc,0}^\infty(t, x) = \partial_t u_{sc}^\infty(t, x).$$

The remaining asymptotic wave functions, corresponding to the space derivatives, are not the derivatives of u_{sc}^∞ but differ from them by terms which converge to zero in $L_2(\Omega)$ when $t \to \infty$. This can be proved by means of Lemma 2.7 of [21].

Exercises for §1.4.

1. Derive the representation (1.4.7) for $h(x)$ from equations (1.4.6) and (1.3.46).

2. Use equations (1.4.4) and (1.3.55) to complete the verification of the representation (1.4.10) for the echo waveform $e(\tau, \theta, \theta_0)$.

1.5 Sonar Echoes from Moving Scatterers. The preceding sections treated scattering from objects that were stationary with respect to the sonar system. In this section the analysis is extended to moving scatterers under the simplifying assumption that their velocities are subsonic and essentially constant during the scattering of a pulse.

The relative velocity of the sonar system and the scatterer, during the short interval when the pulse interacts with the scatterer, will be described by a constant vector $\vec{v}$ with magnitude $v = |\vec{v}|$ such that $0 < v < 1 (=$ speed of sound). The space-time coordinates of a Galilean reference frame in which the sonar system is at rest will be denoted by $(t, x) = (t, x_1, x_2, x_3)$, as in the preceding sections. The coordinates of a second Galilean frame in which the scatterer Γ is at rest will be denoted by $(t', x') = (t', x'_1, x'_2, x'_3)$. The two coordinate systems are related by a Lorentz transformation based on $\vec{v}$. It may be assumed that the spatial axes of the two systems are parallel and that the origin $\mathcal{O}'$ of the second system moves along the x_1-axis in the positive direction with speed v. The Lorentz transformation then takes the form [1]

$$(1.5.1) \qquad t' = \frac{t - vx_1}{\sqrt{1-v^2}}, \; x'_1 = \frac{x_1 - vt}{\sqrt{1-v^2}}, \; x'_2 = x_2, \; x'_3 = x_3,$$

with inverse

$$(1.5.2) \qquad t = \frac{t' + vx'_1}{\sqrt{1-v^2}}, \; x_1 = \frac{x'_1 + vt'}{\sqrt{1-v^2}}, \; x_2 = x'_2, \; x_3 = x'_3.$$

The assumption that Γ is in the far field of the transmitter implies that the sonar signal is effectively a plane wave near Γ, as was shown in §1.1. Thus the signal is described relative to the sonar system frame by a scalar potential

$$(1.5.3) \qquad u_0(t, x) = s(x \cdot \theta_0 - t, \theta_0).$$

The Lorentz transformation (1.5.1) satisfies the relation

$$(1.5.4) \qquad x \cdot \theta_0 - t = \gamma_0(x' \cdot \theta'_0 - t')$$

where

$$(1.5.5) \qquad \gamma_0 = \frac{1 - v\theta_0^1}{\sqrt{1-v^2}} = \frac{1 - \vec{v} \cdot \theta_0}{\sqrt{1-v^2}}$$

and

$$(1.5.6) \qquad \theta'^1_0 = \frac{\theta_0^1 - v}{1 - v\theta_0^1}, \; \theta'^2_0 = \frac{\sqrt{1-v^2}\theta_0^2}{1 - v\theta_0^1}, \; \theta'^3_0 = \frac{\sqrt{1-v^2}\theta_0^3}{1 - v\theta_0^1}.$$

Thus the signal takes the form

$$(1.5.7) \qquad u'_0(t', x') = s'(x' \cdot \theta'_0 - t', \theta'_0),$$

relative to the object frame, where

$$s'(\tau, \theta'_0) = s(\gamma_0 \tau, \theta_0). \tag{1.5.8}$$

It is easy to verify that (1.5.6) defines a mapping of the unit vectors $\theta_0 \in R^3$ into unit vectors $\theta'_0 \in R^3$.

The d'Alembert equation is invariant under the Lorentz transformation (1.5.1). Hence, relative to the object frame coordinates (t', x') the echo produced by the scattering of the signal (1.5.7) from Γ is the solution of the same boundary value problem (1.1.15)-(1.1.18) that was solved in §1.3. It follows from the analysis of §1.4 that the echo is given in the far field by the asymptotic wave function

$$u_{sc}^{\infty\prime}(t', x') = \frac{e'(|x'| - t', \theta', \theta'_0)}{|x'|}, \quad x' = |x'|\theta', \tag{1.5.9}$$

with echo wave form

$$e'(\tau, \theta', \theta'_0) = Re\left\{ \int_0^\infty e^{i\tau\omega} T_+(\omega\theta', \omega\theta'_0)\hat{s}'(\omega, \theta'_0)\, d\omega \right\}. \tag{1.5.10}$$

where $T_+(p, p')$ is the function defined in §1.2, equation (1.2.12).

Equations (1.5.9), (1.5.10) describe the echo relative to the object frame. To describe it relative to the sonar system frame it is necessary to apply the Lorentz transformation again. This may be done by noting that along the ray $x' = |x'|\theta'$ with θ' fixed and $|x'| \geq 0$ one has $|x'| = x' \cdot \theta'$. Hence, on using (1.5.4)-(1.5.6) with θ_0 replaced with θ, one gets

$$\begin{aligned}
e'(|x'|-t', \theta', \theta'_0) &= e'(x' \cdot \theta' - t', \theta', \theta'_0) \\
&= e(x \cdot \theta - t, \theta, \theta_0) = e(|x| - t, \theta, \theta_0),
\end{aligned} \tag{1.5.11}$$

where $e(\tau, \theta, \theta_0)$ is defined by

$$e(\tau, \theta, \theta_0) = e'(\gamma^{-1}\tau, \theta', \theta'_0), \tag{1.5.12}$$

$$\gamma = \frac{1 - v\theta^1}{\sqrt{1 - v^2}} = \frac{1 - \vec{v} \cdot \theta}{\sqrt{1 - v^2}}, \tag{1.5.13}$$

and θ'_0 (respectively θ') is derived from θ_0 (respectively θ) by (1.5.6). Thus the echo is described relative to the sonar system frame by

$$u_{sc}^\infty(t, x) = \frac{e(|x| - t, \theta, \theta_0)}{|x|}, \tag{1.5.14}$$

where the echo waveform $e(\tau, \theta, \theta_0)$ is defined by (1.5.10), (1.5.12) and (1.5.13). The resulting form of $e(\tau, \theta, \theta_0)$ can be written (up to the slowly varying factor $|x'|/|x|$)

$$e(\tau, \theta, \theta_0) = Re\left\{ \int_0^\infty e^{i(\frac{\gamma_0}{\gamma}\tau)\omega} T_+(\omega\gamma_0\theta', \omega\gamma_0\theta'_0)\hat{s}(\omega, \theta_0)\, d\omega \right\}. \tag{1.5.15}$$

If $v = 0$ then $\gamma = \gamma_0 = 1$, $\theta' = \theta$, $\theta_0' = \theta_0$ and (1.5.15) reduces to equation (1.4.10) for the echo waveform of a stationary scatterer. Comparison of (1.4.10) and (1.5.15) shows that the motion of the scatterer relative to the sonar system causes three kinds of distortion in the echo. These are a frequency shift in the signal waveform of amount

$$\text{(1.5.16)} \qquad \Delta = \frac{\gamma}{\gamma_0} - 1,$$

a frequency shift in the scattering amplitude T_+ of amount

$$\text{(1.5.17)} \qquad \Delta'\omega = \gamma_0 - 1,$$

and an angular distortion in the scattering amplitude T_+ due to the replacement of θ and θ_0 by θ' and θ_0'. The frequency shift (1.5.16) is the usual Doppler shift.

Final Approximations. Equation (1.5.14) gives the echo waveform, in the far field of the scatterer, due to the incident pulse $s(x \cdot \theta_0 - t, \theta_0)$. Recall that, in the original formulation of the sonar echo prediction problem, the transmitter was localized near x_0 and produced a signal at the scatterer which was given by

$$\text{(1.5.18)} \qquad u_0(t, x) = \frac{s(x \cdot \theta_0 - t + |x_0|, \theta_0)}{|x_0|},$$

apart from an error of order $1/|x_0|^2$. The echo produced by the signal (1.5.18) may be obtained from (1.5.14) by multiplying by the amplitude factor $1/|x_0|$ and replacing t by $t - |x_0|$. Thus, one has

$$\text{(1.5.19)} \qquad u_{sc}(t, x) \approx \frac{e(|x| - t + |x_0|, \theta, \theta_0)}{|x||x_0|}$$

in the far field of the scatterer. In particular, if the transmitter and receiver coincide, so $x = x_0 = |x_0|(-\theta_0)$, then the echo at the receiver is given by

$$\text{(1.5.20)} \qquad u_{sc}(t, x_0) \approx \frac{e(2|x_0| - t, -\theta_0, \theta_0)}{|x_0|^2},$$

where $e(\tau, \theta, \theta_0)$ is given by (1.5.15).

High Frequency Limit. A. Majda's theorem (1.1.37) implies that if $\hat{s}(\omega, \theta_0)$ is concentrated in a high frequency band $\omega \geq \omega_0 \gg 1$, where T_+ is essentially equal to T_+^∞, then

$$\text{(1.5.21)} \qquad e(\tau, \theta, \theta_0) \approx \left(\frac{\pi}{2}\right)^{1/2} T_+^\infty(\theta', \theta_0') s\left(\frac{\gamma_0}{\gamma}\tau, \theta_0\right),$$

where $\gamma = \gamma(\theta) = (1 - \vec{v} \cdot \theta)/\sqrt{1 - v^2}$ and $\gamma_0 = \gamma(\theta_0)$. In particular, equations (1.5.20) and (1.5.21) give

$$\text{(1.5.22)} \qquad u_{sc}(t, x_0) \approx K(x_0)\, s\left((\gamma(\theta_0)/\gamma(-\theta_0))(2|x_0| - t), \theta_0\right),$$

where $K(x_0)$ is a suitable amplitude factor. Thus, in the high frequency limit, the echo waveform is equal to the signal waveform, with a time delay equal to twice the scatterer's range $|x_0|$ and with a time dilation $\gamma(\theta_0)/\gamma(-\theta_0)$ due to the scatterer's velocity.

Exercises for §1.5.

1. Verify equations (1.5.4)-(1.5.6) for the Lorentz transformation (1.5.1).

2. Show that the transformation $\theta_0 \to \theta_0'$ defined by (1.5.6) carries unit vectors into unit vectors. Verify that the transformation does not preserve the lengths of vector that are not unit vectors.

3. Give the details of the derivation of the representation (1.5.15) for the echo wave form $e(\tau, \theta, \theta_0)$.

Part 2. RADAR ECHO STRUCTURE

2.1 The Radar Echo Prediction Problem. The analysis of this problem presented below closely parallels the analysis of sonar echo prediction. Thus both pulse mode and CW mode operation of radar systems are treated and the construction of pulse mode echoes is derived from the theory of CW mode echoes. Again, the analysis of each of these modes is based of a boundary value problem. Each step in the analysis of the sonar problem has its analogue for the radar problem. In particular, the scalar d'Alembert equation for the sonar field is replaced by a vector wave equation for the radar field. The vector nature of the radar problem necessitates more complex algebraic calculations. Nevertheless, each of the principal steps in the analysis of the sonar problem has its analogue for the radar problem.

Radar echo prediction is analyzed below under the following physical assumptions.

- The radar system (transmitter and receiver) operates in a homogeneous isotropic unlimited medium which is at rest in a suitable Galilean reference frame.
- The radar system is stationary with respect to that frame.
- The radar signals are generated by localized electric currents.
- The scatterers are bounded objects which are perfect electrical conductors.
- The transmitter and receiver are in the far field of the scattering objects.
- The speeds of the scattering objects are much less than the speed of radar waves (i.e., the speed of light) and are essentially constant during the interval required for the radar pulse to sweep over the objects.
- Secondary echoes due to the radar system components are negligible.
- Noise in the medium is negligible.

The notation used below for the radar problem will parallel that used for the sonar problem as closely as possible, in order to stress the analogies between the two problems. In particular, $(t, x) = (t, x_1, x_2, x_3) \in R^4$ denotes the time and space coordinates of a Galilean reference frame in which the medium and the radar system are at rest. $\Gamma \subset R^3$ denotes a closed bounded set such that $\Omega = R^3 - \Gamma$ is connected. As before, the set Γ represents the scattering object or objects.

The medium filling Ω is characterized by a dielectric constant ϵ_0 and a magnetic permeability μ_0. It will be assumed, for simplicity, that $\epsilon_0 = 1$ and $\mu_0 = 1$, since this can be achieved by a suitable choice of units. Maxwell's equations then take the form

(2.1.1)
$$\nabla \times \overrightarrow{H} - \partial_t \overrightarrow{E} = \overrightarrow{J},$$
$$\nabla \times \overrightarrow{E} + \partial_t \overrightarrow{H} = \overrightarrow{0},$$

in the notation of conventional vector analysis, where $\overrightarrow{E}$ and $\overrightarrow{H}$ denote the electric and magnetic field vectors, respectively, and $\overrightarrow{J}$ denotes the electric current density that generates the field. Elimination of the field $\overrightarrow{H}$ yields the second order vector wave equation

$$(2.1.2) \qquad \partial_t^2 \overrightarrow{E} + \nabla \times (\nabla \times \overrightarrow{E}) = -\partial_t \overrightarrow{J}.$$

It is this equation that will be used below to analyze the radar echo prediction problem. Note that if the the electric field has been determined then the magnetic field can be found from the second of Maxwell's equations (2.1.1.) by differentation and quadrature.

Notation. The vector equation (2.1.2) is equivalent to a system of three second order partial equations for the components of the electric field. In what follows the notation and conventions of matrix algebra will be used to write this system and related equations. Thus if

$$(2.1.3) \qquad \overrightarrow{E} = E_1 \hat{\imath}_1 + E_2 \hat{\imath}_2 + E_3 \hat{\imath}_3,$$

where $(\hat{\imath}_1, \hat{\imath}_2, \hat{\imath}_3)$ is the orthonormal basis associated with the coordinates (x_1, x_2, x_3), then (2.1.3) defines a one-to-one correspondence

$$(2.1.4) \qquad \overrightarrow{E} \leftrightarrow u = (E_1, E_2, E_3)^T,$$

where the notation M^T denotes the transpose of a matrix M. Under the mapping defined by

$$(2.1.5) \qquad \overrightarrow{a} = a_1 \hat{\imath}_1 + a_2 \hat{\imath}_2 + a_3 \hat{\imath}_3 \leftrightarrow a = (a_1, a_2, a_3)^T$$

the dot product and the cross product of vector algebra correspond to the following matrix operations:

$$(2.1.6) \qquad \overrightarrow{p} \cdot \overrightarrow{a} \leftrightarrow p^T a = p \cdot a = p_1 a_1 + p_2 a_2 + p_3 a_3,$$

and

$$(2.1.7) \qquad \overrightarrow{p} \times \overrightarrow{a} \leftrightarrow p \times a = M(p)a$$

where $M(p)$ is the matrix defined by

$$(2.1.8) \qquad M(p) = \begin{pmatrix} 0 & -p_3 & p_2 \\ p_3 & 0 & -p_1 \\ -p_2 & p_1 & 0 \end{pmatrix}.$$

Similarly, the vector differential operator "curl" has the representation

$$(2.1.9) \qquad \nabla \times \overrightarrow{E} \leftrightarrow \nabla \times u = M(\partial)u$$

where $\partial = (\partial_1, \partial_2, \partial_3)$. Moreover, iteration gives

$$(2.1.10) \qquad \nabla \times (\nabla \times \vec{E}) \leftrightarrow \nabla \times (\nabla \times u) \equiv Au \equiv A(\partial)u$$

where

$$(2.1.11) \qquad A(p) = M(p)^2 = pp - |p|^2 1.$$

In the last equation pp denotes the 3×3 matrix whose jk^{th} component is $p_j p_k$, and 1 denotes the 3×3 identity matrix. Hence,the differential identity (2.1.10) can also be written

$$(2.1.12) \qquad \nabla \times (\nabla \times u) = (\nabla\nabla - \Delta 1)u,$$

where Δ denotes the three-dimensional Laplacian:

$$(2.1.13) \qquad \Delta = \partial_1^2 + \partial_2^2 + \partial_3^2.$$

Note also that on squaring the matrix $A(p)$ one finds that it has the minimal polynomial $\lambda^2 + |p|^2 \lambda$; see [20]. Thus, one has the identity

$$(2.1.14) \qquad A(p)^2 + |p|^2 A(p) = 0.$$

In the correspondence defined by (2.1.5) every second-order tensor corresponds to a 3×3 matrix C. The tensors pp and 1 of (2.1.11) are instances of this. It will be helpful below to use the dyadic notation of Gibbs for the contraction of a tensor and a vector[27]. Thus,

$$(2.1.15) \qquad C \cdot a = Ca \quad (3 \times 1) \quad \text{and} \quad a \cdot C = a^T C \quad (1 \times 3).$$

The same convention holds for differential operators. In particular, in (2.1.12), one has

$$(2.1.16) \qquad \nabla\nabla u = (\nabla\nabla) \cdot u = \nabla(\nabla \cdot u).$$

The bilinear form corresponding to a tensor C is given by

$$(2.1.17) \qquad a \cdot C \cdot b = a^T C b.$$

Note that the transpose of C satisfies $a \cdot C^T \cdot b = b \cdot C \cdot a$. Moreover, two tensors C_1 and C_2 are equal if they satisfy the identity $a \cdot C_1 \cdot b = a \cdot C_2 \cdot b$ for all vectors a and b, or even for all vectors of a basis.

The electromagnetic field generated by a radar transmitter is characterized by a function

$$(2.1.18) \qquad u = u(t,x) = (u_1(t,x), u_2(t,x), u_3(t,x))^T, \quad t \in R, \quad x \in \Omega,$$

whose values are the components $E_j = u_j$, $j = 1, 2, 3$ of the electric field. u is a solution of the inhomogeneous vector wave equation

(2.1.19)
$$\partial_t^2 u + \nabla \times (\nabla \times u) = f(t, x), \quad t \in R, \quad x \in \Omega,$$

where

(2.1.20)
$$f(t, x) = -\partial_t J(t, x),$$

and $J(t, x)$ is the electric current density that generates the field. As in Part 1, $f(t, x)$ must ultimately be chosen to produce a desired transmitter radiation pattern. For the present, f is assumed to be known.

Pulse Mode Scattering. Consider a single short pulse of duration T which is generated by a transmitter that is localized near a point x_0, so that the support of $f(t, x)$ satisfies (1.1.4), as in the sonar case. The electromagnetic field generated by $f(t, x)$ is characterized by a vector function $u(t, x)$ that satisfies (2.1.19), together with the boundary condition

(2.1.21)
$$M(\nu)u \equiv \nu \times u = 0 \quad \text{for} \quad t \in R, \quad x \in \partial\Omega,$$

where $\nu = \nu(x)$ is a normal vector to $\partial\Omega$. The initial condition (1.1.6) is required, as before. The boundary condition (2.1.21) characterizes objects that are perfect electrical conductors, while the initial condition implies that there is no signal before the sources begin to act.

The Primary Field. The primary field of the transmitter, denoted by $u_0(t, x)$, is the field that is generated by the sources when no scatterers are present. It can again be represented as a retarded potential. This may be shown by noting that (2.1.19),(2.1.20), and the well known identity

(2.1.22)
$$\nabla \cdot (\nabla \times u) = 0,$$

imply that

(2.1.23)
$$\partial_t^2 \nabla \cdot u = -\partial_t \nabla \cdot J.$$

On integrating this twice over the interval $t_0 \leq \tau \leq t$, and using the initial condition (1.1.6), one finds that

(2.1.24)
$$\nabla \cdot u(t, x) = -\int_{t_0}^t \nabla \cdot J(\tau, x) \, d\tau, \quad \text{for} \quad t \in R, \quad x \in \Omega.$$

Equations (2.1.19),(2.1.24) and the identity (2.1.12) imply that u satisfies the alternative wave equation

(2.1.25)
$$\partial_t^2 u - \Delta u = \nabla\nabla \cdot \int_{t_0}^t J(\tau, x) \, d\tau - \partial_t J(t, x).$$

In particular, the primary field $u_0(t, x)$ is determined by the wave equation

$$(2.1.26) \qquad \partial_t^2 u_0 - \Delta u_0 = (\nabla\nabla - \partial_t^2 1) \cdot \int_{t_0}^t J(\tau, x)\, d\tau, \quad \text{for} \quad t \in R, \quad x \in R^3,$$

and the initial condition that $u_0(t, x) = 0$ for all $t \le t_0$. Equation (2.1.26) is equivalent to three scalar wave equations and may be integrated by the retarded potential formula. Thus if the function IJ is defined by

$$(2.1.27) \qquad IJ(t, x) = \int_{t_0}^t J(\tau, x)\, d\tau,$$

then the function $u_1(t, x)$ defined by

$$(2.1.28) \quad u_1(t, x) = \frac{1}{4\pi} \int_{|x' - x_0| \le \delta_0} \frac{IJ(t - |x - x'|, x')}{|x - x'|}\, dx' \quad \text{for} \quad t \in R, \quad x \in R^3$$

satisfies

$$(2.1.29) \qquad \partial_t^2 u_1 - \Delta u_1 = IJ(t, x), \quad \text{for} \quad t \in R, \quad x \in R^3,$$

and the initial condition $u_1(t, x) = 0$ for $t \le t_0$. The linearity of the wave equation then implies that the primary field is given by

$$(2.1.30) \qquad u_0(t, x) = (\nabla\nabla - \partial_t^2 1) \cdot u_1(t, x).$$

The Far Field Approximation. The variation of the primary field $u_0(t, x)$ near the scatterer Γ will be simplified by using the assumption that Γ lies in the far field of the transmitter. As in the sonar case, it is assumed that the origin of coordinates lies in Γ and that

$$(2.1.31) \qquad \Gamma \subset \{x : |x| \le \delta\}.$$

The far field assumption then takes the form

$$(2.1.32) \qquad x_0 = -|x_0|\theta_0, \quad |x_0| >> \delta_0 + \delta,$$

and substitution of the estimate (1.1.11) for $|x - x'|$ into (2.1.28) gives

$$(2.1.33) \qquad u_1(t, x) = \frac{s_1(x \cdot \theta_0 - t + |x_0|, \theta_0)}{|x_0|} + \mathcal{O}(\frac{1}{|x_0|^2}), \quad |x_0| \to \infty,$$

uniformly for $t \in R$, $|x| \le \delta$, where

$$(2.1.34) \quad s_1(\tau, \theta_0) = \frac{1}{4\pi} \int_{|x' - x_0| \le \delta_0} IJ(\theta_0 \cdot (x' - x_0) - \tau, x')\, dx' \quad \text{for} \quad \tau \in R.$$

Application of the operator $(\nabla\nabla - \partial_t^2 1)\cdot$ to (2.1.33) gives

$$(2.1.35) \qquad u_0(t,x) = \frac{s(x\cdot\theta_0 - t + |x_0|,\theta_0)}{|x_0|} + \mathcal{O}(\frac{1}{|x_0|^2}), \qquad |x_0| \to \infty,$$

where

$$(2.1.36) \qquad s(\tau,\theta_0) = (\theta_0\theta_0 - 1)\cdot\frac{1}{4\pi}\int_{|x'-x_0|\leq\delta_0} \partial_t J(\theta_0\cdot(x'-x_0) - \tau, x')\,dx'.$$

Note that the tensor

$$(2.1.37) \qquad P(\theta) = 1 - \theta\theta$$

is the orthogonal projection of R^3 onto the plane through the origin with normal vector θ. In particular, $P(\theta)$ is real, symmetric and idempotent; i.e.,

$$(2.1.38) \qquad P(\theta)^2 = P(\theta).$$

Combining (2.1.20),(2.1.36) and (2.1.37) gives the following alternative representation of the wave profile $s(\tau,\theta_0)$:

$$(2.1.39) \qquad s(\tau,\theta_0) = P(\theta_0)\frac{1}{4\pi}\int_{|x'-x_0|\leq\delta_0} f(\theta_0\cdot(x'-x_0) - \tau, x')\,dx'.$$

Electromagnetic Plane Waves. If the error term in (2.1.35) is dropped then the primary field becomes an electromagnetic plane wave. In the remainder of these lectures the primary field is assumed to have this form, namely,

$$(2.1.40) \qquad u_0(t,x) = s(x\cdot\theta_0 - t,\theta_0), \quad \text{where} \quad \text{supp } s(\cdot,\theta_0) \subset [a,b].$$

Note that (2.1.38),(2.1.39) imply that one has

$$(2.1.41) \qquad P(\theta_0)s(\tau,\theta_0) = s(\tau,\theta_0).$$

It is easy to verify that this the only restriction on the plane wave; i.e., (2.1.40) defines a solution of the homogeneous vector wave equation

$$(2.1.42) \qquad \partial_t^2 u_0 + \nabla\times(\nabla\times u_0) = 0 \quad \text{for} \quad t \in R, \quad x \in R^3,$$

for any profile $s(\tau,\theta_0)$ that satisfies (2.1.41). As in the sonar case, a primary task of the design engineer is to design the transmitter in such a way that the signal waveform is a prescribed function.

Pulse Mode Scattering, continued. The electromagnetic field produced when a plane wave (2.1.40) is scattered by Γ is the solution $u(t,x)$ of the boundary value problem

$$(2.1.43) \qquad \partial_t^2 u + \nabla\times(\nabla\times u) = 0 \quad \text{for} \quad t \in R, \quad x \in \Omega$$

$$(2.1.44) \qquad \nu\times u = 0 \quad \text{for} \quad t \in R, \quad x \in \partial\Omega$$

$$(2.1.45) \qquad u(t,x) \equiv u_0(t,x) \quad \text{for} \quad t + b + \delta < 0.$$

The secondary field, or echo, is defined by

$$(2.1.46) \qquad u_{sc}(t,x) = u(t,x) - u_0(t,x) \quad \text{for} \quad t \in R, x \in \Omega.$$

It is shown below that, as in the sonar case, the echo has the form

$$(2.1.47) \qquad u_{sc}(t,x) \approx \frac{e(|x| - t, \theta, \theta_0)}{|x|}, \qquad x = |x|\theta$$

in the far field. The principal goal of Part 2 of these lectures is to calculate the relationship between the echo waveform $e(\tau, \theta, \theta_0)$ and the signal waveform $s(\tau, \theta_0)$. Again, the calculation will be based on the theory of

CW Mode Scattering. Electric current density functions of the form

$$(2.1.48) \qquad J(t,x) = Re\{J(x)e^{-i\omega t}\}$$

generate CW mode fields of the same form:

$$(2.1.49) \qquad u(t,x) = Re\{w(x)e^{-i\omega t}\},$$

where the vector field $w(x) = (w_1(x), w_2(x), w_3(x))^T$ has complex components. $u(t,x)$ must satisfy the vector wave equation (2.1.19),(2.1.20), with $J(t,x)$ defined by (2.1.48), and the boundary condition (2.1.21). The initial condition (1.1.6) is not appropriate for the CW mode field and is replaced by the Silver-Müller radiation condition [2]. The corresponding boundary value problem for $w(x)$ is

$$(2.1.50) \qquad \nabla \times (\nabla \times w) - \omega^2 w = f(x) \quad \text{for} \quad x \in \Omega,$$
$$(2.1.51) \qquad \nu \times w = 0 \quad \text{for} \quad x \in \partial\Omega,$$
$$(2.1.52) \qquad \theta \times (\nabla \times w) + i\omega w = \mathcal{O}(1/|x|^2) \quad \text{for} \quad |x| \to \infty,$$

where $x = |x|\theta$ and it is assumed that

$$(2.1.53) \qquad f(x) = i\omega J(x),$$

and

$$(2.1.54) \qquad \text{supp} \quad f \subset \{x : |x - x_0| \leq \delta_0\},$$

in agreement with (1.1.4). The Silver-Müller radiation condition, like the Sommerfeld radiation condition for the sonar problem, ensures that w is a pure outgoing wave in the far field and guarantees the uniqueness of the CW mode field; see [2].

The primary CW mode field is the CW mode field $w_0(x)$ generated by $f(x)$ when no scattering object is present. It is given by

$$(2.1.55) \qquad w_0(x) = (1 + \frac{1}{\omega^2}\nabla\nabla)\left\{\frac{1}{4\pi}\int_{|x'-x_0|\leq\delta_0} \frac{e^{i\omega|x-x'|}}{|x-x'|} f(x')\, dx'\right\}$$

for all $x \in R^3$. This result may be derived by putting $IJ(t,x) = (-i\omega)^{-1}e^{-i\omega t}J(x)$ and $u_0(t,x) = e^{-i\omega t}w_0(x)$ in (2.1.28),(2.1.30) and making $t \to \infty$. The details are left for the exercises.

Application of the far field assumption, as expressed by (1.1.11), to the integral

$$(2.1.56) \qquad I(x) = \frac{1}{4\pi} \int_{|x'-x_0|\le\delta_0} \frac{e^{i\omega|x-x'|}}{|x-x'|} f(x')\,dx'$$

gives

$$(2.1.57) \qquad I(x) = \frac{T(\omega\theta_0)}{|x_0|} e^{i\omega\theta_0\cdot x} + \mathcal{O}(1/|x_0|^2), \quad |x_0| \to \infty,$$

uniformly for $|x| \le \delta$ where

$$(2.1.58) \qquad T(\omega\theta_0) = \frac{1}{4\pi} \int_{|x'-x_0|\le\delta_0} e^{-i\omega\theta_0\cdot x'} f(x')\,dx'.$$

On applying the operator $1 + \frac{1}{\omega^2}\nabla\nabla$ to (2.1.57) and using (2.1.55) one finds the estimate

$$(2.1.59) \qquad w_0(x) = P(\theta_0)\left\{ \frac{T(\omega\theta_0)}{|x_0|} e^{i\omega\theta_0\cdot x}\right\} + \mathcal{O}(1/|x_0|^2), \quad |x_0| \to \infty.$$

If the error term in (2.1.59) is dropped then the primary field $w_0(x)$ becomes a CW mode plane wave

$$(2.1.60) \qquad w_0(x,\omega\theta_0) = e^{i\omega\theta_0\cdot x} P(\theta_0)\cdot a,$$

where a is a constant vector. It is easy to verify that, for all choices of the vector a, (2.1.60) defines a solution of Maxwell's equation $\nabla\times(\nabla\times w_0) - \omega^2 w_0 = 0$ for the electric field. A complete set of CW mode plane waves for the direction θ_0 is obtained by choosing a basis of constant vectors a_1, a_2, a_3. As an alternative to this it will be convenient to introduce the matrix plane wave

$$(2.1.61) \qquad \Psi_0(x,\omega\theta_0) = (2\pi)^{-3/2} e^{i\omega\theta_0\cdot x} P(\theta_0).$$

Note that the columns of $\Psi_0(x,\omega\theta_0)$ are vector plane waves of the form (2.1.60).

The CW mode electric field which is produced when $\Psi_0(x,\omega\theta_0)$ is scattered by Γ will be denoted by $\Psi^+(x,\omega\theta_0)$. It is the solution

$$(2.1.62) \qquad \Psi^+(x,\omega\theta_0) = \Psi_0(x,\omega\theta_0) + \Psi_{sc}^+(x,\omega\theta_0), \quad x \in \Omega,$$

of the boundary value problem

$$(2.1.63) \qquad \nabla\times(\nabla\times\Psi^+) - \omega^2\Psi^+ = 0 \quad \text{for} \quad x \in \Omega,$$

$$(2.1.64) \qquad \nu\times\Psi^+ = 0 \quad \text{for} \quad x \in \partial\Omega,$$

$$(2.1.65) \qquad \theta\times(\nabla\times\Psi_{sc}^+) + i\omega\Psi_{sc}^+ = \mathcal{O}(1/|x|^2), \quad |x| \to \infty.$$

It is shown below that in the far field of Γ the secondary field $\Psi_{sc}^{+}(x, \omega\theta_0)$ is a diverging spherical wave

$$(2.1.66) \qquad \Psi_{sc}^{+}(x, \omega\theta_0) \approx \frac{e^{i\omega|x|}}{4\pi|x|} T_{+}(\omega\theta, \omega\theta_0), \qquad x = |x|\theta.$$

Note that the coefficient $T_{+}(\omega\theta, \omega\theta_0)$ is a matrix-valued function in the present case. It will be called the matrix scattering amplitude for the scatterer Γ.

With the above notation and conventions, the solution of the radar echo prediction problem takes the same form as for the sonar case, namely,

$$(2.1.67) \qquad e(\tau, \theta, \theta_0) = Re\{ \int_0^\infty e^{i\tau\omega} T_{+}(\omega\theta, \omega\theta_0)\hat{s}(\omega, \theta_0) \, d\omega \},$$

where $\hat{s}(\omega, \theta_0)$ is the Fourier transform of the signal waveform:

$$(2.1.68) \qquad \hat{s}(\omega, \theta_0) = \frac{1}{(2\pi)^{1/2}} \int_{-\infty}^\infty e^{-i\omega\tau} s(\tau, \theta_0) \, d\tau.$$

The only difference from the sonar case is that $T_{+}(\omega\theta, \omega\theta_0)$ is matrix valued, $\hat{s}(\omega, \theta_0)$ is vector valued and the order of the factors in the integrand of (2.1.67) must be preserved. The derivation of equation (2.1.67) is the principal result of Part 2 of these lectures.

The organization of the remainder of Part 2 parallels that of Part 1. Thus, §2.2 reviews the definition of CW mode electric fields and analyzes their far field structure. §2.3 reviews the definition of pulse mode radar wave fields and develops their representation by means of CW mode fields. §2.4 describes the theory of asymptotic wave fields for vector valued fields, as developed in [22] and [23], and applies it to the derivation of the far field form (2.1.47) of radar echoes and the integral relation (2.1.67). §2.5 contains the generalization of (2.1.67) to the case of moving scatterers.

Exercises for §2.1.

1. Verify in detail equations (2.1.9)-(2.1.11) for the curl operator.

2. Verify identity (2.1.14) where $A(p)$ is defined by (2.1.11).

3. Verify equation (2.1.26) for the primary field $u_0(t, x)$.

4. Verify equation (2.1.39) for the incident wave profile $s(\tau, \theta_0)$.

5. Show that if $s(\tau, \theta_0)$ is any twice-differentiable vector-valued function of τ that satisfies (2.1.41) then $u_0(t, x) = s(x \cdot \theta_0 - t, \theta_0)$ is a solution of the homogeneous vector wave equation (2.1.42).

6. Verify the representation (2.1.55) for the primary CW mode field $w_0(x)$. Suggestion: Straight differentation will not work! Try the method suggested in the text.

2.2   The Structure of CW Mode Radar Echoes. The definition and basic properties of the CW mode fields $\Psi^+(x,p)$ are developed in this section. The definition is based of the boundary value problem (2.1.61)-(2.1.65), but in a formulation appropriate for scatterers Γ with non-smooth boundaries. In the classical theory of the electromagnetic problem, a solution is a function in the class $C^2(\Omega) \cap C^1(\bar{\Omega})$, where $\bar{\Omega} = \Omega \cup \partial\Omega$; see [2]. It is known that if $\partial\Omega$ is a smooth surface then a unique classical solution $\Psi^+(x,p)$ exists for each $p = \omega\theta_0 \in R^3 - \{0\}$. However, if $\partial\Omega$ has edges and/or vertices then at such points the normal vector ν is undefined and the boundary condition (2.1.64) is not meaningful. In this case it is known that if (2.1.64) is enforced only at points where $\partial\Omega$ is smooth then all solutions have first derivatives that are singular at the edges and vertices and no classical solution exists [7,8]. Moreover, if the strength of these singularities is not controlled then the uniqueness of the solution is lost [3,8].

The Edge Condition. Physically, non-uniqueness can occur because line or point sources can be situated in the edges or vertices. This type of non-uniqueness is eliminated by the "edge condition" [3] which requires that the energy in each bounded portion of Ω should be finite. It may be formulated mathematically as the condition that $u(x) = \Psi^+(x,\omega\theta_0) \cdot a$ should satisfy [18]

$$(2.2.1) \qquad \int_{K\cap\Omega} \{|\nabla \times u(x)|^2 + |u(x)|^2\}\, dx < \infty \quad \text{for every cube } K \subset R^3.$$

In the theory developed here the edge condition (2.2.1) is formulated by means of the function classes

$$(2.2.2) \qquad L_2^{loc}(\bar{\Omega}) = \{u : u \in L_2(K \cap \Omega) \text{ for every cube } K \subset R^3\},$$

and

$$(2.2.3) \qquad L_2^{loc}(\nabla\times, \bar{\Omega}) = L_2^{loc}(\bar{\Omega}) \cap \{u : \nabla \times u \in L_2^{loc}(\bar{\Omega})\}.$$

Condition (2.2.1) is equivalent to the condition $u \in L_2^{loc}(\nabla\times, \bar{\Omega})$. Moreover, if $u(x) = \Psi^+(x,\omega\theta_0) \cdot a \in L_2^{loc}(\nabla\times, \bar{\Omega})$ is a weak solution of (2.1.63), in the sense of the theory of distributions, then u is also in the class

$$(2.2.4) \qquad L_2^{loc}(\nabla \times (\nabla\times), \bar{\Omega}) = L_2^{loc}(\nabla\times, \bar{\Omega}) \cap \{u : \nabla \times (\nabla \times u) \in L_2^{loc}(\bar{\Omega})\}.$$

The Generalized Boundary Condition for Perfect Conductors. For general domains $\Omega \subset R^3$ and functions $u \in L_2^{loc}(\nabla \times (\nabla\times), \bar{\Omega})$, the boundary condition (2.1.64) for perfect conductors is replaced by the generalized boundary condition

$$(2.2.5) \qquad \int_\Omega \{(\nabla \times u) \cdot v - u \cdot (\nabla \times v)\}\, dx = 0$$

for all $v \in L_2^{loc}(\nabla\times, \bar{\Omega})$ such that $v(x) = 0$ outside of a bounded set. This formulation is based on a vector version of Green's theorem which states that, for domains with smooth boundaries, one has

$$(2.2.6) \qquad \int_\Omega \{(\nabla \times u) \cdot v - u \cdot (\nabla \times v)\}\, dx = \int_{\partial\Omega} (\nu \times u) \cdot v\, dS.$$

Clearly, this implies that classical solutions of the CW scattering problem satisfy the generalized boundary condition (2.2.5). The great advantage of the generalized boundary condition is that it is meaningful for all domains. Moreover, it implies the classical boundary condition at boundary points near which $\partial\Omega$ is smooth. In particular, the classical and generalized boundary conditions are equivalent for domains with smooth boundaries. An early application of this formulation may be found in [18].

Vector fields which lie in the function class

$$(2.2.7) \qquad L_2^{E,loc}(\nabla \times (\nabla\times), \bar{\Omega}) = L_2^{loc}(\nabla \times (\nabla\times), \bar{\Omega}) \cap \{u : (2.2.5)\text{holds}\}$$

are precisely those fields u for which $u, \nabla \times u$ and $\nabla \times (\nabla \times u)$ are in $L_2^{loc}(\bar{\Omega})$ and the edge condition and generalized boundary condition hold. Thus the boundary value problem (2.2.62)-(2.1.65) may be reformulated as follows.

The CW Mode Field $\Psi^+(x,p)$. $\Psi^+(x,p)$ is defined to be the (matrix) CW electric field that is produced when the matrix plane wave

$$(2.2.8) \qquad \Psi_0(x,p) = (2\pi)^{-3/2} e^{ip\cdot x} P(p), \quad p \in R^3 - \{0\},$$

is scattered by Ω. Here the projection $P(p)$ is defined by

$$(2.2.9) \qquad P(p) = 1 - \theta\theta, \quad p = |p|\theta \neq 0.$$

In particular, $P(p)$ is homogeneous of degree zero. With this notation, the properties that characterize $\Psi^+(x,p)$ may be formulated as follows:

$$(2.2.10) \qquad \Psi^+(x,p) = \Psi_0(x,p) + \Psi_{sc}^+(x,p), \quad x \in \Omega,$$
$$(2.2.11) \qquad \Psi^+(x,p) \in L_2^{E,loc}(\nabla \times (\nabla\times), \bar{\Omega})$$
$$(2.2.12) \qquad \nabla \times (\nabla \times \Psi^+) - |p|^2 \Psi^+ = 0 \quad \text{for} \quad x \in \Omega,$$
$$(2.2.13) \qquad \theta \times (\nabla \times \Psi_{sc}^+) + i|p|\Psi_{sc}^+ = \mathcal{O}(1/|x|^2), \quad |x| \to \infty.$$

The meaning of the vector operations in conditions (2.2.10)-(2.2.13), when applied to the matrix valued function $\Psi^+(x,p)$, is that those conditions hold for the vector valued function $\Psi^+(x,p)\cdot a$ for all choices of the constant vector a. This is equivalent to the requirement that conditions (2.2.10)-(2.2.13) hold for each column of the matrix $\Psi^+(x,p)$. Note also that, in this formulation of the problem, the generalized boundary condition and the edge condition are incorporated into condition (2.2.11). Also, (2.2.12) implies that $\Psi^+(x,p)$ satisfies the vector Helmholtz equation:

$$(2.2.14) \qquad \Delta\Psi^+ + |p|^2 \Psi^+ = 0 \quad \text{for} \quad x \in \Omega.$$

It follows, as in the scalar case, that the components of $\Psi^+(x,p)$ are analytic functions in the interior of Ω. In particular, condition (2.2.13) is meaningful. The existence and uniqueness of the function $\Psi^+(x,p)$ can be proved by the methods introduced for the scalar case in [21].

The remainder of this section presents a calculation of the far field form (2.1.66) of the function $\Psi_{sc}^+(x, p)$. The results include integral representations of the matrix scattering amplitude $T_+(p, p')$ which generalize the representations (1.2.13) and (1.2.15) of the sonar case. The derivation is based on the CW Green's matrix for Maxwell's equations, defined by

$$(2.2.15) \qquad G(x - x', \omega) = \left\{ 1 + \frac{1}{\omega^2} \nabla \nabla \right\} \frac{e^{i\omega|x - x'|}}{4\pi|x - x'|},$$

where 1 denotes the 3×3 unit matrix. $G(x - x', \omega)$ is uniquely characterized by the properties

$$(2.2.16) \qquad \nabla \times (\nabla \times G) - \omega^2 G = \delta(x - x')1 \quad \text{for} \quad x \in R^3,$$

$$(2.2.17) \qquad \theta \times (\nabla \times G) + i\omega G = \mathcal{O}(1/|x|^2) \quad \text{for} \quad |x| \to \infty.$$

The interpretation of the vector operations in (2.2.16),(2.2.17) is the same as in equations (2.2.10)-(2.2.13) above. Equation (2.2.15) can be derived directly from conditions (2.2.16)-(2.2.17). However, it is already implicit in equation (2.1.55), derived above, which implies that

$$(2.2.18) \qquad w_0(x) = \int_{R^3} G(x - x', \omega) f(x') \, dx'$$

is the unique solution of the problem

$$(2.2.19) \qquad \nabla \times (\nabla \times w_0) - \omega^2 w_0 = f(x) \quad \text{for} \quad x \in R^3,$$

$$(2.2.20) \qquad \theta \times (\nabla \times w_0) + i\omega w_0 = \mathcal{O}(1/|x|^2) \quad \text{for} \quad |x| \to \infty.$$

However, care must be used in interpreting (2.2.18) because $G(x - x', \omega)$, when interpreted as a distribution, is not a locally integrable function as is the function $G_0(x - x', \omega)$ in the scalar case. Instead, $G(x - x', \omega)$ contains both a multiple of the delta function and a singular integral operator. Details may be found in [9].

The derivation of the integral representation for Ψ_{sc}^+ will be based on a second vector Green's theorem which states that

$$(2.2.21) \qquad \begin{aligned} &\int_\Omega [(\nabla \times (\nabla \times u)) \cdot v - u \cdot (\nabla \times (\nabla \times v))] \, dx \\ &= \int_{\partial\Omega} [(\nu \times (\nabla \times u)) \cdot v - u \cdot (\nu \times (\nabla \times v))] \, dS \\ &= \int_{\partial\Omega} [(\nu \times u) \cdot (\nabla \times v) - (\nabla \times u) \cdot (\nu \times v)] \, dS \end{aligned}$$

where ν denotes the unit normal field on $\partial\Omega$, pointing out of Ω. Note that the last two integrals are alternative forms of the boundary integral. Equation (2.2.21) may be derived from equation (2.2.6). Alternatively, it is a consequence of the classical scalar divergence theorem and the differential identity

$$(2.2.22) \qquad \begin{aligned} &\nabla \cdot \{(\nabla \times u) \times v - (\nabla \times v) \times u\} \\ &= (\nabla \times (\nabla \times u)) \cdot v - u \cdot (\nabla \times (\nabla \times v)). \end{aligned}$$

The integral representation will be derived by applying (2.2.21) to $G(x - x', |p|)$ and Ψ_{sc}^+. The following notation is used:

- $\cdot$ a and b denote constant vectors.
- $\cdot$ r_0 and R denote real numbers such that $0 < r_0 < R$, $\Gamma \subset \{x' : |x'| < r_0\}$.
- $\cdot$ x denotes a point of R^3 such that $r_0 < |x| < R$.

Moreover, it will be convenient to write

$$
\begin{aligned}
u_0(x') &= \Psi_0(x', p) \cdot a, \\
u_{sc}(x') &= \Psi_{sc}^+(x', p) \cdot a, \\
u(x') &= \Psi^+(x', p) \cdot a, \\
v(x') &= G(x' - x, |p|) \cdot b,
\end{aligned}
$$

(2.2.23)

as well as

$$
\begin{aligned}
S(r) &= \{x' : |x'| = r\}, \\
\nu' = \nu(x') &= \quad \text{exterior unit normal to} \quad S(r), \\
\nabla' &= \quad \text{gradient operator at point} \quad x', \\
\Omega(r_0, R) &= \{x' : r_0 < |x'| < R\}.
\end{aligned}
$$

(2.2.24)

Also, note that

$$
\begin{aligned}
\nabla' \times (\nabla' \times u_{sc}) - |p|^2 u_{sc} = 0 &\quad \text{for all} \quad x' \in \Omega, \\
\nabla' \times (\nabla' \times v) - |p|^2 v = \delta(x' - x) b &\quad \text{for all} \quad x' \in R^3.
\end{aligned}
$$

(2.2.25)

On applying the first equation of (2.2.21) to $u = u_{sc}(x')$ and $v(x')$ in the domain $\Omega(r_0, R)$, one gets

(2.2.26)
$$
\begin{aligned}
&-b \cdot \Psi_{sc}^+(x, p) \cdot a \\
&= \int_{\partial\Omega(r_0, R)} \{\nu' \times (\nabla' \times u_{sc}) \cdot v - u_{sc} \cdot \nu' \times (\nabla' \times v)\} \, dS' \\
&= \int_{S(R)} - \int_{S(r_0)},
\end{aligned}
$$

because $\partial\Omega(r_0, R) = S(R) - S(r_0)$ if $S(R)$ and $S(r_0)$ are oriented by their outward unit normals. Note that equation (2.2.26) implies that the integral over $S(R)$ is actually independent of R, provided that $R > |x|$. Also, both G and Ψ_{sc}^+ satisfy the Silver-Müller radiation condition. It follows easily that

(2.2.27)
$$
\lim_{R \to \infty} \int_{S(R)} = 0.
$$

The verification is left for the exercises. Hence, (2.2.26) implies that

(2.2.28) $\quad b \cdot \Psi_{sc}^+(x, p) \cdot a = \int_{S(r_0)} \{\nu' \times (\nabla' \times u_{sc}) \cdot v - u_{sc} \cdot \nu' \times (\nabla' \times v)\} \, dS'.$

Next, if one applies (2.2.21) to $u = u_0(x')$ and $v(x')$ in the *interior* of $S(r_0)$, where both satisfy the homogeneous vector wave equation, one gets

$$(2.2.29) \qquad 0 = \int_{S(r_0)} \{ \nu' \times (\nabla' \times u_0) \cdot v - u_0 \cdot \nu' \times (\nabla' \times v) \} \, dS'.$$

Adding (2.2.28) and (2.2.29) then gives

$$(2.2.30) \qquad b \cdot \Psi_{sc}^+(x,p) \cdot a = \int_{S(r_0)} \{ \nu' \times (\nabla' \times u) \cdot v - u \cdot \nu' \times (\nabla' \times v) \} \, dS'.$$

On using the definitions (2.2.23) of u, u_{sc} and v and applying the rules of dyadic algebra one obtains the following successive integral representations.

$$b \cdot \Psi_{sc}^+(x,p) \cdot a$$

$$(2.2.31) \quad \begin{aligned} &= \int_{S(r_0)} \{ (G \cdot b) \cdot (\nu' \times \nabla' \times \Psi^+ \cdot a) - (\nu' \times \nabla' \times G \cdot b) \cdot \Psi^+ \cdot a) \} \, dS' \\ &= \int_{S(r_0)} \{ b \cdot G \cdot (\nu' \times \nabla' \times \Psi^+ \cdot a) + (\nabla' \times G \cdot b) \cdot (\nu' \times \Psi^+ \cdot a) \} \, dS' \\ &= \int_{S(r_0)} \{ b \cdot G \cdot (\nu' \times \nabla' \times \Psi^+ \cdot a) - b \cdot (\nabla' \times G) \cdot (\nu' \times \Psi^+ \cdot a) \} \, dS'. \end{aligned}$$

In the calculation indicated above, the symmetry of the dyadic G and the antisymmetry of the dyadic $\nabla' \times G$ were used. The verification of these facts is left for the exercises. Finally, in the first and last members of equations (2.2.31), the arbitrary vectors a and b may be dropped (quotient law for tensors), giving

$$(2.2.32) \qquad \begin{aligned} \Psi_{sc}^+(x,p) = \int_{S(r_0)} &\{ G(x - x', |p|) \cdot (\nu' \times \nabla' \times \Psi^+(x',p)) \\ &- (\nabla' \times G(x - x', |p|)) \cdot (\nu' \times \Psi^+(x',p) \} \, dS'. \end{aligned}$$

This result is the analogue of the representation (1.2.10) for the sonar case.

Far Field Form of G and $\nabla \times G$. Equation (2.2.32) will be used to demonstrate the far field form (2.1.66) of $\Psi_{sc}^+(x,p)$ and to calculate an integral representation for the scattering amplitude $T_+(p,p')$. For this purpose, the far field forms of $G(x - x', \omega)$ and $\nabla \times G(x - x', \omega)$, for x' fixed and $|x| \to \infty$, are needed. Note that the definition (2.2.15) of $G(x - x', \omega)$ implies that one has

$$(2.2.33) \quad G(x - x', \omega) = \left\{ 1 + \frac{1}{\omega^2} \nabla\nabla \right\} \gamma(R) = (\gamma + \frac{1}{\omega^2} \frac{\gamma'}{R}) 1 + \frac{1}{\omega^2} (\gamma'' - \frac{\gamma'}{R}) \eta\eta,$$

where

$$(2.2.34) \qquad \gamma(R) = \frac{e^{i\omega R}}{4\pi R}, \quad R = |x - x'| \quad \text{and} \quad \eta = \frac{x - x'}{R}.$$

On combining (2.2.34) and the relation $|x - x'| = |x| - x' \cdot \theta + \mathcal{O}(1/|x|)$ where x' is restricted to a bounded set and $x = |x|\theta, |x| \to \infty$, one gets

$$\gamma(R) = \gamma(|x|)e^{-i\omega\theta \cdot x'} + \mathcal{O}(\frac{1}{|x|^2}),$$

$$\gamma'(R) = i\omega\gamma(|x|)e^{-i\omega\theta \cdot x'} + \mathcal{O}(\frac{1}{|x|^2}),$$

(2.2.35)

$$\gamma''(R) = -\omega^2\gamma(|x|)e^{-i\omega\theta \cdot x'} + \mathcal{O}(\frac{1}{|x|^2}),$$

$$\eta = \theta + \mathcal{O}(\frac{1}{|x|}).$$

Next, combining (2.2.33) and (2.2.35) gives, after simplification, the relation

(2.2.36) $$G(x - x', \omega) = \frac{e^{i\omega|x|}}{4\pi|x|}e^{-i\omega\theta \cdot x'} P(\theta) + \mathcal{O}(\frac{1}{|x|^2}).$$

A similar calculation gives

(2.2.37) $$\nabla \times G(x - x', \omega) = \frac{e^{i\omega|x|}}{4\pi|x|}ie^{-i\omega\theta \cdot x'} M(\omega\theta) + \mathcal{O}(\frac{1}{|x|^2}),$$

where $M(\theta)$ is defined by (2.1.8). Recall that

(2.2.38) $$\Psi_0(x, p) = (2\pi)^{-3/2}e^{ip \cdot x} P(p).$$

Moreover, a simple calculation gives

(2.2.39) $$\nabla \times \Psi_0(x, p) = i(2\pi)^{-3/2}e^{ip \cdot x} M(p).$$

Thus, if one defines $p' = |p|\theta$, then (2.2.36),(2.2.37) may be written as

(2.2.40) $$G(x - x', |p|) = \frac{e^{i|p||x|}}{4\pi|x|}(2\pi)^{3/2}\Psi_0(x', -p') + \mathcal{O}(\frac{1}{|x|^2}).$$

and

(2.2.41) $$\nabla \times G(x - x', |p|) = \frac{e^{i|p||x|}}{4\pi|x|}(2\pi)^{3/2}(\nabla' \times \Psi_0(x', -p')) + \mathcal{O}(\frac{1}{|x|^2}).$$

Integral Representation of $T_+(p, p')$. The above estimates are uniform for all $x' \in S(r_0)$. Hence they may be substituted into equation (2.2.32) to give

$$\Psi_{sc}^+(x, p) = \frac{e^{i|p||x|}}{4\pi|x|}(2\pi)^{3/2} \int_{S(r_0)} \{\Psi_0(x', -p') \cdot (\nu' \times \nabla' \times \Psi^+(x', p))$$

(2.2.42) $$- (\nabla' \times \Psi_0(x', -p')) \cdot (\nu' \times \Psi^+(x', p))\} \, dS' + \mathcal{O}(\frac{1}{|x|^2}).$$

This has the form (2.1.66) and gives an integral representation for $T_+(p,p')$. If $p' = |p|\theta$, so $|p'| = |p|$, then (2.2.42) implies the representation

$$
\begin{aligned}
T_+(p',p) = (2\pi)^{3/2} \int_{S(r_0)} & \{\Psi_0(x,-p') \cdot (\nu \times \nabla \times \Psi^+(x,p)) \\
& - (\nabla \times \Psi_0(x,-p')) \cdot (\nu \times \Psi^+(x,p))\} \, dS.
\end{aligned}
$$

(2.2.43)

This result is the analogue of the representation (1.2.13) for the sonar case. Note that (2.2.43) implies that $P(p')T_+(p',p) = T_+(p',p)$. This follows from (2.2.38), (2.2.39) and the easily verified relations $P(p)^2 = P(p)$ and $P(p)M(p) = M(p)$.

The Reciprocity Theorem. The analogue of the sonar reciprocity theorem (1.2.14) is the statement that

$$
(2.2.44) \qquad T_+(p,p')^T = T_+(-p',-p) \quad \text{for all} \quad p,p' \quad \text{such that} \quad |p| = |p'|.
$$

This result has a simple, and rather surprising, physical interpretation. To formulate it, note that if a and b are constant vectors then (2.1.61)-(2.1.66) imply that

$$
(2.2.45) \quad
\begin{aligned}
a \cdot T_+(p,p') \cdot b = {} & \text{far field component along } a \text{ in direction } p \\
& \text{due to a plane wave in direction } p', \text{ polarized along } b.
\end{aligned}
$$

Moreover, since $a \cdot C \cdot b = b \cdot C^T \cdot a$, equation (2.2.44) is equivalent to the statement that

$$
(2.2.46) \qquad a \cdot T_+(p,p') \cdot b = b \cdot T_+(-p',-p) \cdot a.
$$

This states that the component along a of the far field amplitude in the direction p due to a plane wave propagating in the direction p' and polarized along b is equal to the component along b of the far field in the direction $-p'$ due to a plane wave propagating in the direction $-p$ and polarized along a.

The reciprocity theorem (2.2.44) may be verified by applying the vector Green's theorem (2.2.21) to the fields $u(x) = \Psi^+(x,p) \cdot a$ and $v(x) = \Psi^+(x,p') \cdot b$ where a and b are arbitrary constant vectors. Of course, if $\partial\Omega$ is not smooth then the calculation must be based on the version of the vector Green's theorem given by the generalized boundary condition (2.2.5). The technique for doing this parallels the calculation for the sonar case which is given in [21].

An Alternative Representation for $T_+(p,p')$. The integral representation (2.2.43) for $T_+(p,p')$ is independent of r_0, provided that $S(r_0)$ contains the scatterer Γ. More generally, $S(r_0)$ can be replaced by any smooth surface S that contains Γ. This is evident from the derivation where the set $\Omega(r_0,R)$ can be replaced by any region that is bounded externally by $S(R)$ and internally by S. In particular, if $\partial\Omega$ is smooth one may take $S = \partial\Omega$. In this case the representation takes the form

$$
(2.2.47) \qquad T_+(p',p) = (2\pi)^{3/2} \int_{\partial\Omega} \Psi_0(x,-p') \cdot (\nu \times \nabla \times \Psi^+(x,p)) \, dS,
$$

because of the boundary condition (2.1.64). This result remains true for scatterers Γ which are piecewise smooth, with edges and vertices, but the use of the vector Green's theorem in the proof must be based on the generalized boundary condition (2.2.5). Note that (2.2.47) is the analogue for the radar problem of equation (1.2.15) for the sonar problem.

An Alternative Representation of $\Psi^+(x,p)$. The integral representation (2.2.32) for $\Psi^+(x,p)$ is also independent of the surface $S(r_0)$. In particular, if $\partial\Omega$ is piecewise smooth one may replace $S(r_0)$ by $\partial\Omega$. The result is

$$(2.2.48) \qquad \Psi^+(x,p) = \Psi_0(x,p) + \int_{\partial\Omega} G(x-x',|p|) \cdot (\nu' \times \nabla' \times \Psi^+(x',p)) \, dS',$$

again because of the boundary condition (2.1.64).

The Induced Surface Current Density $J(x,p)$. If $E(t,x)$, $H(t,x)$ represents the electromagnetic field near a perfect electrical conductor Γ then the field of tangent vectors on $\partial\Gamma = \partial\Omega$ defined by

$$(2.2.49) \qquad J(t,x) = \nu(x) \times H(t,x), \quad x \in \partial\Omega,$$

represents the density of the electric currents induced in the conductor by the field [12]. In particular, for CW fields with frequency ω one has $\nabla \times E = i\omega H$ and hence

$$(2.2.50) \qquad i\omega J = \nu \times i\omega H = \nu \times (\nabla \times E).$$

The matrix CW electric field $\Psi^+(x,p)$ represents the scattered electric fields due to the matrix incident plane wave field $\Psi_0(x,p)$. The corresponding magnetic fields $\Psi_H^+(x,p)$ are defined by Maxwell's equation

$$(2.2.51) \qquad i|p|\Psi_H^+(x,p) = \nabla \times \Psi^+(x,p).$$

Hence, the corresponding matrix current density $J(x,p)$ is defined by

$$(2.2.52) \qquad i|p|J(x,p) = i|p|(\nu \times \Psi_H^+(x,p)) = \nu \times (\nabla \times \Psi^+(x,p)).$$

Note that the columns of $J(x,p)$ are the current densities induced by the scattering by Γ of the plane waves represented by the corresponding columns of Ψ_0.

It is important that the electric fields $\Psi^+(x,p)$ and the magnetic fields $\Psi_H^+(x,p)$ are completely determined by their matrix current density $J(x,p)$. This is immediate from (2.2.48) which can be written

$$(2.2.53) \qquad \Psi^+(x,p) = \Psi_0(x,p) + i|p| \int_{\partial\Omega} G(x-x',|p|) \cdot J(x',p) \, dS'.$$

A corresponding representation of $\Psi_H^+(x,p)$ may be found by applying the $\nabla\times$ operator to (2.2.53) and using the Maxwell equation (2.2.51). The result may be written

$$(2.2.54) \qquad \begin{aligned} \Psi_H^+(x,p) &= \Psi_{0,H}(x,p) + \int_{\partial\Omega} \nabla \times G(x-x',|p|) \cdot J(x',p) \, dS' \\ &= \Psi_{0,H}(x,p) + \int_{\partial\Omega} \nabla G_0(x-x',|p|) \times J(x',p) \, dS', \end{aligned}$$

where $G_0(x - x', \omega)$ is the scalar Green's function (1.2.11) of Part 1. Similarly, (2.2.47) and (2.2.52) imply the representation

$$(2.2.55) \qquad T_+(p', p) = (2\pi)^{3/2} i |p| \int_{\partial\Omega} \Psi_0(x, -p') \cdot J(x', p) \, dS.$$

An Integral Equation for $J(x, p)$. It has been seen that all the field quantities of the CW mode scattering problem can be calculated by quadrature when the induced current density $J(x, p)$ is known. To complete this section it will be shown that $J(x, p)$ satisfies an integral equation. This result provides a basis for calculating all the CW mode fields.

To derive the integral equation, for the case of a smooth scatterer $\partial\Omega$, assume that the normal field $\nu(x)$ on $\partial\Omega$ is extended to a smooth field defined in a neighborhood of $\partial\Omega$. Then, for points x near to $\partial\Omega$, one has

$$(2.2.56) \qquad \begin{aligned} \nu(x) \times \Psi_H^+(x, p) &= \nu(x) \times \Psi_{0,H}(x, p) \\ &+ \nu(x) \times \int_{\partial\Omega} \nabla G_0(x - x', |p|) \times J(x', p) \, dS'. \end{aligned}$$

The limit as $x \to \partial\Omega$ of the last integral can be calculated by means of the standard jump relations of potential theory. The theory may be found in [2]. Specifically, the integral in (2.2.56) is defined for all $x \in R^3$ and defines a function that is smooth except for jump discontinuities at the points of $\partial\Omega$. The amounts of these discontinuities are given in [2], Theorem 2.17 (p.51). Applied to the integral in (2.2.56), this gives

$$(2.2.57) \qquad \begin{aligned} &\left(\int_{\partial\Omega} \nabla G_0(x - x', |p|) \times J(x', p) \, dS' \right)_+ \\ &= \int_{\partial\Omega} \nabla G_0(x - x', |p|) \times J(x', p) \, dS' - \frac{1}{2}\nu(x) \times J(x, p), \end{aligned}$$

where x is an arbitrary point of $\partial\Omega$ and the notation $(f(x))_+$ denotes the limit as x is approached through points of Ω. Passage to the limit in (2.2.56) gives, by equation (2.2.57),

$$(2.2.58) \qquad \begin{aligned} J(x, p) &= J_0(x, p) \\ &+ \nu(x) \times \left\{ \int_{\partial\Omega} \nabla G_0(x - x', |p|) \times J(x', p) \, dS' - \frac{1}{2}\nu(x) \times J(x, p) \right\}, \end{aligned}$$

where $J_0(x, p) = \nu(x) \times \Psi_{0,H}^+(x, p)$. The vector identity $\nu \times (\nu \times J) = -J$ may be used to rewrite this equation as

$$(2.2.59) \qquad \frac{1}{2}J(x, p) = J_0(x, p) + \int_{\partial\Omega} \nu \times \{\nabla G_0(x - x', |p|) \times J(x', p)\} \, dS',$$

or

$$(2.2.60) \qquad J(x, p) = 2J_0(x, p) + \int_{\partial\Omega} K(x, x', |p|) \cdot J(x', p)\} \, dS',$$

where

$$K(x, x', |p|) \cdot J(x', p) = 2\nu(x) \times \{\nabla G_0 \times J(x', p)\}$$
$$(2.2.61) \qquad = 2\left[\nabla G_0(\nu(x) \cdot J(x', p)) - (\nu(x) \cdot \nabla G_0)J(x', p)\right].$$

Thus the kernel $K(x, x', |p|)$, which acts on fields of tangent vectors on $\partial\Omega$, is given by

$$(2.2.62) \qquad K(x, x', |p|) = \left\{\nabla \frac{e^{i|p||x-x'|}}{2\pi|x-x'|}\right\}\nu(x) - \left\{\frac{\partial}{\partial\nu(x)}\frac{e^{i|p||x-x'|}}{2\pi|x-x'|}\right\}1.$$

Equation (2.2.60) is a Fredholm integral equation for the induced electric current density $J(x, p)$. It has been shown to be a necessary condition for the solvability of the CW mode scattering problem. It can also be shown to be a sufficient condition under suitable smoothness conditions on $\partial\Omega$; see, for example, [2]. $J(x, p)$ determines the CW mode field $\Psi^+(x, p)$ uniquely. Moreover, the matrix scattering amplitude $T_+(p, p')$ is determined by (2.2.55) when $J(x, p)$ is known. Thus the scattering problem has been reduced to the problem of solving a Fredholm equation.

It must be remarked that the jump relations that were used at (2.2.57) to derive (2.2.60) are known to be valid only if the components of $J(x, p)$ are Hölder continuous [2]. This will be the case if the surface $\partial\Omega$ is in the class $C^{1,\alpha}$; see [2]. Under this same condition the kernel $K(x, x', |p|)$ has a weak singularity at $x = x'$ [2] and hence defines a compact operator to which the Fredholm alternative theorem applies. This property of $K(x, x', |p|)$ is well known for the second term on the right in (2.2.62); see, for example, [2], p.61. For the first term, note that $\nu(x') \cdot J(x', p) = 0$ because $J(x', p)$ is tangent to $\partial\Omega$ at x'. It follows that

$$\nu(x) \cdot J(x', p) = (\nu(x) - \nu(x')) \cdot J(x', p)$$
$$(2.2.63) \qquad = \mathcal{O}(|x - x'|^\alpha)|J(x', p)| \quad \text{for} \quad x \to x',$$

which implies that the first term on the right in (2.2.62) also has a weak singularity.

Exercises for §2.2.

1. Verify the vector identity

$$\nabla \cdot (u \times v) = (\nabla \times u) \cdot v - u \cdot (\nabla \times v)$$

and use it to verify the vector Green's theorem (2.2.6).

2. Show that the boundary value problem (2.2.10)-(2.2.13) for $\Psi^+(x, p)$ can have at most one solution. Pattern your demonstration on that for the scalar case given in [21].

3. Verify the vector identity (2.2.22) and then use it to verify both of equations (2.2.21).

4. Verify the limit relation (2.2.27).

5. Verify equation (2.2.29) for the primary field u_0.

6. Show the correctness of equations (2.2.31).

7. Show that the CW mode Green's matrix $G(x - x', \omega)$, defined by (2.2.15), is symmetric and $\nabla \times G(x - x', \omega)$ is antisymmetric. Suggestion: Show that $\nabla \times G \cdot a = \nabla G_0 \times a$ for all constant vectors a.

8. Verify the integral representation (2.2.32) for Ψ_{sc}^+.

9. Verify the representation (2.2.33) for $G(x - x', \omega)$.

10. Verify the estimates (2.2.35) and then use them to derive the far field forms of G and $\nabla \times G$: equations (2.2.36) and (2.2.37).

11. Verify the representation (2.2.39) for $\nabla \times \Psi_0$.

12. Verify the integral representation (2.2.43) for $T_+(p', p)$.

13. Verify the reciprocity theorem (2.2.44).

14. Verify the alternative representation (2.2.47) for $T_+(p', p)$ and the alternative representation (2.2.48) for Ψ^+.

15. Verify the integral representation (2.2.54) for the magnetic field Ψ_H^+.

16. Derive the limit relation (2.2.57) from the jump relations of potential theory as formulated in [2], p.51.

2.3   The Structure of Pulse Mode Radar Echoes. In this section the Hilbert space method of §1.3 is extended to radar echo prediction. First, the spectral theorem is used to construct the scattered wave field for essentially arbitrary scatterer geometry and pulse structure. Second, the generalized eigenfunction expansions, based on the CW mode fields $\Psi^{\pm}(x,p)$, are described. Finally, the representation (1.3.56) of pulse mode sonar echoes is extended to the case of radar echoes.

The construction of pulse mode radar echoes given below is based of the fact that the differential operator $A = \nabla \times (\nabla \times)$, acting on vector valued functions that satisfy the edge condition and the generalized boundary condition, defines a selfadjoint operator in the Hilbert space

$$(2.3.1) \qquad\qquad \mathcal{H} = L_2(\Omega, C^3),$$

which consists of all functions $u : \Omega \to C^3$ whose components $u_j(x)$ (j=1,2,3) are complex-valued Lebesgue measurable functions on Ω such that

$$(2.3.2) \qquad \int_\Omega \{|u_1(x)|^2 + |u_2(x)|^2 + |u_3(x)|^2\} \, dx < \infty.$$

It is well known that $\mathcal{H}$ is a Hilbert space with inner product defined by

$$(2.3.3) \qquad (u,v) = \int_\Omega \{\overline{u}_1(x)v_1(x) + \overline{u}_2(x)v_2(x) + \overline{u}_3(x)v_3(x)\} \, dx,$$

where the overbar denotes the complex conjugate.

The edge condition and the generalized boundary condition for functions in $\mathcal{H}$ will be formulated by means of the function classes

$$(2.3.4) \qquad L_2(\nabla\times, \Omega) = \mathcal{H} \cap \{u : \nabla \times u \in \mathcal{H}\},$$

$$(2.3.5) \qquad L_2(\nabla \times (\nabla\times), \Omega) = L_2(\nabla\times, \Omega) \cap \{u : \nabla \times (\nabla \times u) \in \mathcal{H}\},$$

$$(2.3.6) \qquad L_2^E(\nabla \times (\nabla\times), \Omega) = L_2(\nabla \times (\nabla\times), \Omega) \cap \{u : (2.2.5) \text{ holds}\}.$$

The operator A in $\mathcal{H}$ will be defined by

$$(2.3.7) \qquad\qquad D(A) = L_2^E(\nabla \times (\nabla\times), \Omega),$$

$$(2.3.8) \qquad\qquad Au = \nabla \times (\nabla \times u) \quad \text{for all} \quad u \in D(A).$$

The solution of the radar echo prediction problem given below is based on the fact that A is a selfadjoint non-negative operator in the Hilbert space $\mathcal{H}$:

$$(2.3.9) \qquad\qquad A = A^* \geq 0.$$

Moreover, the non-negative square root of A, denoted by $A^{1/2}$, exists and satisfies

(2.3.10) $$D(A^{1/2}) = L_2(\nabla \times, \Omega),$$

(2.3.11) $$\|A^{1/2}u\| = \|\nabla \times u\| \quad \text{for all} \quad u \in D(A^{1/2})$$

where $\|u\| = (u, u)^{1/2}$ denotes the norm in $\mathcal{H}$. A proof of these facts may be based on Kato's theory of sesquilinear forms in Hilbert spaces [4], pp 322-331.

The vector wave equation (2.1.19),(2.1.20) may be interpreted as an ordinary differential equation

(2.3.12) $$\partial_t^2 u + Au = f(t, \cdot), \quad t \in R,$$

for a function $t \to u(t, \cdot) \in \mathcal{H}$, where

(2.3.13) $$f(t, x) = -\partial_t J(t, x), \quad \text{supp } J \subset \{t : t_0 \leq t \leq t_0 + T\},$$

and a solution of (2.3.12) is sought which satisfies

(2.3.14) $$u(t, \cdot) = 0 \quad \text{for} \quad t < t_0,$$

just as in §1.3.

The formal solution of (2.3.12)-(2.3.14) is exactly the same as for the sonar problem. In particular, note that here the condition

(2.3.15) $$\int_{t_0}^{t_0+T} f(t, x) \, dt = 0 \quad \text{for all} \quad x \in \Omega$$

is implied by hypothesis (2.3.13). It follows, as in §1.3, that

(2.3.16) $$u(t, x) = Re\{v(t, x)\} \quad \text{for} \quad t \geq t_0 + T,$$

where

(2.3.17) $$v(t, \cdot) = e^{-itA^{1/2}} h$$

and

(2.3.18) $$h = \int_{t_0}^{t_0+T} e^{i\tau A^{1/2}} \int_{t_0}^{\tau} f(t, \cdot) \, dt \, d\tau.$$

Spectral Properties of A. Property (2.3.9) implies that the operator $\mathcal{H}$ has a spectrum $\sigma(A)$ which is real and non-negative: $\sigma(A) \subset [0, \infty)$. The spectral representation described below implies that one actually has $\sigma(A) = [0, \infty)$. It is also known that A has no positive eigenvalues [2]. However, 0 is known to be an eigenvalue of A of infinite multiplicity. This property will be illustrated in the simple case that corresponds to $\Omega = R^3$ (no scatterer). The eigenfunctions of A

for the eigenvalue 0 are the *static* states of the electromagnetic field. They have no analogue in the acoustic case. The Hilbert space $\mathcal{H}$ has a direct sum decomposition

$$(2.3.19) \qquad\qquad \mathcal{H} = \mathcal{H}^0 \oplus \mathcal{H}^c,$$

where $\mathcal{H}^0$ is the null space of A; i.e., the eigenspace of A for the eigenvalue 0, and $\mathcal{H}^c$ is its orthogonal complement. It is known that this decomposition reduces A and that the part of A in $\mathcal{H}^c$ has a pure continuous spectrum [4]. Clearly, the propagating states that produce radar signals and echoes all lie in $\mathcal{H}^c$. For this reason $\mathcal{H}^c$ will be called the *scattering subspace* of A.

Spectral Analysis of A_0. Let A_0 denote the operator that corresponds to taking $\Omega = R^3$ in (2.3.7),(2.3.8). A spectral analysis of A_0 can be based on the Plancherel theory of the Fourier transform; cf. [20]. The Plancherel theory implies that the formulas

$$(2.3.20) \qquad \hat{f}(p) = L_2(R^3)\text{-}\lim_{M\to\infty} \frac{1}{(2\pi)^{3/2}} \int_{|x|\leq M} e^{-ip\cdot x} h(x)\, dx,$$

$$(2.3.21) \qquad f(x) = L_2(R^3)\text{-}\lim_{M\to\infty} \frac{1}{(2\pi)^{3/2}} \int_{|p|\leq M} e^{ip\cdot x} \hat{h}(p)\, dp,$$

define a unitary operator $\Phi : L_2(R^3) \to L_2(R^3)$ and its inverse Φ^*:

$$(2.3.22) \qquad\qquad \Phi f = \hat{f}, \quad \Phi^* \Phi = \Phi\Phi^* = 1.$$

In what follows the notation $L_2(R^3)$-lim will be suppressed, for simplicity. However, all Fourier integrals are to be understood in the sense of (2.3.20),(2.3.21). The Fourier transform has the characteristic property that $(\Phi \partial_j f)(p) = ip_j \hat{f}(p)$. In particular, one has

$$(2.3.23) \qquad (\Phi A_0 u)(p) = A_0(p)\hat{u}(p), \quad \text{for all} \quad p \in R^3,$$

where $A_0(p)$ is related to the symbol $A(p)$ of (2.1.11) by

$$(2.3.24) \qquad\qquad A_0(p) = A(ip) = -A(p) = |p|^2 1 - pp.$$

Thus the minimal polynomial of $A_0(p)$ is $Q(\lambda, p) = \lambda^2 - |p|^2 \lambda$, so that

$$(2.3.25) \qquad\qquad A_0(p)^2 - |p|^2 A_0(p) = 0;$$

see (2.1.14). Hence, $A_0(p)$ has the eigenvalues $\lambda_0(p) \equiv 0$, $\lambda_1(p) = |p|^2$. The corresponding orthogonal projections

$$(2.3.26) \qquad\qquad P_0(p) = \eta\eta,$$

$$(2.3.27) \qquad\qquad P(p) = 1 - \eta\eta, \quad \text{where} \quad p = |p|\eta,$$

may be found by direct calculation or by a general procedure given in [20]. The bounded operators in $\mathcal{H}_0 = L_2(R^3, C^3)$ defined by

$$(2.3.28) \qquad P_0 f = \Phi^*(P_0(p)\hat{f}(p)), \quad Pf = \Phi^*(P(p)\hat{f}(p)),$$

may be shown to be the orthogonal projections onto $\mathcal{H}_0^0$, the nullspace of A_0, and the subspace $\mathcal{H}_0^c$ for the continuous spectrum, respectively;

$$(2.3.29) \qquad h = P_0 h + Ph, \quad \text{for all} \quad h \in \mathcal{H}_0.$$

This may be interpreted as an eigenfunction expansion for A_0. Indeed, if $h_c = Ph$ then

$$(2.3.30) \qquad \hat{h}_c(p) = P(p)\hat{h}(p) = \frac{1}{(2\pi)^{3/2}} \int_{R^3} e^{-ip\cdot x} P(p)h(x)\,dx,$$

and

$$(2.3.31) \qquad h_c(x) = \frac{1}{(2\pi)^{3/2}} \int_{R^3} e^{ip\cdot x} P(p)\hat{h}_c(p)\,dp,$$

because $P(p) = P(p)^2$. Note that the CW mode plane wave $\Psi_0(x,p)$ satisfies

$$(2.3.32) \qquad \Psi_0(x,p) = (2\pi)^{-3/2} e^{ix\cdot p} P(p), \quad \Psi_0(x,p)^* = (2\pi)^{-3/2} e^{-ix\cdot p} P(p),$$

where M^* denotes the Hermitian adjoint (conjugate transpose) of M, because $P(p)$ is real and symmetric. These equations imply that equations $(2.3.30), (2.3.31)$ may be written

$$(2.3.33) \qquad \hat{h}_c(p) = \int_{R^3} \Psi_0(x,p)^* \cdot h(x)\,dx,$$

$$(2.3.34) \qquad h_c(x) = \int_{R^3} \Psi_0(x,p) \cdot \hat{h}_c(p)\,dp.$$

A spectral representation for A_0 may be based on $(2.3.23)$. It was shown in [20] that if $F(\lambda)$ is any function that is defined, continuous and bounded on $[0, \infty)$ then one has

$$(2.3.35) \qquad \begin{aligned} F(A_0)u(x) &= (2\pi)^{-3/2} \int_{R^3} e^{ix\cdot p} \left\{ F(0)P_0(p) + F(|p|^2)P(p) \right\} \hat{u}(p)\,dp \\ &= F(0)P_0 u(x) + (2\pi)^{-3/2} \int_{R^3} e^{ix\cdot p} F(|p|^2)P(p)\hat{u}(p)\,dp, \end{aligned}$$

which may be written

$$(2.3.36) \qquad F(A_0)u(x) = F(0)P_0 u(x) + \int_{R^3} \Psi_0(x,p)F(|p|^2)\hat{u}_c(p)\,dp.$$

In particular, taking $F(\lambda) = exp(-it\lambda^{1/2})$ provides a representation of the solution

$$(2.3.37) \qquad v(t,x) = e^{-itA_0^{1/2}} h(x),$$

of the initial value problem for Maxwell's equations in R^3, namely,

$$(2.3.38) \qquad v(t, x) = P_0 h(x) + \int_{R^3} e^{-it|p|} \Psi_0(x, p) \cdot \hat{h}_c(p) \, dp.$$

Note that (2.3.38) divides the solution into the static part $P_0 h$ and the propagating part, given by the integral.

Spectral Analysis of A. The key to extending the sonar echo structure results of Part 1 to the case of radar echoes is a spectral analysis of the electromagnetic operator $A = \nabla \times (\nabla \times)$, based on the CW mode fields $\Psi^+(x, p)$ and the associated eigenfunction expansion. The expansion theorem states that every function h in $\mathcal{H}^c$, the scattering subspace, has a generalized Fourier transform

$$(2.3.39) \qquad \hat{h}_+(p) = L_2(R^3)\text{-}\lim_{M \to \infty} \int_{\Omega(M)} \Psi^+(x, p)^* \cdot h(x) \, dx,$$

where $\Omega(M) = \Omega \cap \{x : |x| < M\}$, and a corresponding eigenfunction representation

$$(2.3.40) \qquad h(x) = L_2(\Omega)\text{-}\lim_{M \to \infty} \int_{|p| \leq M} \Psi^+(x, p) \cdot \hat{h}_+(p) \, dp.$$

Moreover, the transformation Φ_+ defined by

$$(2.3.41) \qquad \Phi_+ h = \hat{h}_+, \quad h \in \mathcal{H}^c,$$

is a partial isometry [4] and diagonalizes the part of A in $\mathcal{H}^c$ in the sense that

$$(2.3.42) \qquad (\Phi_+ F(A) h)(p) = F(|p|^2)(\Phi_+ h)(p)$$

for every bounded measurable function $F(\lambda)$ which is defined for $\lambda \geq 0$. In what follows the relations (2.3.39), (2.3.40) will be written in the symbolic form

$$(2.3.43) \qquad \hat{h}_+(p) = \int_\Omega \Psi^+(x, p)^* \cdot h(x) \, dx$$

$$(2.3.44) \qquad h(x) = \int_{R^3} \Psi^+(x, p) \cdot \hat{h}_+(p) \, dp.$$

However, they will always be interpreted in the mean square sense of (2.3.39), (2.3.40).

The eigenfunction expansion described above may be derived from a construction of the spectral family $\{\Pi(\lambda) : \lambda \geq 0\}$ for the operator A. Details may be found, for the acoustic case and for general first order symmetric hyperbolic systems, in [21] and [10]. The proofs require a "local compactness theorem" which, for the electromagnetic case, is available in the work of N. Weck [14] and C. Weber [13]. The details are too lengthy and technical to be presented here.

As in the sonar case, there is a second family of generalized eigenfunctions associated with A. Its definition is based on the observation that the matrix plane waves satisfy the identity

$$(2.3.45) \qquad \overline{\Psi_0(x, -p)} = \Psi_0(x, p).$$

It follows that the family defined by

(2.3.46) $$\Psi^-(x,p) = \overline{\Psi^+(x,-p)}$$

satisfies the same defining condition (2.2.10)-(2.2.13) as $\Psi^+(x,p)$, but with the Silver-Müller radiation condition (2.2.13) replaced by the *incoming* Silver- Müller radiation condition

(2.3.47) $$\theta \times (\nabla \times \Psi^-_{sc}) - i|p|\Psi^-_{sc} = \mathcal{O}(1/|x|^2), \quad |x| \to \infty.$$

It can be verified by direct calculation that relations (2.3.39)-(2.3.42) hold with $\Psi^+(x,p), \hat{h}_+(p)$ and Φ_+ replaced by $\Psi^-(x,p), \hat{h}_-(p)$ and Φ_-, respectively. The far field behavior of $\Psi^-(x,p)$ follows from (2.2.42), (2.2.43) and (2.3.46) which give

(2.3.48) $$\Psi^-_{sc}(x,p) = \frac{e^{-i|p||x|}}{4\pi|x|}T_-(p',p) + \mathcal{O}(\frac{1}{|x|^2}),$$

where $x = |x|\theta$, $p' = |p|\theta$ and

(2.3.49) $$T_-(p',p) = \overline{T_+(p',-p)}.$$

Construction of Pulse Mode Radar Echoes. The eigenfunction expansions will now be used to construct the radar echo wave function defined by (2.3.17), (2.3.18). Equations (2.3.42)-(2.3.44), and their analogues for the second expansion, imply that one has

(2.3.50) $$(\Phi_\pm \, e^{-itA^{1/2}}h)(p) = e^{-it|p|} \, \hat{h}_\pm(p),$$

and hence

(2.3.51) $$v(t,x) = \int_{R^3} e^{-it|p|}\Psi^\pm(x,p) \cdot \hat{h}_\pm(p) \, dp,$$

where it is assumed that $h \in \mathcal{H}^c$ and

(2.3.52) $$\hat{h}_\pm(p) = \int_\Omega \Psi^\pm(x,p)^* \cdot h(x) \, dx.$$

If h is defined by (2.3.18) where f satisfies (2.3.13) then the continuity of Φ_+ and Φ_-, together with (2.3.42), imply that

(2.3.53) $$\hat{h}_\pm(p) = \int_{t_0}^{t_0+T} e^{i\tau|p|} \int_{t_0}^{\tau} \Phi_\pm f(t,p) \, dt d\tau.$$

Interchange of the t and τ integrations and use of (2.3.15) gives

(2.3.54) $$\hat{h}_\pm(p) = i|p|^{-1} \int_{t_0}^{t_0+T} \int_\Omega e^{i|p|t}\Psi^\pm(x,p)^* \cdot f(t,x) \, dx dt,$$

or

(2.3.55) $$\hat{h}_{\pm}(p) = (2\pi)^{1/2} i |p|^{-1} \hat{f}_{\pm}(-|p|, p),$$

where

(2.3.56) $$\hat{f}_{\pm}(\omega, p) = \frac{1}{(2\pi)^{1/2}} \int_{t_0}^{t_0+T} e^{-i\omega t} \int_{\Omega} \Psi^{\pm}(x, p)^* \cdot f(t, x)\, dx dt.$$

The total field produced by the scattering of electromagnetic waves with a source function $f(t, x)$ is represented by (2.3.51) where $\hat{h}_+(p)$ and $\hat{h}_-(p)$ are defined by (2.3.55) and (2.3.56). These equations are not directly applicable to the boundary value problem (2.1.40) and (2.1.43)-(2.1.45) for the scattering of plane waves. However, this problem may be reduced to the preceding one by the method used for the sonar case. Again, let r_0 and R be radii such that $\delta < r_0 < R$, where $\Gamma \subset \{x : |x| \leq \delta\}$, and let $j(x) \in C^{\infty}(R^3)$ be a function with the properties

(2.3.57) $$j(x) = 0 \quad \text{for} \quad |x| \leq r_0, \quad j(x) = 1 \quad \text{for} \quad |x| \geq R.$$

The solution of the boundary value problem (2.1.43)-(2.1.45) will be constructed as a sum

(2.3.58) $$u(t, x) = j(x) u_0(t, x) + \tilde{u}_{sc}(t, x).$$

Note that $\tilde{u}_{sc}(t, x)$ coincides with the echo (2.1.46) for $|x| \geq R$ and all $t \in R$, by (2.3.57).

The solution of (2.1.43)-(2.1.45) is given by (2.3.58) if $\tilde{u}_{sc}(t, x)$ satisfies

(2.3.59) $$\partial_t^2 \tilde{u}_{sc} + \nabla \times (\nabla \times \tilde{u}_{sc}) = f(t, x) \quad \text{for} \quad t \in R, x \in \Omega,$$
(2.3.60) $$\nu \times \tilde{u}_{sc} = 0 \quad \text{for} \quad t \in R, x \in \partial\Omega,$$
(2.3.61) $$\tilde{u}_{sc}(t, x) = 0 \quad \text{for} \quad t + b + R < 0, x \in \Omega,$$

where

(2.3.62) $$f(t, x) = -\left(\partial_t^2 + \nabla \times (\nabla \times)\right) j(x) u_0(t, x).$$

Note that

(2.3.63) $$\operatorname{supp} f \subset \{(t, x) : -b - R \leq t \leq -a + R \quad \text{and} \quad r_0 \leq |x| \leq R\}$$

by (2.1.40) and (2.3.57). In particular, (2.3.13) holds with $t_0 = -b - R$ and $T = 2R + b - a$. Of course, if $\partial\Omega$ is not smooth then $\tilde{u}_{sc}$ must satisfy the edge condition and generalized boundary condition; i.e., $\tilde{u}_{sc}(t, \cdot) \in D(A) = L_2^E(\nabla \times (\nabla \times), \Omega)$ for all $t \in R$. The solution is given by

(2.3.64) $$\tilde{u}_{sc}(t, x) = Re\{\tilde{v}_{sc}(t, x)\},$$

where $\tilde{v}_{sc}$ has two representations,

$$(2.3.65) \qquad \tilde{v}_{sc}(t,x) = \int_{R^3} e^{-it|p|}\Psi^\pm \hat{h}_\pm(p)\,dp,$$

corresponding to Ψ^+ and Ψ^-, and $\hat{h}_+(p)$ and $\hat{h}_-(p)$ are defined by (2.3.55), (2.3.56) and (2.3.62). This section is concluded with a calculation of $\hat{h}_-(p)$.

The support of f in space-time is bounded, by (2.3.63), and a simple calculation gives

$$(2.3.66) \qquad \frac{1}{(2\pi)^{1/2}} \int_{-\infty}^{\infty} e^{-i\omega t}\, f(t,x)\,dt = (\omega^2 1 - \nabla \times (\nabla\times))j(x)e^{-i\omega\theta_0 \cdot x}\hat{s}(-\omega,\theta_0)$$

where $\hat{s}(\omega,\theta_0)$ is defined by (2.1.68). Thus, by (2.3.54), (2.3.55) and (2.3.66) one has, after some algebra,

$$(2.3.67) \quad \hat{f}_\pm(\omega,p) = \int_\Omega \Psi^\pm(x,p)^* \cdot (\omega^2 1 - \nabla \times (\nabla\times))j(x)e^{-i\omega\theta_0 \cdot x}P(\theta_0)\hat{s}(-\omega,\theta_0)\,dx.$$

The factor $P(\theta_0)$ could be dropped, by (2.1.41), but it will be needed in subsequent steps. The remainder of the calculation of $\hat{f}_\pm(\omega,p)$ is algebraically rather complicated and it will be convenient to introduce the following additional notation.

$$(2.3.68) \qquad \omega = -|p|,$$
$$(2.3.69) \qquad a = \hat{s}(-\omega,\theta_0),$$
$$(2.3.70) \qquad b = \text{arbitrary constant vector},$$
$$(2.3.71) \qquad A(x) = \overline{\Psi^-(x,p)} \cdot b = \Psi^+(x,-p) \cdot b,$$
$$(2.3.72) \qquad B(x) = j(x)e^{-i\omega\theta_0 \cdot x}P(\theta_0)a.$$

Note that the second equation in (2.3.71) follows from the definition (2.3.46). Moreover, by (2.3.57), $B(x)$ is zero for $|x| \le r_0$ and concides with a vector plane wave for $|x| \ge R$. It follows that one has

$$(2.3.73) \qquad \text{supp}\,(\nabla \times (\nabla\times) - \omega^2 1)B(x) \subset \{x : r_0 \le |x| \le R\} = \Omega(r_0,R).$$

Thus, (2.3.67) implies that

$$(2.3.74) \qquad \begin{aligned} b \cdot \hat{f}_-(\omega,p) &= \int_\Omega (\overline{\Psi^-(x,p)} \cdot b)^T \cdot (\omega^2 1 - \nabla \times (\nabla\times))B(x)\,dx \\ &= -\int_{\Omega(r_0,R)} A^T(\nabla \times (\nabla\times) - \omega^2 1)B(x)\,dx. \end{aligned}$$

Next, the vector Green's theorem (2.2.21), in the form

$$(2.3.75) \qquad \begin{aligned} \int_\Omega \{A^T(\nabla \times \nabla \times B - \omega^2 B) - (\nabla \times \nabla \times A - \omega^2 A)^T B\}\,dx \\ = \int_{\partial\Omega} \{(\nabla \times A)^T(\nu \times B) - (\nu \times A)^T(\nabla \times B)\}\,dS, \end{aligned}$$

will be applied to the fields A and B defined by (2.3.71), (2.3.72) and the domain $\Omega(r_0, R)$. Then, since A satisfies $\nabla \times \nabla \times A - \omega^2 A = 0$ and $j(x) \equiv 0$ on $|x| = r_0$ while $j(x) \equiv 1$ on $|x| = R$, Green's theorem and (2.3.74) give

$$(2.3.76) \qquad b \cdot \hat{f}_-(\omega, p) = - \int_{S(R)} \{ (\nabla \times A)^T (\nu \times B) - (\nu \times A)^T (\nabla \times B) \} \, dS$$

whence

$$
\begin{aligned}
(2.3.77) \qquad & b \cdot \hat{f}_-(-|p|, p) \\
& = - \int_{S(R)} \{ (\nabla \times \Psi^+(x, -p) \cdot b)^T (\nu \times \{ e^{i|p|\theta_0 \cdot x} P(\theta_0) a \}) \\
& \qquad - (\nu \times \Psi^+(x, -p) \cdot b)^T (\nabla \times \{ e^{i|p|\theta_0 \cdot x} P(\theta_0) a \}) \, dS.
\end{aligned}
$$

On noting the definition (2.1.61) of the matrix plane wave $\Psi_0(x, p)$ and writing $p_0 = |p|\theta_0$, one can rewrite this as

$$
\begin{aligned}
(2.3.78) \qquad & b \cdot \hat{f}_-(-|p|, p) \\
& = (2\pi)^{3/2} \int_{S(R)} \{ (\nu \times \Psi^+(x, -p) \cdot b) \cdot \nabla \times \Psi_0(x, p_0) a \\
& \qquad - (\nabla \times \Psi^+(x, -p) \cdot b) \cdot (\nu \times \Psi_0(x, p_0) a) \} \, dS.
\end{aligned}
$$

Finally, this may be related to the integral representation (2.2.43) for the matrix scattering amplitude by applying the rules of vector algebra to the integrand in (2.3.78) and noting that the matrix plane wave and its curl are symmetric and antisymmetric, respectively: $(\Psi_0)^T = \Psi_0$ and $(\nabla \times \Psi_0)^T = -\nabla \times \Psi_0$. Using these facts, one can write

$$
\begin{aligned}
(2.3.79) \qquad & b \cdot \hat{f}_-(-|p|, p) \\
& = (2\pi)^{3/2} \int_{S(R)} \{ (\Psi_0(x, p_0) \cdot a) \cdot (\nu \times \nabla \times \Psi^+(x, -p) \cdot b) \\
& \qquad - (\nu \times \nabla \times \Psi_0(x, p_0) \cdot a) \cdot (\Psi^+(x, -p) \cdot b) \} \, dS.
\end{aligned}
$$

On comparing this result with (2.2.43) and using the reciprocity theorem, one gets

$$
\begin{aligned}
(2.3.80) \qquad & b \cdot \hat{f}_-(-|p|, p) = a \cdot T_+(-p_0, -p) \cdot b \\
& \qquad\qquad = b \cdot T_+(-p_0, -p)^T \cdot a = b \cdot T_+(p, p_0) \cdot a.
\end{aligned}
$$

This holds for arbitrary b and hence b may be cancelled. On recalling the definition of a, (2.3.69), one gets

$$(2.3.81) \qquad \hat{f}_-(-|p|, p) = T_+(p, |p|\theta_0) \hat{s}(|p|, \theta_0),$$

and hence, by (2.3.55),

$$(2.3.82) \qquad \hat{h}_-(p) = (2\pi)^{1/2} i |p|^{-1} T_+(p, |p|\theta_0) \hat{s}(|p|, \theta_0).$$

Combining (2.1.46), (2.3.58), (2.3.64) and (2.3.65) gives the representation (2.3.83)

$$u_{sc}(t,x) = Re\left\{(2\pi)^{1/2}\, i \int_{R^3} e^{-it|p|}\Psi^-(x,p)|p|^{-1}T_+(p,|p|\theta_0)\hat{s}(|p|,\theta_0)\,dp\right\},$$

for $|x| \geq R$. This result is actually valid for all $x \in \Omega$, by the argument given following (1.3.56). Note that the property $P(p)T_+(p,p') = T_+(p,p')$, which follows from (2.2.43), implies that $P(p)\hat{h}_-(p) = \hat{h}_-(p)$ and hence that $h(x) \in \mathcal{H}^c$. Thus the scattering of a plane electromagnetic pulse by a perfectly conducting object does not generate an electrostatic field component.

Exercises for §2.3.

1. Use Kato's theorem on sesquilinear forms in Hilbert spaces to prove that the operator A defined by (2.3.8) satisfies (2.3.9), (2.3.10) and (2.3.11).

2. Show that the operator $A_0(p)$ defined by (2.3.24) satisfies $A_0(p)P_0(p) = 0$ and $A_0(p)P(p) = |p|^2 P(p)$.

3. Verify the representation (2.3.36) for $F(A_0)$.

4. Assuming the correctness of the generalized eigenfunction expansion (2.3.39)-(2.3.42), prove that the second family $\{\Psi^-(x,p) : p \in R^3 - \{0\}\}$ satisfies the analogous relations.

5. Verify the representation (2.3.55) for $\hat{h}_-(p)$.

6. Verify the relations (2.3.66) and (2.3.67).

2.4 Radar Echoes in the Far Field. The radar echo wave form that arrives at the receiver is the far field form of $u_{sc}(t, x)$, because of the basic assumption that the receiver lies in the far field of the scatterers. As in the sonar case, the far field form of $u_{sc}(t, x)$ coincides with its asymptotic form for large t. It was shown above that

$$(2.4.1) \qquad u_{sc}(t, x) = Re\{v_{sc}(t, x)\},$$

near the receiver, where

$$(2.4.2) \qquad v_{sc}(t, x) = \int_{R^3} e^{-it|p|} \Psi^-(x, p) \cdot \hat{h}_-(p)\, dp,$$

and $\hat{h}_-(p)$ is given by (2.3.82):

$$(2.4.3) \qquad \hat{h}_-(p) = (2\pi)^{1/2} i|p|^{-1} T_+(p, |p|\theta_0)\hat{s}(|p|, \theta_0).$$

Of course, $v_{sc}(t, x)$ has a similar representation which is based on $\Psi^+(x, p)$. However, the representation (2.4.2), based on $\Psi^-(x, p)$, is better suited to the study of the asymptotic behavior of $v_{sc}(t, x)$ for $t \to +\infty$. (The other representation is better suited to study the behavior for $t \to -\infty$.) This is because of the discovery that, for $t \to +\infty$, the wave function (2.4.2) is asymptotically equal to the free wave

$$(2.4.4) \qquad v_{sc}^0(t, x) = \int_{R^3} e^{-it|p|} \Psi_0(x, p) \cdot \hat{h}_-(p)\, dp$$

in the sense that

$$(2.4.5) \qquad \lim_{t \to +\infty} \|v_{sc}(t, \cdot) - v_{sc}^0(t, \cdot)\|_{\mathcal{H}} = 0,$$

where $\mathcal{H} = L_2(\Omega, C^3)$. This was proved for the sonar case in [21], and the proof for the radar case is entirely similar.

On combining (2.4.4) and definition (2.1.61) of $\Psi_0(x, p)$, one has the explicit representation

$$(2.4.6) \qquad v_{sc}^0(t, x) = (2\pi)^{-3/2} \int_{R^3} e^{i(x \cdot p - t|p|)} P(p) \cdot \hat{h}_-(p)\, dp,$$

where $\hat{h}_-(p)$ is given by (2.4.3). It was shown in [21] that the asymptotic behavior for $t \to +\infty$ of (2.4.6) is given by

$$(2.4.7) \qquad \lim_{t \to +\infty} \|v_{sc}^0(t, \cdot) - v_{sc}^\infty(t, \cdot)\|_{\mathcal{H}_0} = 0,$$

where $\mathcal{H}_0 = L_2(R^3, C^3)$

$$(2.4.8) \qquad v_{sc}^\infty(t, x) = \frac{F(|x| - t, \theta)}{|x|}, \qquad x = |x|\theta,$$

and

(2.4.9)
$$F(\tau, \theta) = \frac{1}{(2\pi)^{1/2}} \int_0^\infty e^{i\tau\omega}(-i\omega)P(\theta)\hat{h}_-(\omega\theta)\,d\omega.$$

On combining (2.4.5) and (2.4.7) and using the triangle inequality one gets

(2.4.10)
$$\lim_{t\to+\infty} \|v_{sc}(t, \cdot) - v_{sc}^\infty(t, \cdot)\|_{\mathcal{H}} = 0,$$

and, a fortiori,

(2.4.11)
$$\lim_{t\to+\infty} \|u_{sc}(t, \cdot) - u_{sc}^\infty(t, \cdot)\|_{\mathcal{H}} = 0,$$

where

(2.4.12)
$$u_{sc}^\infty(t, x) = \frac{e(|x| - t, \theta, \theta_0)}{|x|}, \quad x = |x|\theta,$$

and $e(\tau, \theta, \theta_0) = Re\{F(\tau, \theta)\}$. Thus,

(2.4.13)
$$e(\tau, \theta, \theta_0) = \frac{1}{(2\pi)^{1/2}} Re\left\{\int_0^\infty e^{i\tau\omega}(-i\omega)P(\theta)\hat{h}_-(\omega\theta)\,d\omega\right\}.$$

Finally, combining (2.4.13) and (2.4.3) gives (2.1.67):

(2.4.14)
$$e(\tau, \theta, \theta_0) = Re\{\int_0^\infty e^{i\tau\omega}T_+(\omega\theta, \omega\theta_0)\hat{s}(\omega, \theta_0)\,d\omega\}.$$

Of course, the derivatives of $u_{sc}(t, x)$ also have asymptotic wave functions, just as in the sonar case, provided that the incident wave profile $\hat{s}(\omega, \theta_0)$ has sufficiently many derivatives.

Exercises for §2.4.

1. Prove that the wave $v_{sc}(t, x)$ defined by (2.4.2) is asymptotically equal to the free wave $v_{sc}^0(t, x)$ defined by (2.4.4) in the sense of equation (2.4.5). Suggestion: Use the method of [21], Theorem 7.1.

2. Show how the methods of [21], Lecture 2, can be used to prove (2.4.7).

2.5 Radar Echoes from Moving Scatterers. In this section the radar echoes from moving scatterers are analyzed under the simplifying assumptions that

(2.5.1) The speeds v of the scattering objects are much smaller than the speed $c = 1$ of the radar signals.

(2.5.2) The speeds v are essentially constant during the interval required for the radar pulse to sweep over the object.

It will be shown that under these hypotheses the results of §1.5 extend fully to radar echoes.

As in §1.5, the relative velocity of the radar system and the scatterer, during the short interval in which they interact, will be described by a constant vector $\vec{v}$ with magnitude $v = |\vec{v}|$. To see the appropriateness of assumption (2.5.1), note that a maximum credible speed of scatterers for an earth-based radar system is the escape velocity for earth: $v_{\text{metric}} = 11.2 \ km/sec$. In the same units, the speed of electromagnetic waves is $c_{\text{metric}} = 3 \times 10^5 \ km/sec$. Thus the dimensionless ratio

$$(2.5.3) \qquad v = \frac{v}{c} = \frac{v_{metric}}{c_{metric}} = \frac{11.2}{3 \times 10^5} = 3.7 \times 10^{-5} \ << \ 1.$$

Assumption (2.5.2) will be regarded as essentially a definition of what is meant by a short pulse.

The radar echo calculation will make use of the two Galilean frames and the Lorentz transformation of §1.5, equations (1.5.1), (1.5.2). A major difference between the sonar and radar problems is that sonar echoes transform like scalars under the Lorentz transformation, whereas radar echoes are measured by electric and magnetic field components. The latter make up the six independent components of a second order antisymmetric tensor on the four dimensional space-time manifold [1]. Let the electric and magnetic field components on the system frame (t, x) and the object frame (t', x') be denoted by $(E_1, E_2, E_3), (H_1, H_2, H_3)$ and $(E_1', E_2', E_3'), (H_1', H_2', H_3',)$, respectively. Then application of the transformation rules of tensor algebra leads to the relations [1,p.114]

$$(2.5.4) \qquad \begin{aligned} E_1' &= E_1, \\ E_2' &= \frac{E_2 - vH_3}{\sqrt{1 - v^2}}, \\ E_3' &= \frac{E_3 + vH_2}{\sqrt{1 - v^2}}, \end{aligned}$$

and

$$(2.5.5) \qquad \begin{aligned} H_1' &= H_1, \\ H_2' &= \frac{H_2 + vE_3}{\sqrt{1 - v^2}}, \\ H_3' &= \frac{H_3 - vE_2}{\sqrt{1 - v^2}}. \end{aligned}$$

It is clear from (2.5.4), (2.5.5) with $v << 1$ that, for the radar applications studied here, the electric and magnetic field components transform essentially like scalars.

With the above simplification, it is obvious that the remainder of the calculation of the radar echoes from moving scatterers is identical with that for the sonar case. In particular, near the receiver the echo is given by

$$(2.5.6) \qquad u_{sc}(t, x) \approx u_{sc}^{\infty}(t, x) = \frac{e(|x| - t, \theta, \theta_0)}{|x|},$$

where

$$(2.5.7) \qquad e(\tau, \theta, \theta_0) = Re\left\{ \int_0^{\infty} e^{i(\frac{\gamma_0}{\gamma}\tau)\omega} T_+(\omega\gamma_0\theta', \omega\gamma_0\theta_0')\hat{s}(\omega, \theta_0)\, d\omega \right\}.$$

It has been emphasized that the calculation of radar echoes given here requires not only $0 \leq v < 1$, as in the sonar case, but $0 \leq v << 1$. Notice that, in this approximation, the Lorentz transformation of unit vectors defined by (1.5.6) reduces to the identity. Morever, γ_0 is very close to 1. Thus, for consistency, (2.5.7) should be replaced by

$$(2.5.8) \qquad e(\tau, \theta, \theta_0) = Re\left\{ \int_0^{\infty} e^{i(\frac{\gamma_0}{\gamma}\tau)\omega} T_+(\omega\theta, \omega\theta_0)\hat{s}(\omega, \theta_0)\, d\omega \right\}.$$

Note that one should not put $\gamma/\gamma_0 = 1$ in the exponential because it is this time-dilatation which produces the Doppler shift that is characteristic of radar echoes from moving objects.

Final Approximations. Equation (2.5.6) gives the echo waveform, in the far field of the scatterer, due to the incident pulse $s(x \cdot \theta_0 - t, \theta_0)$. Recall that, in the original formulation of the radar echo prediction problem, the transmitter was localized near x_0 and produced a signal at the scatterer which was given by

$$(2.5.9) \qquad u_0(t, x) = \frac{s(x \cdot \theta_0 - t + |x_0|, \theta_0)}{|x_0|},$$

apart from an error of order $1/|x_0|^2$. The echo produced by the signal (2.5.9) may be obtained from (2.5.6) by multiplying by the amplitude factor $1/|x_0|$ and replacing t by $t - |x_0|$. Thus, as in the sonar case, one has

$$(2.5.10) \qquad u_{sc}(t, x) \approx \frac{e(|x| - t + |x_0|, \theta, \theta_0)}{|x||x_0|}$$

in the far field of the scatterer. In particular, if the transmitter and receiver coincide, so $x = x_0 = |x_0|(-\theta_0)$, then the echo at the receiver is given by

$$(2.5.11) \qquad u_{sc}(t, x_0) \approx \frac{e(2|x_0| - t, -\theta_0, \theta_0)}{|x_0|^2},$$

where $e(\tau, \theta, \theta_0)$ is given by (2.5.8).

High Frequency Limit. If A. Majda's theorem (1.1.37) holds for the electro-magnetic case, then it implies that if $\hat{s}(\omega, \theta_0)$ is concentrated in a high frequency band $\omega \geq \omega_0 >> 1$, where T_+ is essentially equal to T_+^{∞}, then one has

$$(2.5.12) \qquad e(\tau, \theta, \theta_0) \approx \left(\frac{\pi}{2}\right)^{1/2} T_+^{\infty}(\theta, \theta_0)s(\frac{\gamma_0}{\gamma}\tau, \theta_0),$$

where $\gamma = \gamma(\theta) = (1 - \vec{v} \cdot \theta)/\sqrt{1 - v^2}$ and $\gamma_0 = \gamma(\theta_0)$. In particular, equations (2.5.11) and (2.5.12) give

$$(2.5.13) \qquad u_{sc}(t, x_0) \approx K(x_0)\, s\left((\gamma(\theta_0)/\gamma(-\theta_0))(2|x_0| - t), \theta_0\right),$$

where $K(x_0)$ is a suitable amplitude factor, in complete analogy with the sonar case.

Exercises for §2.5.

1. Use the method of [1], or another reference on special relativity theory, to derive equations (2.5.4), (2.5.5).

REFERENCES

[1]   P. G. BERGMANN, *Introduction to the Theory of Relativity*, Prentice-Hall, New York, 1942.

[2]   D. COLTON AND R. KRESS, *Integral Equation Methods in Scattering Theory*, J. Wiley and Sons, New York, 1983.

[3]   A. E. HEINS AND S. SILVER, *The Edge Condition and Field Representation Theorems in the Theory of Electromagnetic Diffraction*, Proc. Camb. phil. Soc., 51 (1955), pp. 149–169.

[4]   T. KATO, *Perturbation Theory for Linear Operators*, Springer Verlag, New York, 1966.

[5]   V. D.KUPRADZE, *The Method of Integral Equations in Diffraction Theory*, Mat. Sbornik, 41 (1934), pp. 561–581.

[6]   A. MAJDA, *High Frequency Asymptotics for the Scattering Matrix and the Inverse Problem of Acoustical Scattering*, Comm. Pure Appl. Math., 29 (1976), pp. 261–291.

[7]   A. MAUE, *Zur Formulierung eines allgemeinen Beugungsproblem durch eine Integralgleichung*, Zeit. Physik, 126 (1949), pp. 601–618.

[8]   J. MEIXNER, *Strenge Theorie der Beugung elektromagnetischer Wellen an vollkommen leitenden Kreisscheibe*, Z. Naturforsch., 3a (1948), pp. 506–518.

[9]   J. R. SCHULENBERGER AND C. H. WILCOX, *The Singularities of the Green's Matrix in Anisotropic Wave Motion*, Indiana J. Math., 20 (1971), pp. 1093–1117.

[10] J. R. SCHULENBERGER AND C. H. WILCOX, *Eigenfunction Expansions and Scattering Theory for Wave Propagation Problems of Classical Physics*, Arch. Rational Mech. Anal., 46 (1972), pp. 280–320.

[11] E. SKUDRZYK, *The Foundations of Acoustics*, Springer Verlag, New York, 1971.

[12] J. A. STRATTON, *Electromagnetic Theory*, McGraw-Hill, New York, 1941.

[13] C. WEBER, *Hilbertraummethoden zur Untersuchung der Beugung elektromagnetischer Wellen an Dielektra*, U. Stuttgart doctoral thesis (1977).

[14] N. WECK, *Maxwell Boundary Value Problem on Riemannian Manifolds with Nonsmooth Boundaries*, J. Math. Anal. Appl., 46 (1974), pp. 410–437.

[15] H. WEYL, *Kapazität von Strahlungsfeldern*, Math. Zeit., 55 (1952), pp. 187–198.

[16] C. H. WILCOX, *A Generalization of Theorems of Rellich and Atkinson*, Proc.AMS, 7 (1956), pp. 271–276.

[17] C.H. WILCOX, *Spherical Means and Radiation Conditions*, Arch. Rational Mech. Anal., 3 (1959), pp. 133–148.

[18] C.H. WILCOX, *The Mathematical Foundations of Diffraction Theory*, in Electromagnetic Waves, U. Wisconsin Press, Madison, Wisconsin, 1962.

[19] C. H. WILCOX, *Initial-Boundary Value Problems for Linear Hyperbolic Partial Differential Equations of the Second Order*, Arch. Rational Mech. Anal., 10 (1962), pp. 361–400.

[20] C. H. WILCOX, *Transient Wave Propagation in Homogeneous Anisotropic Media*, Arch. Rational Mech. Anal., 37 (1970), pp. 323–343.

[21] C. H. WILCOX, *Scattering Theory for the d'Alembert Equation in Exterior Domains*, Lecture Notes in Mathematics V. 442, Springer Verlag, New York, 1975.

[22] C. H. WILCOX, *Asymptotic Wave Functions and Energy Distributions in Strongly Propagative Media*, J. Math. pures et appl., 57 (1978), pp. 275–321.

[23] C. H. WILCOX, *Electromagnetic Signal Propagation in Crystals*, Applicable Anal., 8 (1978), pp. 83–94.

[24] C. H. WILCOX, *Sonar Echo Analysis*, Math. Meth. in the Appl. Sci., 1 (1979), pp. 70–88.

[25] C.H. WILCOX, *Radar Echo Analysis by the Singularity Expansion Method*, Electromagnetics, 1 (1981), pp. 481–491.

[26] C. H. WILCOX, *Sound Propagation in Stratified Fluids: Applied Mathematical Sciences V. 50*, Springer Verlag, New York, 1984.

[27] E. B. WILSON, *Vector Analysis*, Yale Univ. Press, 1929.

THE SYNTHESIS PROBLEM FOR RADAR AMBIGUITY FUNCTIONS*

CALVIN H. WILCOX†

Introduction. The ambiguity function $A(x, y)$ corresponding to a function $u(t)$ is defined by

$$A(x, y) = \int_{-\infty}^{\infty} u\left(t - \frac{x}{2}\right) \overline{u}\left(t + \frac{x}{2}\right) e^{-2\pi i y t} dt \, ,$$

where $\overline{u}$ denotes the complex conjugate of u. This paper deals with the problem of finding the functions $u(t)$ that gives rise to prescribed functions, or classes of functions, $A(x, y)$. The problem plays a central rôle in the analysis of one of the fundamental problems of radar. This problem is formulated, and reduced to the study of the correspondence $u(t) \rightarrow A(x, y)$, in §1. In the succeeding sections several approaches to the problem of synthesizing prescribed ambiguity functions are developed and applied to answering questions suggested by the radar problem.

§1. The Radar Problem. Let us consider a radar system X that is fixed in space. We assume that a number of objects $\mathcal{O}_1, \mathcal{O}_2, \ldots, \mathcal{O}_N$ are moving in the environment of X. The function of the radar system is to determine, for some preassigned instant of time which we take to be $t = 0$, the number N of objects, and their ranges $R_0^1, R_0^2, \ldots, R_0^N$ and velocities $v_1, v_2, \ldots, v_N$. This is to be done by transmitting, during some interval of time $-T < t < T$, an electromagnetic wave, or pulse, and then receiving and analyzing the echo produced by the objects. The transmitted pulse and the echo can be described by real valued functions

$$s(t) = \text{Amplitude of Transmitted Pulse at } X$$

and

$$e(t) = \text{Amplitude of Echo at } X \, .$$

The echo $e(t)$ does not determine completely the number of objects and their ranges and velocities, and the problem arises of finding which pulses $s(t)$ provide the most information about these quantities. To formulate the question mathematically, we must describe the process by which information is extracted from $e(t)$. This process depends on a representation for $s(t)$ which we now proceed to derive.

*Sponsored by the U.S. Army under Contract No. :DA–11–022–ORD–2059, Mathematics Research Center, U.S. Army

†University of Utah, Department of Mathematics, Salt Lake City, Utah 84112

229

In what follows, extensive use is made of the theory of the Fourier integral in the form due to Plancherel; see, e.g., [1, 10, 12]* We write

$$U(f) = \underset{A \to \infty}{\text{l. i. m}} \int_{-A}^{A} u(t)e^{-2\pi i f t}dt \ , \quad u(t) = \underset{A \to \infty}{\text{l. i. m}} \int_{-A}^{A} U(f)e^{2\pi i t f}df$$

and indicate the relationship between u and U by the notation

$$u(t) \sim U(f) \ .$$

Frequent use is made of Parseval's formula which reads, in our notation,

$$\int_{-\infty}^{\infty} u(t)\overline{v}(t)dt = \int_{-\infty}^{\infty} U(f)\overline{V}(f)df \ .$$

A function
$$s(t) \sim S(f)$$

is real valued if and only if
$$S(-f) = \overline{S}(f) \ .$$

Thus $s(t)$ is determined by the positive frequencies in its spectrum or, equivalently, by the function

$$\psi(t) = \underset{A \to \infty}{\text{l. i. m}} \ 2 \int_{0}^{A} S(f)e^{2\pi i t f}df \ .$$

It is known [10, p. 128] that

$$\psi(t) = s(t) + i\sigma(t)$$

where s and σ are a pair of Hilbert transforms, i.e.,

$$\sigma(t) = \frac{1}{\pi} \, p \int_{-\infty}^{\infty} \frac{s(\tau)}{t - \tau} \, d\tau \ , \quad s(t) = -\frac{1}{\pi} \, p \int_{-\infty}^{\infty} \frac{\sigma(\tau)}{t - \tau} \, d\tau$$

and

$$\int_{-\infty}^{\infty} s^2(t)dt = \int_{-\infty}^{\infty} \sigma^2(t)dt \ .$$

The p denotes the Cauchy principal value of the integrals. It follows that

$$\int_{-\infty}^{\infty} |\psi(t)|^2 dt = \int_{-\infty}^{\infty} \{s^2(t) + \sigma^2(t)\}dt = 2 \int_{-\infty}^{\infty} s^2(t)dt \ ,$$

*Numbers in square brackets denote references from the list at the end of the paper.

a quantity which is proportional to the total energy carried by the pulse $s(t)$. We assume $s(t)$ is normalized so that

$$\int_{-\infty}^{\infty} |\psi(t)|^2 dt = 1 \ .$$

Next we define

$$t_0 = \int_{-\infty}^{\infty} t|\psi(t)|^2 dt$$

and

$$f_0 = \int_{-\infty}^{\infty} f|\Psi(f)^2 df$$

and assume these quantities are finite. t_0 is called the epoch and f_0 the carrier frequency of the pulse $s(t)$. Notice that t_0 can have any value, but $f_0 > 0$ because

$$\cdot \quad \Psi(f) = \begin{cases} 2S(f) & , \quad f \geq 0 \ , \\ 0 & , \quad f < 0 \ . \end{cases}$$

Now consider the function

$$u(t) = \psi(t + t_0)e^{-2\pi i f_0(t+t_0)}$$

and its transform

$$U(f) = \Psi(f + f_0)e^{2\pi i t_0 f} \ .$$

We have

$$\int_{-\infty}^{\infty} |u(t)|^2 dt = 1 \ , \quad \int_{-\infty}^{\infty} t|u(t)|^2 dt = 0 \ , \quad \int_{-\infty}^{\infty} f|U(f)|^2 df = 0$$

and

$$s(t) = \text{Re}\{\psi(t)\} = \text{Re}\{u(t - t_0)e^{2\pi i f_0 t}\} \ .$$

The normalized function $u(t)$ is called the waveform for $s(t)$. It is "slowly varying" in the sense that its spectrum is centered around the zero frequency. In what follows, we write all pulses $s(t)$ in this way and deal with the waveforms $u(t)$.

Returning now to the radar problem, we assume that the echo produced by objects $O_1, O_2, \ldots$, having ranges $R_0^1, R_0^2, \ldots$ and velocities $v_1, v_2, \ldots$ at $t = 0$, has the form

$$e(t) = K_1 \ s\left(t - \frac{2R^1}{c + v_1}\right) + K_2 \ s\left(t - \frac{2R^2}{c + v_2}\right) + \cdots$$

where

$$R^k = R_0^k + v_k t$$

is the range of $\mathcal{O}_k$ at time t and c is the speed of electromagnetic waves. The physical assumptions underlying this approximation are that the radar cross sections of the $\mathcal{O}_k$ are independent of frequency (true for all high frequencies), that all the $\mathcal{O}_k$ are in the far field of the radar system X, that multiply reflected waves among the $\mathcal{O}_k$ are negligible, and that R^k is approximately linear for $-T < t < T$. In particular, the echo from a single object with range

$$R = R_0 + vt$$

is assumed to be

$$e(t) = Ks\left(t - \frac{2R}{c+v}\right) = Ks\left(at - \frac{2R_0}{c+v}\right)$$

where

$$a = \frac{c-v}{c+v} = 1 - \frac{2v}{c+v} \ .$$

On using the representation for $s(t)$ in terms of the waveform $u(t)$ we find that

$$e(t) = \text{Re}\{\psi_e(t)\} = \text{Re}\left\{\sqrt{a}\ e^{-2\pi i f_0 x_0} u(at - t_0 - x_0) e^{-2\pi i y_0 t} e^{2\pi i f_0 t}\right\}$$

where

$$x_0 = \frac{2R_0}{c+v} \ , \ y_0 = \frac{2 f_0 v}{c+v} \ .$$

We have taken $K = \sqrt{a}$ to normalize $e(t)$. We further assume that v is small compared with c so that v/c is negligible and a is approximately 1. To this approximation

$$e(t) = \text{Re}\{\psi_e(t)\} = \text{Re}\left\{[e^{-2\pi i f_0 x_0} u(t - t_0 - x_0) e^{-2\pi i y_0 t}] e^{2\pi i f_0 t}\right\} \ ,$$

where

$$x_0 = \frac{2}{c} R_0 \ , \ y_0 = \frac{2 f_0}{c} v \ .$$

The quantities x_0 and y_0 are, respectively, the time delay in the echo, due to the range of the object, and the Doppler, or frequency, shift in the echo, due to the object's velocity. They determine the desired quantities R_0 and v uniquely. x_0 and y_0 are estimated by comparing the actual echo $e(t)$ with the possible echoes

$$e_{x,y}(t) = \text{Re}\{\psi_{x,y}(t)\} = \text{Re}\left\{[e^{-2\pi i f_0 x} u(t - t_0 - x) e^{-2\pi i y t}] e^{2\pi i f_0 t}\right\}$$

for various values of x and y. This is done by forming the function

$$I(x,y) = \left| \int_{-\infty}^{\infty} \psi_e(t) \overline{\psi}_{x,y}(t) dt \right|^2 \ ,$$

electronically, and displaying it on a screen as a brightness pattern. Observe that, since $u(t)$ is normalized,

$$I(x_0, y_0) = 1 \ , \ I(x,y) \leq 1 \quad \text{for all} \quad x, y \ ,$$

by Schwarz's inequality. Thus, for a single object, range and velocity are identified by the brightest spot in the pattern.

Notice that

$$\int\limits_{-\infty}^{\infty} \psi_e(t)\overline{\psi}_{x,y}(t)dt = e^{2\pi i[f_0(x-x_0)+t_0(t-y_0)]+\pi i(y-y_0)(x+x_0)} \int\limits_{-\infty}^{\infty} u\left(t+\frac{x_0-x}{2}\right)e^{-2\pi i(y_0-y)t}dt ,$$

whence

$$I(x,y) = |A(x_0-x, y_0-y)|^2$$

where

$$A(x,y) = \int\limits_{-\infty}^{\infty} u\left(t-\frac{x}{2}\right)\overline{u}\left(t+\frac{x}{2}\right)e^{-2\pi iyt}dt .$$

When there are several objects $\mathcal{O}_1, \mathcal{O}_2, \ldots$ we assume the ranges $x_1, x_2, \ldots$ and velocities $v_1, v_2, \ldots$ fluctuate in a random way so that the echoes arrive at X incoherently. Then the cross terms in the product $I(x,y)$ are small and, approximately,

$$I(x,y) = M_1^2|A(x_1-x, y_1-y)|^2 + M_2^2|A(x_2-x, y_2-y)|^2 + \cdots$$

where $M_1, M_2 \ldots$ depend on the ranges and radar cross sections of $\mathcal{O}_1, \mathcal{O}_2, \ldots$. Thus, in all cases, $I(x,y)$ is determined by the function $A(x,y)$. The number of objects and their various ranges and velocities obviously are not completely determined by the pattern $I(x,y)$, and the ambiguity in the determination is described by the structure of $A(x,y)$. Hence, it is natural to call $A(x,y)$ the ambiguity function for the waveform $u(t)$.

It is obviously desirable to choose the waveform $u(t)$ so as to minimize the ambiguity in the determination of the number of objects, their ranges, and velocities; i.e., so $A(x,y)$ is a prescribed function or has a certain prescribed properties. This is the synthesis problem. The waveforms $u(t)$ are subject to certain restrictions. They must be square-integrable and normalized as described above. In addition, we shall require that bandwidth β defined by

$$\beta^2 = \int\limits_{-\infty}^{\infty} f^2|u(f)|^2 df$$

and the time duration τ defined by

$$\tau^2 = \int\limits_{-\infty}^{\infty} t^2|u(t)|^2 dt$$

are finite. The only other restriction is the condition

$$U(f) = \Psi(f+f_0)e^{2\pi it_0f} = 0 \quad \text{for} \quad f \leq -f_0 .$$

It is shown below to have no practical significance for the radar problem, and we shall ignore it in what follows.

§2. Some Properties of Ambiguity Functions. In this section, precise definitions are given for the classes of functions $u(t)$ and $A(x, y)$ suggested by the radar problem, and a number of their properties are derived.

DEFINITION. A function $u(t)$ is in class W (and is called a waveform) $\Longleftrightarrow$ $u(t)$, $tu(t)$, and $f\, U(f)$ are in $L^2(-\infty, \infty)$ and

$$(2.1) \qquad \int_{-\infty}^{\infty} |u(t)|^2 dt = 1 \, , \quad \int_{-\infty}^{\infty} t|u(t)|^2 dt = 0 \, , \quad \int_{-\infty}^{\infty} f|U(f)|^2 df = 0 \, .$$

Notice that the existence of the integrals in (2.1) follows from the integrability assumptions. It is also evident that $u(t) \in W \Longleftrightarrow U(f) \in W$. Additional consequences of the definition are contained in the following lemmas.

LEMMA 2.1. $u(t) \in W \Longrightarrow u(t)$ and $U(f)$ are in $L^1(-\infty, \infty)$ and

$$(2.2) \qquad U(f) = \int_{-\infty}^{\infty} u(t)e^{-2\pi i f t} dt \, , \quad u(t) = \int_{-\infty}^{\infty} U(f)e^{2\pi i t f} df \, .$$

Proof. Since $u(t) = (1/t)(tu(t))$ and $1/t$ and $tu(t) \in L^2$ on any set excluding the origin we have $u(t) \in L^1$. Similarly $U(f) \in L^1$ and the two integrals in (2.2) exist. It follows that their values are $U(f)$ and $u(t)$, respectively [1, p. 19].

LEMMA 2.2. $u(t) \in W \Longrightarrow u'(t)$ and $U'(f)$ exist almost everywhere and are in $L^2(-\infty, \infty)$.

This is contained in a theorem of Titchmarsh [10, p. 92].

LEMMA 2.3. $u(t) \in W \Longrightarrow u(t)$ and $U(f)$ are continuous and bounded on $(-\infty, \infty)$ and

$$(2.3) \qquad \lim_{|t| \to \infty} u(t) = 0 \, , \quad \lim_{|f| \to \infty} U(f) = 0 \, .$$

Proof. The continuity follows from (2.2)[12, p. 72], and (2.3) follows from (2.2) and the Riemann–Lebesgue theorem [12, p. 14]. The boundedness is a consequence of the continuity and (2.3).

DEFINITION. A function $A(x, y)$ is in class X (and is called ambiguity function) $\Longleftrightarrow$

$$(2.4) \qquad A(x, y) = A_u(x, y) \equiv \int_{-\infty}^{\infty} u\left(t - \frac{x}{2}\right) \overline{u}\left(t + \frac{x}{2}\right) e^{-2\pi i y t} dt$$

where $u(t) \in W$. (The function A_u is defined by (2.4).)

It is convenient to introduce the functionals

$$F_{x,y}[u, v] = \int_{-\infty}^{\infty} u\left(t - \frac{x}{2}\right) \overline{v}\left(t + \frac{x}{2}\right) e^{-2\pi i y t} dt$$

and

$$\mathcal{F}_{x,y}[u,v] = \int\limits_{-\infty}^{\infty} u(t)\overline{v}(t+x)e^{-2\pi iyt}dt \ .$$

Obviously,

(2.5)
$$F_{x,y}[u,v] = e^{-\pi ixy}\mathcal{F}_{x,y}[u,v]$$

and

$$A_u(x,y) = F_{x,y}[u,u] \ .$$

We will now derive some properties of ambiguity functions. A number of our derivations depend on

Parseval's Theorem [12, p. 70]: If $u(t)$ and $v(t)$ are in $L^2(-\infty,\infty)$ then

$$\int\limits_{-\infty}^{\infty} u(t)\overline{v}(t)dt = \int\limits_{-\infty}^{\infty} U(f)\overline{V}(f)df \ .$$

This result implies

LEMMA 2.4. If $u(t)$ and $v(t)$ are in W, then

$$F_{x,y}[u,v] = F_{y,x}[\overline{V},\overline{U}] \ .$$

Proof. Since

$$u(t)e^{-2\pi iyt} \sim U(f+y) \ , \quad v(t+x) \sim V(f)e^{2\pi ixf} \ ,$$

Parseval's theorem implies

$$\int\limits_{-\infty}^{\infty} u(t)e^{-2\pi iyt}\overline{v}(t+x)dt = \int\limits_{-\infty}^{\infty} U(f+y)\overline{V}(f)e^{-2\pi ixf}df \ ,$$

or

$$\mathcal{F}_{x,y}[u,v] = \mathcal{F}_{y,x}[\overline{V},\overline{U}] \ .$$

On multiplying this by $e^{-\pi ixy}$ we get the same result for $F_{x,y}[u,v]$.

COROLLARY 2.4. $u(t) \in W \Longrightarrow A_u(x,y) = A_{\overline{U}}(y,x)$, or

(2.6)
$$A(x,y) = A_u(x,y) = \int\limits_{-\infty}^{\infty} \overline{U}\left(f-\frac{y}{2}\right)U\left(f+\frac{y}{2}\right)e^{-2\pi ixf}df \ .$$

This follows from Lemma 2.4 with $v = u$. In what follows, we make use of the inner product notation

$$(u,v) = \int\limits_{-\infty}^{\infty} u(t)\overline{v}(t)dt \ , \quad \|u\|^2 = (u,u) \ .$$

Our next result is

LEMMA 2.5. *If $u(t)$ and $v(t)$ are in W, then $F_{x,y}[u,v]$ is in $L^2(E)$, where E denotes the Euclidean plane, and*

$$\int_E F_{x,y}[u_1,v_1]\overline{F}_{x,y}[u_2,v_2]dxdy = (u_1,u_2)\overline{(v_1,v_2)}$$

for all u_1, u_2, v_1, v_2 in W.

Proof. Since the functions in W are bounded and in $L^2(-\infty, \infty)$, the products $u(t)\overline{v}(t+x)$ are in L^2 and

$$u(t)\overline{v}(t+x) \sim \mathfrak{F}_{x,y}[u,v]$$

for each fixed x. Hence by Parseval's theorem

$$\int_{-\infty}^{\infty} \mathfrak{F}_{x,y}[u_1,v_1]\overline{\mathfrak{F}}_{x,y}[u_2,v_2]dy = \int_{-\infty}^{\infty} u_1(t)\overline{v}_1(t+x)\overline{u}_2(t)v_2(t+x)dt \ .$$

Now the iterated integral

$$\int_{-\infty}^{\infty} dt \int_{-\infty}^{\infty} |u_1(t)\overline{v}_1(t+x)\overline{u}_2(t)v_2(t+x)|dx$$

converges. Hence by Fubini's theorem (as stated, e.g., in [6, p. 207]),

$$(u_1,u_2)\overline{(v_1,v_2)} = \int_{-\infty}^{\infty} dt \int_{-\infty}^{\infty} u_1(t)\overline{v}_1(t+x)\overline{u}_2(t)v_2(t+x)dx$$

$$= \int_{-\infty}^{\infty} dx \int_{-\infty}^{\infty} \mathfrak{F}_{x,y}[u_1,v_1]\overline{\mathfrak{F}}_{x,y}[u_2,v_2]dy = \int_E \mathfrak{F}_{x,y}[u_1,v_1]\overline{\mathfrak{F}}_{xy}[u_2,v_2]dxdy \ .$$

The same result for $F_{x,y}[u,v]$ follows from (2.5).

COROLLARY 2.5. $A(x,y) \in X \implies A(x,y) \in L^2(E)$ and

$$\int_E |A(x,y)|^2 dxdy = 1 \ .$$

This result has been called the "law of conservation of ambiguity" and the "radar uncertainty principle". It means that the total ambiguity for a waveform is a constant determined by the total energy, and cannot be diminished by varying the waveform.

LEMMA 2.6. *Let $A(x,y) \in X$. Then for each fixed x, $A(x,y) \in L^1(-\infty, \infty)$ as a function of y. Similarly, for each fixed y, $A(x,y) \in L^1(-\infty, \infty)$ as a function of x.*

This result follows from (2.4) and (2.6) and the theorem that $L^1(-\infty, \infty)$ is closed under the convolution operation. For a proof, see, e.g., [1, p. 6].

We can now verify the assertion, made at the end of §1, that the condition $U(f) = 0$ for $f \le -f_0$ is of no practical significance for the radar problem. Suppose that $A(x,y) = A_{\overline{U}}(x,y)$ is a prescribed function, but $U(f) \ne 0$ for $f \le -f_0$. Let

$$
U_0(f) = \begin{cases} U(f) & , \quad f > -f_0 \\ \\ 0 & , \quad f \le -f_0 \end{cases}
$$

and write $A_0(x,y) = A_{\overline{U}_0}(x,y)$. Then, if M is a bound for U, (2.6) implies

$$
|A(x,y) - A_0(x,y)| \le 2M \int_{-\infty}^{-f_0} |U(f)| df .
$$

Hence the difference $A - A_0$ is uniformly small for large f_0. The hyphothesis that the carrier frequency f_0 is large is one of the primary physical assumptions underlying our treatment of the radar problem.

§3. The Construction of a Waveform from Its Ambiguity Function.

It is a theorem that if $u(t)$ and its transform $U(f)$ are both in $L^1(-\infty, \infty)$ then

$$
u(t) = \int_{-\infty}^{\infty} U(f)e^{2\pi itf} df
$$

almost everywhere [1, p. 19]. Hence Lemma 2.6 implies that

$$
u\left(\xi - \frac{x}{2}\right) \overline{u}\left(\xi + \frac{x}{2}\right) = \int_{-\infty}^{\infty} A(x,y)e^{2\pi i\xi y} dy
$$

for all ξ and x. (It holds for all ξ because both sides are continuous.) On introducing the variables

$$
t = \xi - \frac{x}{2} \quad , \quad \tau = \xi + \frac{x}{2} \quad \text{so} \quad \xi = \frac{1}{2}(t + \tau) \quad , \quad x = \tau - t
$$

(3.2)

$$
f = \eta + \frac{y}{2} \quad , \quad \phi = \eta - \frac{y}{2} \quad \text{so} \quad \eta = \frac{1}{2}(f + \phi) \quad , \quad y = f - \phi
$$

this result takes the following form.

THEOREM 3.1. *Let $A(x, y) = A_u(x, y)$ be an ambiguity function, corresponding to $u(t) \in W$. Then*

$$u(t)\overline{u}(\tau) = \int_{-\infty}^{\infty} A(\tau - t, y)e^{\pi i(t+\tau)y} dy$$

and

$$U(f)\overline{U}(\phi) = \int_{-\infty}^{\infty} A(x, f - \phi)e^{\pi i(f+\phi)x} dx .$$

The last equation follows from (2.6). Theorem 3.1 provides a uniqueness theorem for the representation (2.4) of ambiguity functions. We state it as

COROLLARY 3.1. *If $A(x, y) = A_u(x, y) = A_v(x, y)$ where u and v are in W then*

$$v(t) = c\, u(t) , \quad c \quad \text{a constant} , \quad |c| = 1 .$$

Proof. By Theorem 3.1, we have $u(t)\overline{u}(\tau) = v(t)\overline{v}(\tau)$, whence $v(t) = c\, u(t)$. But then $(c\overline{c} - 1)u(t)\overline{u}(\tau) = 0$, whence $|c|^2 = 1$ since $u(t)$ is not identically zero.

It is significant for our work that the operations used above to produce $u(t)\overline{u}(\tau)$ from $A(x, y)$ can be generalized to apply to an arbitrary $F(x, y) \in L^2(E)$. Indeed, for almost all x such an $F(x, y)$ is in $L^2(-\infty, \infty)$ as a function of y. Hence,

(3.3)
$$G(x, \xi) = \underset{R \to \infty}{\text{l.i.m}} \int_{-R}^{R} F(x, y)e^{2\pi i \xi y} dy$$

exists and is in $L^2(-\infty, \infty)$, and by Parseval's theorem

$$\int_{-\infty}^{\infty} |G(x, \xi)|^2 d\xi = \int_{-\infty}^{\infty} |F(x, y)|^2 dy .$$

By Fubini's theorem, this equation can be integrated with respect to x and implies

$$\int_{E} |G(x, \xi)|^2 dx d\xi = \int_{E} |F(x, y)|^2 dx dy .$$

Let us define t and τ by (3.2) and put

(3.4)
$$H(t, \tau) = G(x, \xi) .$$

Then since $\left| \dfrac{\partial(t, \tau)}{\partial(x, \xi)} \right| = 1$ we have

$$\int_{E} |H(t, \tau)|^2 dt d\tau = \int_{E} |G(x, \xi)|^2 dx d\xi. \ .$$

Similarly, we can define

(3.5)
$$G'(\eta, y) = \operatorname*{l.i.m}_{R \to \infty} \int_{-R}^{R} F(x, y) e^{2\pi i \eta x} dx$$

and

(3.6)
$$H'(f, \phi) = G'(\eta, y),$$

where f and ϕ are defined by (3.2), and conclude that

$$\int_E |H'(f, \phi)|^2 df d\phi = \int_E |G'(\eta, y)| d\eta dy = \int_E |F(x, y)|^2 dx dy .$$

Since the operations defined by (3.3), (3.4), (3.5), and (3.6) are all reversible, we have proved

LEMMA 3.1. *The correspondences Ω and Ω' defined by*

$$\Omega F = H , \ \Omega' F = H'$$

are unitary mappings of $L^2(E)$ onto itself, i.e.,

$$\int_E |H(t, \tau)|^2 dt d\tau = \int_E |H'(f, \phi)|^2 df d\phi = \int_E |F(x, y)|^2 dx dy .$$

The composite mappings $\Omega'\Omega^{-1}$ and $\Omega\Omega'^{-1}$ are also unitary mappings of $L^2(E)$ onto itself. We shall show that they are essentially the Fourier transform for $L^2(E)$. More precisely, we have

LEMMA 3.2. *If $H \in L^2(E)$ and $H' = \Omega'\Omega^{-1}H$, then*

$$H'(f, \phi) = \operatorname*{l.i.m}_{N \to \infty} \int_{t^2 + \tau^2 \leq N^2} H(t, \tau) e^{-2\pi i(ft - \phi\tau)} dt d\tau .$$

Proof. Formally, we have

$$\int_E H(t, \tau) e^{-2\pi i(ft - \phi\tau)} dt d\tau = \int_E G(x, \xi) e^{-2\pi i(y\xi - \eta x)} d\xi dx$$

where the variables are related as in (3.2). The last integral equals

$$\int_{-\infty}^{\infty} F(x, y) e^{2\pi i \eta x} dx = G'(\eta, y) = H'(f, \phi) .$$

The manipulations used in this deduction can be justified if $H \in L^1(E) \cap L^2(E)$ and is bounded. Since the set of such H is dense in $L^2(E)$, the result follows [1, p. 118].

DEFINITION. A function $F(x, y)$ is in class $Y \iff F(x, y)$, $tH(t, \tau)$, and $fH'(f, \phi)$ are in $L^2(E)$ and

$$(3.7) \qquad\qquad F(-x, -y) = \overline{F}(x, y) .$$

Condition (3.7) is satisfied if $F \in X$ (see (2.4)). Moreover, Theorem (3.1) states that if $A = A_u \in X$ then

$$\Omega A(t, \tau) = u(t)\overline{u}(\tau) , \quad \Omega' A(f, \phi) = U(f)\overline{U}(\phi) ,$$

so that the remaining conditions of the definition are necessary for $F \in X$; i.e., $X \subset Y$. Conversely, if $F \in Y$, the conditions

$$(3.8) \qquad\qquad \Omega F(t, \tau) = u(t)\overline{u}(\tau) \quad \text{or} \quad \Omega' F(f, \phi) = U(f)\overline{U}(\phi)$$

are equivalent, by Lemma 3.2, and imply that $u(t) \in W$ and $F = A_u \in X$. This proves

THEOREM 3.2. $F(x, y) \in Y$ is an ambiguity function if and only if one of the conditions (3.8) holds.

Finally we observe that (3.8) can be formulated without explicit reference to $u(t)$ as follows.

COROLLARY 3.2. $F(x, y) \in Y$ is an ambiguity function $\iff H = \Omega F$ satisfies

$$(3.9) \qquad\qquad H(\xi, \xi)H(t, \tau) = H(t, \xi)H(\xi, \tau)$$

$$(3.10) \qquad\qquad H(\xi, \xi) \geq 0 .$$

Proof. The necessity of (3.9) and (3.10) is obvious from (3.8). To prove their sufficiency, notice that (3.7) is equivalent to the condition that H is Hermitian;

$$(3.11) \qquad\qquad H(\tau, \tau) = \overline{H}(\tau, t) .$$

Hence (3.9) implies that

$$H(t, t)H(\tau, \tau) = |H(t, \tau)|^2 .$$

Integrating this over t and τ gives

$$\left(\int_{-\infty}^{\infty} H(t, t)dt \right)^2 = 1 ,$$

whence

$$\int_{-\infty}^{\infty} H(t, t)dt = 1$$

by (3.10). Hence there is a ξ for which $H(\xi, \xi) > 0$. Using such a ξ we define

$$u(t) = \frac{H(t, \xi)}{\sqrt{H(\xi, \xi)}} .$$

Then

$$u(t)\overline{u}(\tau) = \frac{H(t, \xi)\overline{H}(\tau, \xi)}{H(\xi, \xi)} = \frac{H(t, \xi)H(\xi, \tau)}{H(\xi, \xi)} = H(t, \tau)$$

by (3.11) and (3.9). Thus $F \in X$, by Theorem 3.2.

§4. **The Use of Expansions in Series of Orthogonal Functions.** A sequence of functions

$$(4.1) \qquad \phi_0(t) , \quad \phi_1(t) , \ldots, \phi_m(t) , \ldots$$

in $L^2(-\infty, \infty)$ is orthonormal if $(\phi_m, \phi_n) = \delta_{nm}$, the usual Kronecker delta. It is complete, in the mean square or L^2 sense, if for each $\phi(t)$ in $L^2(-\infty, \infty)$ and each $\varepsilon > 0$ there exist coefficients $a_0, a_1, \ldots, a_N$ such that

$$\int_{-\infty}^{\infty} |\phi(t) - \sum_{m=0}^{N} a_m \phi_m(t)|^2 dt < \varepsilon .$$

We shall consider complete orthonormal sequences whose members $\phi_m(t) \varepsilon W$. Their utility for our work follows from

LEMMA 4.1. *Let (4.1) be a complete orthornormal sequence with members $\phi_m(t) \in W$. Then the sequence*

$$\psi_{mn}(x, y) = \int_{-\infty}^{\infty} \phi_m \left(t - \frac{x}{2} \right) \overline{\phi}_n \left(t + \frac{x}{2} \right) e^{-2\pi i y t} dt$$

is orthonormal and complete in $L^2(E)$.

Proof. By Lemma 2.5

$$\int_E \psi_{mn}(x, y) \overline{\psi}_{m'n'}(x, y) dx dy = (\phi_m, \phi_{m,})(\overline{\phi_n, \phi'_n}) = \delta_{mm'} \delta_{nn'} ,$$

which proves the orthonormality. Next, if $F(x, y) \in L^2(E)$, then by Lemma 3.1

$$(4.2) \qquad \begin{aligned} &\int_E |F(x, y) - \sum_{m=0}^{N} \sum_{n=0}^{N} c_{mn} \psi_{mn}(x, y)|^2 dx dy = \\ &\int_E |H(t, \tau) - \sum_{m=0}^{N} \sum_{n=0}^{N} c_{mn} \phi_m(t) \overline{\phi}_n(\tau)|^2 dt d\tau \end{aligned}$$

where $H = \Omega F \in L^2$. But, the sequence $\phi_m(t) \overline{\phi}_n(\tau)$ is known to be orthonormal and complete in $L^2(E)$ [3, p. 56]. Hence (4.2) implies that the ψ_{mn} are complete.

The Riesz–Fischer theorem [8, pp. 69–70] implies that if $u(t) \in L^2(-\infty, \infty)$ then

$$(4.3) \qquad u(t) = \underset{N \to \infty}{\text{l.i.m}} \sum_{m=0}^{N} a_m \phi_m(t) , \quad a_m = (u, \phi_m) ,$$

and

(4.4)
$$\sum_{m=0}^{N} |a_m|^2 = \|u\|^2 < \infty ,$$

and that, conversely, every sequence $a_0, a_1, \ldots$ satisfying (4.4) defines a unique $u(t) \in L^2$ satisfying (4.3). Similarly, by Lemma 4.1, every $F(x,y) \in L^2(E)$ satisfies

(4.5)
$$F(x,y) = \underset{N \to \infty}{\mathrm{l.i.m}} \sum_{m=0}^{N} \sum_{n=0}^{N} c_{mn} \psi_{mn}(x,y) ,$$

$$c_{mn} = \int_E F(x,y) \overline{\psi}_{mn}(x,y) dx dy$$

and

(4.6)
$$\sum_{m=0}^{N} \sum_{n=0}^{N} |c_{mn}|^2 < \infty$$

and, conversely, every sequence c_{mn} satisfying (4.6) defines an F satisfying (4.5).

If $A = A_u \in X$, then, by Lemma 2.5,

$$\int_E A(x,y) \overline{\psi}_{mn}(x,y) dx dy = (u, \phi_m) \overline{(u, \phi_n)} = a_m \overline{a}_n ,$$

where u is given by (4.3). Conversely, if $F(x,y) \in Y$ and

(4.7)
$$c_{mn} = a_m \overline{a}_n$$

then (4.6) implies (4.4) and (4.3) defines a $u \in L^2(-\infty, \infty)$. Moreover, (4.7), (4.5) and (4.2) imply $H(t, \tau) = u(t) \overline{u}(\tau)$. Thus Theorem 3.2 implies

THEOREM 4.1. $F(x,y) \in Y$ is an ambiguity function if and only if $c_{mn} = a_m \overline{a}_n$.

The condition that c_{mn} should factor can be formulated without explicit reference to the sequence a_m, in analogy with Corollary 3.2, as follows.

COROLLARY 4.2. $F(x,y) \in Y$ is an ambiguity function $\iff$ the coefficients c_{mn} defined by (4.5) satisfy

$$c_{kk} c_{mn} = c_{mk} c_{kn} \quad \text{and} \quad c_{kk} \geq 0 \quad \text{for all} \quad k , m , \text{and } n .$$

The proof parallels that of Corollary 3.2 and is omitted. In it we show that $F = A_u$ where $u \in W$ is defined by

$$a_m = \frac{c_{mk}}{\sqrt{c_{kk}}} , c_{kk} > 0$$

and (4.3). Hence we may state

COROLLARY 4.3. *If $F(x,y) \in L^2(E)$ is an ambiguity function, then*

$$u(t) = \underset{M \to \infty}{\mathrm{l.i.m}} \frac{1}{\sqrt{c_{kk}}} \sum_{m=0}^{M} c_{mk} \phi_m(t) \ , \ c_{kk} > 0$$

defines a corresponding waveform.

§5. **Approximation of Arbitrary Functions by Ambiguity Functions.** In this section, we determine the ambiguity function $A = A_u$ that is the best approximation to a given function $F \in Y$ in the mean square sense; i.e., we solve the minimum problem

(5.1) $$\int_E |F(x,y) - A_u(x,y)|^2 dx dy = \text{ minimum } , u \in W \ .$$

If $u \in W$ is any waveform, we have

(5.2)
$$\int_E |F(x,y) - A_u(x,y)|^2 dx dy = \int_E |H(t,\tau) - u(t)\overline{u}(\tau)|^2 dt d\tau$$
$$= 2(1 - \int_E H(t,\tau)\overline{u}(t)u(\tau)dt d\tau)$$

by Lemma 3.1 and (3.11). Thus (5.1) is equivalent to

(5.3) $$J[u] \equiv \int_E H(t,\tau)\overline{u}(t)u(\tau)dt d\tau = \text{ maximum } , u \in W \ .$$

It is evident from (5.2) that $J[u]$ is real, not greater than 1, and equals 1 if and only if $F \in X$.

It is well known that the problem

(5.4) $$|J[u]| = \text{ maximum } , \ u \in L^2(-\infty, \infty) \ , \ \|u\| = 1$$

is related to the eigenvalue problem for the kernel $H(t,\tau)$, i.e., the problem of determining the numbers μ and functions $u \in L^2(-\infty, \infty)$ that satisfy

(5.5) $$\int_{-\infty}^{\infty} H(t,\tau)u(\tau)d\tau = \mu \, u(t) \ .$$

The following facts are known [8, pp. 232–234 and pp. 242–244].

I. (5.5) has solutions if and only if μ belongs to a sequence $\mu_0, \mu_1, \mu_2, \ldots$ (possibly finite) of values. If the sequence is infinite, then zero is its only limit point and we can always assume

$$|\mu_0| > |\mu_1| > |\mu_2| > \cdots$$

II. Each eigenvalue μ_i has finite multiplicity; i.e., the linear space of solutions of (5.5) corresponding to $\mu = \mu_i$ has finite dimension.

III. maximum $|J(u)| = |\mu_0|$ and (5.4) has as solutions the eigenfunctions u corresponding to μ_0.

If $\mu_0 < 0$, it is evident that the maximum in (5.3) can be increased (and therefore a minimum in (5.1) can be reduced) by considering $-H$ instead of H (i.e., approximating $-F$ instead of F). We shall suppose this has been done, so that $\mu_0 > 0$. Then

$$\text{maximum} \quad J[u] = \mu_0$$

and is attained by an eigenfunction u corresponding to μ_0.

It follows easily from Lemma 3.2 and Parseval's theorem that (5.5) is equivalent to

$$(5.6) \qquad \int_{-\infty}^{\infty} H'(f, \phi) U(\phi) d\phi = \mu \, U(f) \ .$$

when $F \in Y$, (5.5) and (5.6) imply that any eigenfunction u is in W; i.e., $tu(t)$ and $fU(f)$ are in $L^2(-\infty, \infty)$. Combining these facts, we arrive at the following solution of (5.1).

THEOREM 5.1. *If $F(x, y) \in Y$, the minimum problem (5.1) always has solutions $u \in W$. The minimum is $2(1 - \mu_0)$ where μ_0 is the largest eigenvalue of (5.5) and $H = \Omega F$. The minimum is attained for any $u \in W$ that is an eigenfunction corresponding to μ_0. (The latter form a finite dimensional linear space.)*

A variant of (5.1) that is of interest in connection with the practical problem of approximating prescribed F by ambiguity functions arises when we restrict u to a finite dimensional space, say,

$$(5.7) \qquad u(t) = \sum_{m=0}^{N} a_m \phi_m(t) \ , \quad \phi_m \in W \ .$$

We assume that the ϕ_m are orthonormal, so that

$$\|u\|^2 = \sum_{m=0}^{N} |a_m|^2 = 1 \ .$$

Then (5.2) implies

$$(5.8) \qquad \int_E |F(x, y) - A_u(x, y)|^2 dx \, dy = 2(1 - \sum_{m=0}^{N} \sum_{n=0}^{N} c_{mn} \overline{a}_m a_n)$$

where c_{mn} is defined by (4.5). The matrix (c_{mn}) is Hermitian, $c_{mn} = \overline{c}_{nm}$, by (3.7) or (3.11). It follows easily that u minimizes (5.8) if and only if

$$\sum_{n=0}^{N} c_{mn} a_n = \mu a_m \ , \quad m = 0, 1, \ldots, N \ .$$

These considerations imply

COROLLARY 5.1. *The minimum of (5.8) is $2(1 - \mu_0)$ where μ_0 is the largest eigenvalue of the Hermitian matrix (c_{mn}). It is attained by the waveform (5.7) where $(a_0, a_1, \ldots, a_N)$ is any eigenvector for (c_{mn}) corresponding to μ_0.*

§6. The Behavior of Ambiguity Functions Near the Origin.

We have seen that if $A \in X$, then

$$|A(x,y)| \leq 1 , \quad |A(0,0)| = 1 .$$

In this section, we determine the behavior of $|A(x,y)|$ near the origin and its dependence on the waveform u, and discuss the associated questions of range-velocity accuracy and resolution for the radar problem. We begin by proving

THEOREM 6.1. *Every $A(x,y) \in X$ is in class C^2, i.e., has continuous first and second derivatives throughout the plane.*

Proof. It is convenient to consider the associated function

$$(6.1) \qquad B(x,y) = e^{\pi i x y} A(x,y) = \int_{-\infty}^{\infty} u(t)\overline{u}(t+x)e^{-2\pi i y t}dt .$$

Obviously, $A \in C^2$ if and only if $B \in C^2$. From (6.1) and the representation

$$B(x,y) = \int_{-\infty}^{\infty} \overline{U}(f)U(f+y)e^{-2\pi i x f}df ,$$

obtained from Corollary 2.4, we get by formal differentiation under the integral sign the integrals

$$I_1(x,y) = \int_{-\infty}^{\infty} u(t)\overline{u}'(t+x)e^{-2\pi i y t}dt = -2\pi i \int_{-\infty}^{\infty} f\overline{U}(f)U(f+y)e^{-2\pi i x f}df$$

$$I_2(x,y) = -2\pi i \int_{-\infty}^{\infty} tu(t)\overline{u}(t+x)e^{-2\pi i y t}dt = \int_{-\infty}^{\infty} \overline{U}(f)U'(f)e^{-2\pi i x f}df$$

$$I_{11}(x,y) = -4\pi^2 \int_{-\infty}^{\infty} f^2\overline{U}(f)U(f+y)e^{-2\pi i x f}df$$

$$I_{12}(x,y) = -2\pi i \int_{-\infty}^{\infty} tu(t)\overline{u}'(t+x)e^{-2\pi i y t}dt = -2\pi i \int_{-\infty}^{\infty} f\overline{U}(f)U'(f+y)e^{-2\pi i x f}df$$

$$I_{22}(x,y) = -4\pi^2 \int_{-\infty}^{\infty} t^2 u(t)\overline{u}(t+x)e^{-2\pi i y t}dt .$$

By assumption, $u(t) \in W$. Hence the functions $u'(t)$, $tu(t)$, $U'(f)$, $fU(f)$ are in $L^2(-\infty, \infty)$, and it follows that all the integrals exist. Moreover, each defines a continuous function in the plane. Thus, for example,

$$I_{22}(x+\xi, y+\eta) - I_{22}(x,y) = -4\pi^2 \int_{-\infty}^{\infty} tu(t) \left\{ t\overline{u}(t+x+\xi)e^{-2\pi i(y+\eta)t} - t\overline{u}(t+x)e^{-2\pi i y t} \right\} dt$$

whence, by Schwarz's inequality,

$$|I_{22}(x+\xi, y+\eta) - I_{22}(x,y)|^2 \le 16\pi^4 \int_{-\infty}^{\infty} t^2 |u(t)|^2 dt \int_{-\infty}^{\infty} |tu(t+x+\xi)e^{2\pi i(y+\eta)t} - tu(t+x)e^{2\pi i y t}|^2 dt$$

$$= 16\pi^4 \tau^2 \int_{-\infty}^{\infty} |\{tu(t+x+\xi) - tu(t+x)\}e^{2\pi i(y+\eta)t} + tu(t+x)\{e^{2\pi i(y+\eta)t} - e^{2\pi i y t}\}|^2 dt$$

Hence, by the triangle inequality,

$$|I_{22}(x+\xi, y+\eta) - I_{22}(x,y)| \le 4\pi^2 \tau \left(\int_{-\infty}^{\infty} |tu(t+x+\xi) - tu(t+x)|^2 dt \right)^{\frac{1}{2}}$$

$$+ 4\pi^2 \tau \left(\int_{-\infty}^{\infty} t^2 |u(t+x)|^2 |e^{2\pi i \eta t} - 1|^2 dt \right)^{\frac{1}{2}}.$$

The first term on the right tends to zero when $\xi \to 0$, see [12, p. 24]. The second tends to zero when $\eta \to 0$, by Lebesgue's dominated convergence theorem, since $tu(t) \in L^2(-\infty, \infty)$. Thus $I_{22}(x,y)$ is continuous and a similar proof applies to the other integrals. Next

$$\int_{x_1}^{x_2} I_1(\xi, y) d\xi = \int_{-\infty}^{\infty} \overline{U}(f) U(f+y) \{e^{-2\pi i x_2 f} - e^{-2\pi i x_1 f}\} df = B(x_2, y) - B(x_1, y) ,$$

by Fubini's theorem, whence $I_1 = \dfrac{\partial}{\partial x} B$. Similarly, $I_{11} = \dfrac{\partial^2}{\partial x^2} B$, etc.

The nature of the maximum of $|A(x,y)|$ at the origin can be described in terms of the level curves $|A(x,y)|^2 = c$ with c near 1. To determine these, we use the Taylor series for $|A(x,y)|^2$. Notice that the partial derivatives of $B(x,y)$ satisfy

$$B_1(0,0) = -2\pi i \int_{-\infty}^{\infty} f|U(f)|^2 df = 0 , \quad B_2(0,0) = -2\pi i \int_{-\infty}^{\infty} t|u(t)|^2 dt = 0$$

$$B_{11}(0,0) = -4\pi^2 \int_{-\infty}^{\infty} f^2 |U(f)|^2 df = -4\pi^2 \beta^2 , \quad B_{22}(0,0) = -4\pi^2 \int_{-\infty}^{\infty} t^2 |u(t)|^2 dt = -4\pi^2 \tau^2 ,$$

$$B_{12}(0,0) = -2\pi i \int_{-\infty}^{\infty} tu(t)\overline{u}'(t) dt \equiv -4\pi^2 \chi$$

where β and τ are the bandwidth and time duration defined in §1 and the last equation defines χ. Thus, since $B(x,y) \in C^2$ and $B(0,0) = 1$, we have

$$B(x,y) = 1 - 2\pi^2(\beta^2 x^2 + 2\chi xy + \tau^2 y^2) + o(x^2 + y^2)$$

near the origin. Multiplying this by its complex conjugate

$$\overline{B}(x,y) = 1 - 2\pi^2(\beta^2 x^2 + 2\overline{\chi} xy + \tau^2 y^2) + o(x^2 + y^2)$$

gives

$$|A(x,y)|^2 = |B(x,y)|^2 = B(x,y)\overline{B}(x,y)$$

$$= 1 - 4\pi^2(\beta^2 x^2 + \{\chi + \overline{\chi}\}xy + \tau^2 y^2) + o(x^2 + y^2) \ .$$

Thus, except for a negligible term, the level curves $|A(x,y)|^2 = 1 - 4\pi^2\delta^2$ for δ small are given by

(6.2) $$\beta^2 x^2 + 2\gamma xy + \tau^2 y^2 = \delta^2$$

where

$$2\gamma = \chi + \overline{\chi} = \frac{1}{\pi} \, Im\left\{ \int_{-\infty}^{\infty} t\overline{u}(t)u'(t)dt \right\} \ .$$

It can be verified that when $u(t) \in W$, then

$$2Re\left\{ \int_{-\infty}^{\infty} t\overline{u}(t)u'(t)dt \right\} = \int_{-\infty}^{\infty} t(\overline{u}u' + u\overline{u}')dt = \int_{-\infty}^{\infty} t\frac{d}{dt}|u|^2 dt = -\int_{-\infty}^{\infty} |u|^2 dt = -1 \ .$$

Hence,

(6.3) $$\frac{1}{4} + 4\pi^2\gamma^2 = \left| \int_{-\infty}^{\infty} t\overline{u}(t)u'(t)dt \right|^2 \leq \int_{-\infty}^{\infty} |u'|^2 dt \int_{-\infty}^{\infty} t^2|u|^2 dt$$

by Schwarz's inequality. Since $u'(t) \sim 2\pi i f U(f)$,

$$\int_{-\infty}^{\infty} |u'|^2 dt = 4\pi^2 \int_{-\infty}^{\infty} f^2 |U(f)|^2 df = 4\pi^2 \beta^2$$

by Parseval's formula, and (6.3) implies

$$\beta^2 \tau^2 - \gamma^2 \geq 1/16\pi^2 > 0 \ .$$

Thus (6.2) always represents an ellipse and an elementary calculation gives for its major semi-axis a, minor semi-axis b, and area πab the values

$$a^2 = \frac{2\delta^2}{\tau^2 + \beta^2 - \sqrt{(\tau^2 - \beta^2)^2 + 4\gamma^2}} \ ,$$

$$b^2 = \frac{2\delta^2}{\tau^2 + \beta^2 + \sqrt{(\tau^2 - \beta^2)^2 + 4\gamma^2}} \ ,$$

$$\pi ab = \frac{\pi \delta^2}{\sqrt{\beta^2 \tau^2 - \gamma^2}} \ .$$

We can now discuss the accuracy with which the range and velocity of an object can be estimated. An object with range and velocity characterized by the coordinates (x_0, y_0) is detected as a bright spot on the x, y-plane described by

$$I(x,y) = |A(x_0 - x, y_0 - y)|^2 \geq 1 - 4\pi^2 \delta^2 \ ;$$

or

(6.4) $$\beta^2(x_0 - x)^2 + 2\gamma(x_0 - x)(y_0 - y) + \tau^2(y_0 - y)^2 \leq \delta^2 \ .$$

This is an elliptical disc with center at (x_0, y_0). It is evident that, for given τ and β, the area of the disc is least when $\gamma = 0$. In this case, the width of the disc along the x-axis (range axis) is $2\delta/\beta$ and along the y-axis (velocity axis) is $2\delta/\tau$. Thus high accuracy in range (velocity) is attained with waveforms having large bandwidth β (time duration τ).

In our formulation of the radar problem, it appears that we can determine (x_0, y_0) exactly by finding the center of the disc, i.e., the maximum of $I(x, y)$. This is because we have ignored the presence of noise in the radar system. If the noise is taken into account, it can be shown [14, p. 105] that the position of the maximum of $I(x, y)$ is a random variable whose standard deviation is measured by the sharpness of the maximum. More precisely, the standard deviations along the range and velocity axes are $1/\beta\sqrt{N/S}$ and $1/\tau\sqrt{N/S}$, respectively, where S/N is the signal-to-noise ratio. Thus the disc (6.4) accurately indicates the ambiguity in (x_0, y_0) provided

$$\delta^2 = \frac{N}{S} \ .$$

Now let us consider the problem of resolving the images of two objects in the range-velocity plane. If the objects have equal radar cross sections, they will be represented in the range-velocity plane by congruent elliptic discs, centered at points (x_1, y_1) and (x_2, y_2) and similarly oriented (see §1). Evidently, the most difficult situation to resolve occurs when the ellipses have a common major axis. Let us agree that, in this situation, two objects are just resolved when the centers of the discs are one major semi-axis part. Then the smallest resolvable separation R in the range-velocity plane is given by

$$(6.5) \qquad R^2 = \frac{2\delta^2}{\tau^2 + \beta^2 - \sqrt{(\tau^2 - \beta^2)^2 + 4\gamma^2}} \ .$$

We shall call R the resolution constant of the waveform u.

In practice, there are limitations on the bandwidths and time durations that can be used, say,

$$(6.6) \qquad \beta \leq \beta_0 \ , \ \tau \leq \tau_0 \ .$$

Hence, in designing a waveform for optimum resolution we seek to minimize R subject to these bounds. Notice that for any fixed β and τ , R is smallest when $\gamma = 0$. Thus we need to consider only those waveforms for which

$$(6.7) \qquad 2\pi\gamma = Im \left\{ \int\limits_{-\infty}^{\infty} t\bar{u}(t)u'(t)dt \right\} = 0 \ .$$

These include all waveforms $u(t) = c\, v(t)$ where c is a complex constant and $v(t)$ is real. Writing $Min(\beta, \tau)$ for the smaller of the numbers β and τ we have

$$R = \frac{\delta}{Min(\beta, \tau)} \quad \text{if} \quad \gamma = 0 \ .$$

Thus the best resolution constant obtainable subject to (6.6) is

$$R = \frac{\delta}{Min(\beta_0, \tau_0)}$$

and is achieved by any waveform for which $\gamma = 0$ and $\beta = \beta_0$, $\beta_0 \leq \tau \leq \tau_0$ (Case 1) or $\tau = \tau_0$, $\tau_0 \leq \beta \leq \beta_0$ (Case 2). Moreover, we may as well assume that $\tau = \beta = \beta_0$ (Case 1) or $\beta = \tau = \tau_0$ (Case 2) since no more resolution is obtained by increasing $\tau(\beta)$.

§7. **The Hermite Waveforms and Their Ambiguity Functions.** In this final section, we develop another approach to the synthesis problem based on a particular orthonormal sequence of waveforms, the Hermite waveforms. Their introduction can be motivated as follows. It is natural to require that both the waveform $u(t)$ and its spectrum $U(f)$ be "small at infinity" in some sense, as well as having finite bandwidth β and time duration τ. Indeed, it might seem desirable from an engineering viewpoint to require the duration and bandwidth to be strictly limited in the sense that

$$(7.1) \qquad u(t) = 0 \quad \text{for} \quad |t| > T \quad \text{and} \quad U(f) = 0 \quad \text{for} \quad |f| > B \ .$$

Any such conditions on $u(t)$ must be considered in the light of the following fundamental theorem due to G. H. Hardy [7, p. 64].

THEOREM OF HARDY. *If for an integer $n \geq 0$,*

$$(7.2) \qquad u(t) = O(t^n e^{-\pi t^2}) , \quad |t| \to \infty ,$$

and

$$(7.3) \qquad U(f) = O(f^n e^{-\pi f^2}) , \quad |f| \to \infty ,$$

then

$$(7.4) \qquad u(t) = P_n(t) e^{-\pi t^2}$$

where $P_n(t)$ is a polynomial of degree not greater than n.

It is evident from Hardy's theorem that there are no waveforms satisfying (7.1). For (7.1) implies (7.2) and (7.3) with $n = 0$, whence $u(t) = K e^{-\pi t^2}$. Then (7.1) implies $K = 0$; i.e., $u(t) \equiv 0$. More generally, if $u(t)$ is smaller at infinity than (7.2), say,

$$(7.5) \qquad u(t) = o(e^{-\pi t^2}) , \quad |t| \to \infty ,$$

then $U(f)$ must be larger at infinity than (7.3) for every n. Indeed, if (7.5) holds and (7.3) is true for some n, then (7.4) holds. Then $P_n(t) = o(1) , \quad |t| \to \infty$, by (7.5); i.e., $P_n(t) \equiv 0$.

Let us denote by Ω_n the set of all waveforms (7.4) with the degree of P_n not greater than n. Obviously, Ω_n is a linear space of dimension $n + 1$ and $\Omega_m \subset \Omega_n$ for $m \leq n$. Hardy's theorem shows that the waveforms of the class

$$\Omega = \Omega_0 \cup \Omega_1 \cup \cdots \cup \Omega_n \cup \cdots$$

are the only ones that have the same order at infinity as their Fourier transforms. If a waveform $u(t)$ is smaller at infinity than every waveform in Ω, then $U(f)$ is larger at infinity than every waveform in Ω, and conversely. Thus the waveforms Ω present themselves naturally as the waveforms that are "small at infinity".

We shall define the nth Hermite waveform to be

$$(7.6) \qquad u_n(t) = \frac{2^{1/4}}{\sqrt{n!}} \, H_n(2\sqrt{\pi} \, t) e^{-\pi t^2} , \quad n = 0, 1, 2, \ldots,$$

where $H_n(x)$, the nth Hermite polynomial, is defined

$$(7.7) \qquad H_n(x) = (-1)^n e^{\frac{x^2}{2}} \left(\frac{d}{dx} \right)^n e^{-\frac{x^2}{2}} , \quad n = 0, 1, 2, \cdots .$$

The waveforms $u_0(t)$ through $u_5(t)$ are tabulated for $0 \leq 2\sqrt{\pi t} \leq 3$ in [4, pp. 26–31]. The Hermite waveforms are known to define a complete orthornormal sequence in $L^2(-\infty, \infty)$. Moreover, they have the interesting property of being proportional to their Fourier transforms. More precisely,

$$(7.8) \qquad u_n(t) \sim U_n(f) = (-i)^n u_n(f) .$$

Proofs of these properties, as well as others used below, can be found in a number of reference books, including [9, Ch. V; 10, Ch. III; 11, Ch. VI; 12, Ch. I].

Since the degree of $H_n(x)$ is n we see that $u_0(t), u_1(t), \ldots, u_n(t)$ are in Ω_n and are linearly independent, whence

$$u(t) \in \Omega_n \iff u(t) = \sum_{m=0}^{n} a_m u_m(t) .$$

Thus Ω is the set of all finite linear combinations of Hermite waveforms. It is obvious from (7.6) and (7.8) that $u_n(t)$, $t u_n(t)$ and $f U_n(f)$ are in $L^2(-\infty, \infty)$. Moreover, the conditions (2.1) are satisfied since $u_n(t)$ is normalized and $u_n(-t) = (-1)^n u_n(t)$. Thus $u_n(t) \in W$.

To find the bandwidth β_n and time duration τ_n of u_n, observe that

$$\tau_n^2 = \int_{-\infty}^{\infty} t^2 |u_n(t)|^2 dt = \int_{-\infty}^{\infty} f^2 |u_n(f)|^2 df = \beta_n^2$$

whence

$$(7.9) \qquad 2\tau_n^2 = 2\beta_n^2 = \tau_n^2 + \beta_n^2 = -\frac{1}{4\pi^2} \int_{-\infty}^{\infty} \{u_n''(t) - 4\pi^2 t^2 u_n(t)\} \overline{u}_n(t) dt .$$

Now $u_n(t)$ is known to satisfy the differential equation

$$(7.10) \qquad u_n''(t) - r\pi^2 t^2 u_n(t) + 2\pi(2n+1)u_n(t) = 0 , \quad n = 0, 1, 2, \ldots .$$

Hence (7.9) implies

$$\tau_n^2 = \beta_n^2 = \frac{2n+1}{4\pi} .$$

Since $u_n(t)$ is real, its constant γ is zero (see 6.7) and, therefore, its resolution constant

$$(7.11) \qquad R_n = \frac{\delta}{\beta_n} = \frac{2\sqrt{\pi}\,\delta}{\sqrt{2n+1}} ,$$

showing that we can get arbitrarily small resolution constants with waveforms $u \in \Omega$. We shall now that (7.11) is the best resolution constant attainable with waveforms $u \in \Omega_n$. More precisely, we have

THEOREM 7.1. *Among all waveforms $u(t) \in \Omega_n$, the Hermite waveform $u_n(t)$ has the smallest resolution constant.*

Proof. Our proof is based on the fact that, for all $u \in \Omega_n$,

$$\tau^2 + \beta^2 = -\frac{1}{4\pi^2} \int_{-\infty}^{\infty} \{u''(t) - 4\pi^2 t^2 u(t)\}\overline{u}(t)dt$$

is largest for $u_n(t)$. To see this, let

$$u(t) = \sum_{m=0}^{n} a_m u_m(t) .$$

Then, by (7.10),

$$u''(t) - 4\pi^2 t^2 u(t) = -2\pi \sum_{m=0}^{n}(2m+1)a_m u_m(t) ,$$

whence

(7.12) $$\tau^2 + \beta^2 = \frac{1}{2\pi} \sum_{m=0}^{n}(2m+1)|a_m|^2 .$$

We must maximize this subject to

(7.13) $$\int_{-\infty}^{\infty} |u(t)|^2 dt = \sum_{m=0}^{n} |a_m|^2 = 1 .$$

Eliminating $|a_n|^2$ between (7.12) and (7.13) gives

$$\tau^2 + \beta^2 = \frac{2n+1}{2\pi} - \frac{1}{\pi} \sum_{m=0}^{n-1}(n-m)|a_m|^2 .$$

Since the terms in the last sum are non-negative, we see that $\tau^2 + \beta^2$ is a maximum if and only if $a_0 = \cdots = a_{n-1} = 0$, whence $|a_n| = 1$ and $u(t) = cu_n(t)$, $|c| = 1$.

To minimize R, we see from the definition (6.5) that it suffices to maximize $\tau^2 + \beta^2$ and simultaneously minimize $(\tau^2 - \beta^2)^2 + 4\gamma^2$. Of course, these two aims may be in competition. However, in the present case u_n maximizes $\tau^2 + \beta^2$ and makes $(\tau^2 - \beta^2)^2 + 4\gamma^2 = 0$, its absolute minimum, since $\tau_n^2 = \beta_n^2$ and $\gamma_n = 0$. Thus, among all $u \in \Omega_n$, u_n minimizes R.

The functions

$$A_{mn}(x,y) = \int_{-\infty}^{\infty} u_m\left(t - \frac{x}{2}\right)\overline{u}_n\left(t + \frac{x}{2}\right) e^{-2\pi i y t} dt$$

form a complete orthonormal sequence in $L^2(E)$ (Lemma 4.1). We show next that A_{mn} can be expressed in terms of the generalized Laguerre polynomials $L_n^{(\alpha)}(x)$ defined by [9, Ch. V]

$$(7.14) \qquad L_n^{(\alpha)}(x) = \frac{1}{n!} e^x x^{-\alpha} \left(\frac{d}{dx}\right)^n (x^{n+\alpha} e^{-x}) \quad \text{for} \quad x \geq 0.$$

THEOREM 7.2.

$$A_{mn}\left(\frac{x}{\sqrt{\pi}}, \frac{y}{\sqrt{\pi}}\right) = (-1)^{m+n} \sqrt{\frac{n!}{m!}} \, e^{-\frac{1}{2}(x^2+y^2)}(x+iy)^{m-n} L_n^{(m-n)}(x^2 + y^2).$$

Proof. We start by finding a generating function for the A_{mn}. To this end, notice that the known generating relation for the Hermite polynomials implies the relation

$$e^{\frac{z^2}{2}-(\sqrt{\pi}\,t-z)^2} = 2^{-\frac{1}{4}} \sum_{n=0}^{\infty} u_n(t)\frac{z^n}{\sqrt{n!}}.$$

Hence,

$$\frac{1}{\sqrt{2}} \sum_{m=0}^{\infty}\sum_{n=0}^{\infty} u_m(t)\overline{u}_n(\tau) \, \frac{z^m w^n}{\sqrt{m!n!}} = e^{\frac{z^2}{2}-(z-\sqrt{\pi}\,t)^2+\frac{w^2}{2}-(w-\sqrt{\pi}\,\tau)^2},$$

and therefore,

$$(7.15)$$
$$\frac{1}{\sqrt{2}} \sum_{m=0}^{\infty}\sum_{n=0}^{\infty} u_m\left(t-\frac{x}{2}\right)\overline{u}_n\left(t+\frac{x}{2}\right) \frac{z^m w^n}{\sqrt{m!n!}} = e^{\frac{z^2}{2}-\left(z-\sqrt{\pi}\,t+\frac{\sqrt{\pi}}{2}x\right)^2+\frac{w^2}{2}-\left(w-\sqrt{\pi}\,t-\frac{\sqrt{\pi}}{2}x\right)^2}.$$

Indeed, since the right hand side of (7.15) is an analytic function of the complex variables z and w, it follows that the double series converges absolutely and the terms may be summed in any order (see, e.g. [2]). Moreover, it is easy to verify that the convergence is uniform with respect to t on any finite interval $-M \leq t \leq M$, whence

$$\frac{1}{\sqrt{2}} \sum_{m,n=0}^{\infty} \left(\int_{-M}^{M} u_m\left(t-\frac{x}{2}\right)\overline{u}_n\left(t+\frac{x}{2}\right) e^{-2\pi iyt}dt\right) \frac{z^m w^n}{\sqrt{m!n!}} =$$

$$(7.16)$$

$$\int_{-M}^{M} e^{\frac{z^2}{2}-(z-\sqrt{\pi}\,t+\frac{\sqrt{\pi}}{2}x)^2+\frac{w^2}{2}-\left(w-\sqrt{\pi}\,t-\frac{\sqrt{\pi}}{2}x\right)^2-2\pi iyt} dt.$$

But

$$\left|\int_{-M}^{M} u_m\left(t-\frac{x}{2}\right)\overline{u}_n\left(t+\frac{x}{2}\right) e^{-2\pi iyt}dt\right|^2 \leq \int_{-M}^{M} \left|u_m\left(t-\frac{x}{2}\right)\right|^2 dt \int_{-M}^{M}\left|u_n\left(t+\frac{x}{2}\right)\right|^2 dt \leq 1.$$

Hence the series in (7.16) converges uniformly in M, x, y, z and w for (say) $|z| \leq 1$, $|w| \leq 1$ and we may make $M \to \infty$. The integral in (7.16) can be calculated explicitly. Omitting the details, we state the result as

LEMMA 7.1. $\sum_{m,n=0}^{\infty} A_{mn}\left(\frac{x}{\sqrt{\pi}}, \frac{y}{\sqrt{\pi}}\right) \frac{z^m w^n}{\sqrt{m!n!}} = e^{-\frac{1}{2}(x^2+y^2)} e^{w(x-iy)-z(x+iy)+wz}$.

Turning to the Laguerre polynomials, we shall prove

LEMMA 7.2. $\sum_{m,n=0}^{\infty} L_n^{(m-n)}(\xi) \frac{\eta^m \zeta^n}{m!} = e^{-\xi\zeta+\eta+\eta\zeta}$.

Proof. The definition (7.14) implies that

$$L_n^{(\alpha)}(x) = \sum_{k=0}^{n} \binom{n+\alpha}{k} \frac{(-x)^{n-k}}{(n-k)!}$$

and, in particular,

$$L_n^{(m-n)}(\xi) = \sum_{k=0}^{n} \binom{m}{k} \frac{(-\xi)^{n-k}}{(n-k)!} \ .$$

On multiplying this by $\eta^m \zeta^n /m!$ and summing the terms in the proper order, we get Lemma 7.2. The necessary changes in order of the summations are easily justified.

Returning to the proof of Theorem 7.2, let us put

$$\xi = x^2 + y^2 \ , \quad \eta = -z(x+iy) \ , \quad \zeta = -\frac{w}{x+iy}$$

in Lemma 7.2. Then $-\xi\zeta + \eta + \eta\zeta = w(x-iy) - z(x+iy) + wz$ and therefore Lemmas 7.1 and 7.2 imply that

$$\sum_{m,n=0}^{\infty} A_{mn}\left(\frac{x}{\sqrt{\pi}}, \frac{y}{\sqrt{\pi}}\right) \frac{z^m w^n}{\sqrt{m!n!}} = e^{-\frac{1}{2}(x^2+y^2)} \sum_{m,n=0}^{\infty} L_n^{(m-n)}(x^2+y^2)[-z(x+iy)]^m \left[\frac{-w}{x+iy}\right]^n /m!$$

$$= e^{-\frac{1}{2}(x^2+y^2)} \sum_{m,n=0}^{\infty} (-1)^{m+n}(x+iy)^{m-n} L_n^{(m-n)}(x^2+y^2) \frac{z^m w^n}{m!}$$

for all z and w, which is equivalent to Theorem 7.2.

Theorem 7.2 is equivalent to the relation

$$\frac{1}{\sqrt{2\pi}} \int_{-\infty}^{\infty} H_m(\tau-x) H_n(\tau+x) e^{-\tau^2/2} e^{-iy\tau} d\tau = (-1)^{m+n} n! e^{-y^2/2}(x+iy)^{m-n} L_n^{(m-n)}(x^2+y^2)$$

between Hermite polynomials and Laguerre polynomials. This relation is known for the case $x = 0$; see, e.g., [5, p. 120]. In the general case it appears to be new.

Theorem 7.2 enables us to write explicitly in terms of elementary functions the ambiguity functions for a large class of waveforms, the waveforms $u \in \Omega$ that are "small at infinity". More precisely, we have

COROLLARY 7.1. *If $u(t) \in \Omega$, then*

$$A(x,y) = e^{-\frac{\pi}{2}(x^2+y^2)} P(x,y)$$

where $P(x,y)$ is a polynomial in x and y.

This is evident from Theorem 7.2 and the relation

$$A(x,y) = \sum_{m=0}^{N} \sum_{n=0}^{N} a_m \bar{a}_n A_{mn}(x,y) \quad \text{if} \quad u(t) = \sum_{m=0}^{N} a_m u_m(t) \ ,$$

which also provides an explicit expression for $P(x,y)$ in terms of $u(t)$. In particualar, we have

COROLLARY 7.2. *The ambiguity function corresponding to the nth Hermite waveform $u_n(t)$ is*

$$A_n(x,y) = e^{-\frac{\pi}{2}(x^2+y^2)} L_n(\pi\{x^2+y^2\})$$

where $L_n(x) \equiv L_n^{(0)}(x)$.

It is evident from Theorem 7.2 that $A_{mn}(x/\sqrt{\pi}, y/\sqrt{\pi})$ has a particularly simple form as a function of the polar coordinates (r, θ) defined by

$$x + iy = re^{i\theta} \ .$$

Indeed, we have

$$A_{mn}\left(\frac{x}{\sqrt{\pi}}, \frac{y}{\sqrt{\pi}}\right) = (-1)^{m-n} \sqrt{\frac{n!}{m!}} \ e^{-\frac{1}{2}r^2} r^{m-n} L_n^{(m-n)}(r^2) e^{i(m-n)\theta} \ , \quad m \geq n \ ,$$

and

$$A_{mn}\left(\frac{x}{\sqrt{\pi}}, \frac{y}{\sqrt{\pi}}\right) = \bar{A}_{nm}\left(\frac{-x}{\sqrt{\pi}}, \frac{-y}{\sqrt{\pi}}\right) = \sqrt{\frac{m!}{n!}} \ e^{-\frac{1}{2}r^2} r^{n-m} L_m^{(n-m)}(r^2) e^{i(m-n)\theta} \ , \quad m \leq n \ .$$

Hence, if we define

$$(7.17) \quad s = \left\{ \begin{array}{l} n, m \geq n \\[2mm] m, m \leq n \end{array} \right\} = \frac{m+n-|m-n|}{2} \\[4mm] k = m - n \quad\quad \right\} \Longleftrightarrow \left\{ \begin{array}{l} m = s + \dfrac{|k|+k}{2} \\[4mm] n = s + \dfrac{|k|-k}{2} \end{array} \right.$$

then

$$(7.18) \quad \frac{1}{\sqrt{\pi}} A_{mn}\left(\frac{x}{\sqrt{\pi}}, \frac{y}{\sqrt{\pi}}\right) = \Psi_{sk}(r) e^{ik\theta}$$

for all $m, n \geq 0$ where

$$\Psi_{sk}(r) = (-1)^{\frac{|k|+k}{2}} \sqrt{\frac{s!}{\pi(s+|k|)!}} e^{-\frac{1}{2}r^2} r^{|k|} L_s^{(|k|)}(r^2) .$$

Moreover, (7.17) establishes a one-to-one correspondence between the pairs (m, n) of integers for which $m \geq 0$, $n \geq 0$ and the pairs (s, k) for which $s \geq 0$, $-\infty < k < \infty$. We know that $(2\pi)^{-\frac{1}{2}} e^{ik\theta}$ is a complete orthornormal sequence with respect to $d\theta$ on $0 \leq \theta \leq 2\pi$. Also $(2\pi)^{\frac{1}{2}} \Psi_{sk}(r)$, $s = 0, 1, 2, \ldots$, is for each k a complete orthornormal sequence with respect to $r dr$ on $0 \leq r < \infty$. This follows from the known fact that $L_s^{(\alpha)}(x)$, $s = 0, 1, 2, \ldots$, is for each α a complete orthogonal sequence with respect to $x^\alpha e^{-x} dx$ on $0 \leq x < \infty$ and satisfies [9, p. 96]

$$\int_0^\infty L_s^{(\alpha)}(x) L_t^{(\alpha)}(x) x^\alpha e^{-x} dx = \Gamma(\alpha+1) \binom{s+\alpha}{s} \delta_{st} .$$

Hence (7.18) provides a new proof that $A_{mn}(x, y)$ is a complete orthornormal sequence with respect to $dx dy = r dr d\theta$ in the plane.

Now consider the correspondence defined by

$$u(t) = \sum_{m=0}^\infty a_m u_m(t) \Longrightarrow R_\phi u(t) = u_\phi(t) = \sum_{m=0}^\infty e^{-im\phi} a_m u_m(t) .$$

R_ϕ is a linear transformation of $L^2(-\infty, \infty)$ onto itself which is unitary, i.e.,

$$\int_{-\infty}^\infty |R_\phi u(t)|^2 dt = \int_{-\infty}^\infty |u(t)|^2 dt = \sum_{m=0}^\infty |a_m|^2 ,$$

since $|e^{-im\phi} a_m| = |a_m|$. If A_ϕ is the ambiguity function for u_ϕ we have, by (7.18),

$$A_\phi(x, y) = \sum_{m=0}^\infty \sum_{n=0}^\infty a_m \overline{a}_n e^{-i(m-n)\phi} A_{mn}(x, y) = \sqrt{\pi} \sum_{s=0}^\infty \sum_{k=-\infty}^\infty a_m \overline{a}_n e^{-ik\phi} \Psi_{sk}(\sqrt{\pi} \, r) e^{ik\theta}$$

(7.19)

$$= \sqrt{\pi} \sum_{s=0}^\infty \sum_{k=-\infty}^\infty a_m \overline{a}_n \Psi_{sk}(\sqrt{\pi}) e^{ik(\theta-\phi)} .$$

This proves

THEOREM 7.3. $A_\phi(x, y) = A(x \cos\phi + y \sin\phi, -x \sin\phi + y \cos\phi)$, i.e., the ambiguity function for u_ϕ is obtained by rotating $A(x, y)$ through the angle ϕ.

COROLLARY 7.3. $F(x, y)$ is an ambiguity function if and only if the rotated functions $F(x \cos\phi + y \sin\phi, -x \sin\phi + y \cos\phi)$ are ambiguity functions.

We showed that the ambiguity functions for the Hermite waveforms have level curves which are circles about the origin, i.e., $A(x, y) = f(x^2 + y^2)$. We now show, using Theorem 7.3, that these are the only such ambiguity functions. Indeed, if A has this property, then $A = A_\phi$ for all ϕ and (7.19) implies that

$$\sum_{m=0}^{\infty} a_m \bar{a}_n \Psi_{sk}(r) = 0 \ , \ k \neq 0 \ .$$

Hence, since the Ψ_{sk} are orthogonal, we have

$$a_m \bar{a}_n = 0 \quad \text{for} \quad k \neq 0 \ , \quad \text{i.e.,} \quad m \neq n \ .$$

This is possible if and only if $|a_n| = 1$ for one fixed n and $a_m = 0$ for $m \neq n$. This proves

COROLLARY 7.4. *An ambiguity function $A = A_u$ satisfies*

$$A_u(x, y) = f(x^2 + y^2)$$

if and only if $u = u_n$ for some n.

An analogous result holds for ambiguity functions $A = A_u$ whose level curves are similar ellipses with centers at the origin and parallel axes. Indeed, if A_u is such a function and ϕ is the angle from the x-axis to the major axis of its level curves, then u_ϕ has the ambiguity function A_ϕ with level curves

$$\frac{x^2}{a^2} + \frac{y^2}{b^2} = \text{ constant } .$$

Notice that $S_\alpha u(t) = \sqrt{\alpha}\, u(\alpha t)$ has the ambiguity function

$$A'(x, y) = \alpha \int_{-\infty}^{\infty} u\left(\alpha t - \frac{\alpha x}{2}\right) \bar{u}\left(\alpha t + \frac{\alpha x}{2}\right) e^{-2\pi i y t} dt$$

$$= \int_{-\infty}^{\infty} u\left(s - \frac{\alpha x}{2}\right) \bar{u}\left(s + \frac{\alpha x}{2}\right) e^{-2\pi i \frac{y}{\alpha} s} ds = A_u\left(\alpha x, \frac{y}{\alpha}\right) \ .$$

Thus A' has the level curves

$$\left(\frac{\alpha x^2}{a}\right) + \left(\frac{y^2}{\alpha b}\right) = \text{ constant } \qquad (\alpha^2 = ab)$$

which are circles if $\alpha^2 = a/b$. Hence the ambiguity function for $S_\alpha R_\phi u(t)$ has level curves which are circles about the origin and, therefore, by Corollary 7.4, $S_\alpha R_\phi u(f) = u_n(t)$ for some n. Thus

(7.20) $$R_\phi u(t) = \frac{1}{\sqrt{\alpha}} u_n\left(\frac{t}{\alpha}\right) = \sum_{m=0}^{\infty} c_m^{(n)} u_m(t)$$

and

(7.21) $$u(t) = \sum_{m=0}^{\infty} e^{im\phi} c_m^{(r)} u_m(t) = R_{-\phi}\left(\frac{1}{\sqrt{\alpha}} u_n\left(\frac{t}{\alpha}\right)\right) \ .$$

Conversely, (7.20) and (7.21) define a waveform whose level curves are elliptic which proves

COROLLARY 7.5. $A_u(x,y)$ has level curves which are similar ellipses with centers at the origin and parallel axes if and only if u is one of the sequence of waveforms defined by (7.20), (7.21).

In conclusion, we point out that considerable information concerning the ambiguity functions

$$A_n(x,y) = e^{-u/2}L_n(u) \, , \, u = \pi r^2 \, ,$$

is available. Thus if

$$0 < u_{1n} < u_{2n} < \cdots < u_{nn}$$

are the zeros of $L_n(u)$ and $u_{kn} = \pi r_{kn}^2$ then

$$0 < r_{1n} < r_{2n} < \cdots < r_{nn}$$

define the zeros of $A_n(x,y)$. It is known that if

$$0 < j_1 < j_2 < \cdots < j_n < \cdots$$

are the zeros of the Bessel function $J_0(x)$ then [9, pp. 123–129]

$$\frac{j_k^2}{4(n+1/2)} < u_{kn} < (k+1/2)\frac{2k+1+\{(2k+1)^2+1/4\}^{\frac{1}{2}}}{n+1/2} \, ,$$

whence

$$\frac{j_k}{2\sqrt{\pi\left(n+\frac{1}{2}\right)}} < r_{kn} < \sqrt{\frac{(k+\frac{1}{2})[2k+1+\{(2k+1)^2+1/4\}^{\frac{1}{2}}]}{\pi(n+\frac{1}{2})}} \, .$$

The smallest zero satisfies

$$\frac{1.202}{\sqrt{\pi\left(n+\frac{1}{2}\right)}} < \frac{j_1}{2\sqrt{\pi\left(n+\frac{1}{2}\right)}} < r_{1n} \le \frac{\sqrt{3}}{\sqrt{2\pi\left(n+\frac{1}{2}\right)}} < \frac{1.225}{\sqrt{\pi\left(n+\frac{1}{2}\right)}} \, .$$

Another estimate which is cruder, but simpler in form, is [9, p. 125]

$$\frac{3\pi}{16}\frac{k+1}{\sqrt{\pi(n+1)}} < r_{kn} < \frac{2(k+1)}{\sqrt{\pi(n+1)}} \, .$$

The zeros j_n satisfy $j_n \sim \pi n$, $n \to \infty$. They are tabulated in [4, p. 166] through $j_{40} = 124.8793$.

Recalling that

$$\beta_n = \tau_n = \sqrt{\frac{(n+\frac{1}{2})}{2\pi}}$$

for u_n, we see that the first zero satisfies

$$\frac{3\sqrt{3}}{16\sqrt{2}}\frac{1}{\beta_n} < r_{1n} < \frac{\sqrt{3}}{2\pi}\frac{1}{\beta_n} \, ,$$

while the last zero satisfies

$$\frac{3\sqrt{2\pi}}{16} \beta_n < r_{nn} < \frac{4\sqrt{2}}{3} \beta_n .$$

These facts can be used to obtain a qualitative picture of the function $|A_n(x,y)|^2 = e^{-u} L_n^2(u)$, $u = \pi r^2$. It is positive except for n zeros which are approximately equally spaced between

$$r_{1n} = c\frac{1}{\sqrt{n+1}} \quad \text{and} \quad r_{nn} = c\sqrt{n+1} ,$$

with $\frac{3\sqrt{\pi}}{16} < c < \frac{2}{\sqrt{\pi}}$. It has an absolute maximum of 1 at $r = 0$ and has n other maxima between the zeros r_{kn}. Beyond the last maximum, it tends monotonically to zero. We have no theoretical information about the magnitudes of the maxima, but tables indicate that they decrease with increasing r and also with increasing n.

N. Wiener has defined and tabulated a function that is closely related to $A_n(x,y)$ [13, pp. 36 and 124–128]. His function, which we call $\mathcal{L}_n(x)$ to distinguish it from $L_n(x)$, is related to $L_n(x)$ by

$$\mathcal{L}_n(x) = \sqrt{2}\, (-1)^n e^{-x} L_n(2x) ,$$

whence

$$A_n(x,y) = \frac{(-1)^n}{\sqrt{2}}\, \mathcal{L}_n\left(\frac{\pi}{2}\, r^2\right)$$

and

$$|A_n(x,y)|^2 = \frac{1}{2}\, \mathcal{L}_n^2\left(\frac{\pi}{2}\, r^2\right) .$$

Thus Wiener's table, which covers $0 \leq n \leq 5$ and $0 \leq \frac{\pi}{2}\, r^2 \leq 30.0$ in steps of $\Delta \frac{\pi}{2}\, r^2 = 0.1$ (roughly, $0 \leq r \leq 4.4$), may be used to graph the first six ambiguity functions.

This paper contains the results of research performed during the summer of 1957 for the Advance Engineering Section, HMEE Department, of the General Electric Company. The preparation of the paper was supported by the United States Army through the Mathematics Research Center, U.S. Army, under contract DA–11–022–ORD–2059.

260

REFERENCES

[1] BOCHNER, S. AND CHANDRESAKHARAN, K, *Fourier Transforms*, Annals of Mathematics Studies, No. 19, Princeton University Press, Princeton, 1949.

[2] BOCHNER, S. AND MARTIN, W.T., *Several Complex Variables*, Princeton University Press, Princeton, 1948.

[3] COURANT, R. AND HILBERT, D., *Methods of Mathematical Physics*, V.I, Interscience Publishers, Inc., New York, 1953.

[4] JAHNKE, E. AND EMDE, F., *Tables of Functions with Formulae and Curves*, Dover Publications, New York, 1945.

[5] MAGNUS, W. AND OBERHETTINGER, F, *Formulas and Theorems for the Special Functions of Mathematical Physics*, Chelsea Publishing Co., New York, 1949.

[6] MUNROE, M.E., *Introduction to measure and integration*, Addison–Wesley, Cambridge, Mass., 1953.

[7] PALEY, R. AND WIENER, N., *Fourier Transforms in the Complex Domain*, American Mathematical Society Colloguium Publications, V. XIX, New York, 1934.

[8] RIESZ, F. AND SZ.–NAGY, B, *Functional Analysis*, F. Ungar Publishing Co., New York, 1955.

[9] SZEGÖ, G., *Orthogonal Polynomials*, American Mathematical Society Colloguium Publications, V. XXIII, New York, 1939.

[10] TITCHMARSH, E.C., *Introduction to the Theory of Fourier Integrals*, Oxford University Press, Oxford, 1937.

[11] TRICOMI, F., *Vorlesungen über Orthogonalreihen*, Springer-Verlag, Berlin, 1955.

[12] WIENER, N., *The Fourier Integral*, Cambridge University Press, Cambridge, 1933.

[13] WIENER, N., *Extrapolation, Interpolation, and Smoothing of Stationary Time Series*, J. Wiley, New York, 1949.

[14] WOODWARD, P.M., *Probability and Information Theory, with Applications to Radar*, Pergamon Press, London, 1953.